O P E N

A NEUROSPIRITUAL EXPLORATION OF THE
SELF-HEALING POWER OF YOUR BRAIN

CHRISTOPHE MORIN, PH.D.
AUTHOR OF THE PERSUASION CODE AND THE SERENITY CODE

ABSTRACT

OPEN is an invaluable resource for individuals seeking self-healing and health practitioners, psychologists, psychiatrists, and medical professionals looking to expand their therapeutic toolkit with innovative, research-backed approaches for complex psychological conditions.

OPEN presents a groundbreaking neurospiritual exploration of the role of consciousness and neuroplasticity in overcoming stress, anxiety, depression, addictions, and trauma (SADAT). Dr. Christophe Morin blends modern science with ancient wisdom in a comprehensive guide drawing on two decades of personal healing and spiritual discovery. Through his transformative experiences with psychedelic-assisted therapy and ancient wellness practices, Morin offers unique perspectives on healing complex psychological conditions affecting hundreds of millions of people worldwide.

At its core, *OPEN* explains the science behind the brain's capacity for change, offering strategies to experience fewer struggles and more joy. Through a mix of personal stories and rigorous research, Morin shows how these approaches can effectively address the maladaptive behaviors that perpetuate chronic suffering in SADAT conditions. By integrating insights from neurobiology, behavioral genetics, personality neuroscience, psychedelic science, quantum mechanics, and ancient spiritual systems, *OPEN* presents a step-by-step approach to raising consciousness, rewriting personal narratives, and reprogramming habits into rituals that cultivate openness, flow, and happiness in your daily life.

More than just a self-healing book, *OPEN* provides a proven roadmap, combining cutting-edge science with time-honored practices to guide readers toward mindfulness, cognitive flexibility, and optimal health. Morin's journey, along with the experiences of those who have followed a similar path, illustrates the remarkable potential of neurospirituality. This emerging field reveals how rituals impact brain function and well-being by examining the biological and spiritual foundations of human experience.

The claims of *OPEN* are bold but speak to the urgency of finding new ways to overcome the despair that plagues so many. *OPEN* your Mind, *OPEN* your Self, *OPEN* your Life, and you will overcome your life's deepest challenges.

*I am very grateful to Dr. Bonnie Bright for her help,
guidance, and exceptional editing contribution.*

*I dedicate this book to the love of my life, my sons, mother, father, brother,
sisters, and all the beautiful souls that have supported my splendid life journey.*

ISBN 978-0-9979550-5-7

Praises for OPEN

In this timely and useful book, Dr. Christophe Morin, a psychologist and neuroscientist, explains in simple terms how you can use neuroplasticity to overcome mental problems such as stress, anxiety, depression, addictions, and trauma, raise your consciousness, and experience peace, joy, and a sense of purpose. The innovative healing approach presented in OPEN is based on extensive scientific evidence and the author's own personal journey. I highly recommend this work.
Dr. Mario Beauregard, Ph.D., Neuroscientist, co-author of *The Spiritual Brain*, author of *Expanding Reality*.

Read this book with an open mind and curiosity. Suspend judgment and watch what happens as Dr. Morin takes you on a journey that will resonate and articulate how you can enhance your daily life. The insights, raw vulnerability, and research behind this book will enlighten any reader.
Dr. Russ Riendeau, Ph.D., Mayo Clinic Certified Wellness Coach

This is Dr. Morin's best work, unlike anything I have read. He is open about his trauma in a way that is both honest and respectful. The way he weaves his personal story with an intense psychedelic-cultural experience and then explains that science is pure genius. You will not be able to stop this intellectual feast.
Dr. Ian McCulloh. Ph.D., LTC (Retired) Associate Professor, Johns Hopkins University

This book is a fascinating journey into the inner healing potential of our brain and soul. The title says it all: Life is much better when we embrace our ability to reprogram toxic habits and live free of stress and anxiety.
Dr. Steve Feinberg, Ph.D. Neurostrategist and Executive Coach, author of *Do what others say can't be done* and *The Advantage Maker*

Dr. Morin offers a new way of being for us all. Sharing his personal journey and rigorous scientific insights in such a compelling manner is more than a creative act. He reminds us that humanity needs plasticity more than ever.
Angela Cretu
Board Member, Ex-CEO of Avon, Angel Investor in Female Founders

This is a very special book, well-informed, articulate, and heartfelt. Dr. Christophe Morin takes us through his healing journey and educates us about our potential to raise our consciousness, rewire our stories, and reprogram our lives.
Joe Tafur MD, author of *The Fellowship of the River: A Medical Doctor's Exploration into Traditional Amazonian Plant Medicine*

Table of Contents

INTRODUCTION

The Power of Being OPEN

"Your vision will become clear only when you look into your heart. Who looks outside, dreams; who looks inside, awakens."

--Carl Jung, Psychologist

Iquitos, Peru. Wednesday, November 08

After thinking about this trip for so long, I am finally in Iquitos. The moment I step off the plane, I am enveloped by the humid embrace of this vibrant place. The air is alive—moist and warm, almost like the jungle is breathing, welcoming me into its folds. My senses are overwhelmed by the rustic charm of the airport, a time capsule that transports me back eight years to my first visit to Peru. A torrent of memories floods my mind—vivid and visceral, the kind that lingers in your bones. Yet, amidst the enchantment, an unwelcome companion resurfaces—fear.

Am I truly prepared for this? I have committed to a Shamanic initiation program that will last 26 grueling days deep in the heart of the Amazonian jungle. The anticipation is a whirlwind of exhilaration and dread. I will be disconnected entirely—not just from my clients, who rely on me daily, but from my 90-year-old mother, whose gentle face I see every week on our video calls. Soon, the comforting hum of emails and news alerts, the endless stimulation of my digital lifeline, will fade into silence. I've always thrived on this constant connectivity, but I wonder now if it's an addiction I've mistaken for a purpose.

This journey is not merely an escape but a confrontational reckoning with the overwhelming flood of information that has numbed my senses and obscured my ability to feel and capture the essence of the world around me. It is my chance to sever the threads of distraction and reclaim the fragments of myself that have been scattered by years of overwork and overstimulation. The promise of rest feels like a distant dream finally coming true. For the first time in four decades, I can exhale fully. I can pause. I can listen—to the jungle, myself, and the wisdom I hope Ayahuasca will reveal.

The stakes feel impossibly high. This isn't just a retreat; it's a pilgrimage. Since my first encounter with Ayahuasca in 2010, I've longed for her guidance, her ability to illuminate the hidden pathways of my soul and soothe the scars of my past. Her voice has echoed in my life ever since, calling me back to a place of serenity and joy that I've only glimpsed. I am not the same person I was when I first took her medicine. My struggles and growth have shaped me into someone who can hear her call more clearly. But I also carry questions—a scientist's hunger for evidence, for understanding the ancient practices of the Shipibo people who have mastered this art of healing over millennia.

As I stand here, the jungle looms, a vast unknown filled with promises and challenges. My heart pounds, not just with excitement but with a deep, primal fear. Can I surrender to what lies ahead? Can I let go of the person I think

I am to discover who I am meant to be?

The Son of a Brilliant Inventor

Figure 1: Old Parthenay, France

I was born in France on April 21, 1960, in a small medieval town called Parthenay. Seeing pictures of my hometown looks charming and romantic. However, during my childhood, I resented being trapped away from the action of a big city like Paris. I was bored to death. My father had settled in Parthenay to start a small body shop in the mid-fifties. At the time, automobile repair shops would not replace bumpers and doors for a small dent. Highly skilled artisans would rework the shape of your vehicle's body to restore it to its original condition.

Figure 2: Andre Morin (1949)

My father's business thrived because he performed these metal surgeries flawlessly. However, he had much more ambitious plans than fixing dents. His big idea was to redesign the shape of the entire body of a car (Figure 2) to make it more stylish or to add more cargo space. Eventually, he transformed trucks into full-blown retail stores (called *camions magasins*) and marketed them under *Etalmobil* in Europe. They would eventually become part of the tapestry of major French and European open markets.

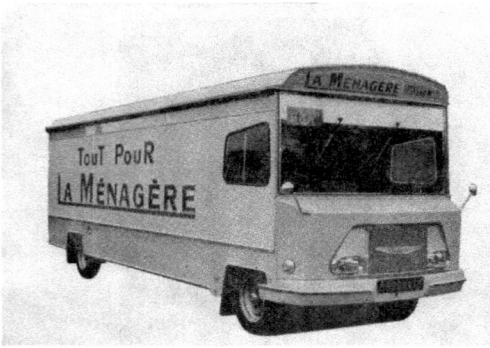

Figure 3: Etalmobil (1961)

My father's ingenuity had no limits. Neither did his ambition. By the time I was born in 1960, he was rapidly expanding his production capacity with my mother's help by building a large factory on the outskirts of my hometown. My father and mother married relatively young after he completed his requisite military service in Algeria. He spent nearly two years apart from his beloved Nicole, during which he wrote to her almost every day. Each letter was enclosed in an envelope adorned with beautifully detailed illustrations. She wrote back, sharing glimpses of her life while they were separated. The cherished collection of these letters, a testament to their deep connection, is now carefully preserved by my older sister as one of our family's most treasured keepsakes.

My father was romantic. He met Nicole and her twin sister Jocelyne at a dance party. He fell in love with the twin he clicked the most with. When we grew up, my father often joked that he was unsure if he married Nicole or Jocelyne since the sisters frequently tricked him by switching their identities. My mother was primarily raised by her mother, a hard-working woman I adored. My whole life, I thought that my grandmother had to take care of her twin daughters alone because her

Figure 4: Letter from Andre to Nicole (1951)

husband abandoned her. It was only recently that I learned that she had refused to live with a cheating husband and kicked him out of their lives. I am not sure at this point what the circumstances behind this dramatic situation were. My mother never talked about it. It remained a family secret for most of my life. I am sure my mother and her sister suffered a great deal from this loss since they both talked fondly of their father, but they chose not to share or discuss the sensitive topic until recently when I started to ask more questions.

Meanwhile, few education opportunities were offered to my mother as she lived in a small village near Parthenay, where I was born. Also, her mother could not afford to pay for her daughter's higher education as she ran a small hairdressing business alone. Nonetheless, Nicole and Jocelyne earned a technical degree at 17 to operate a typewriter at the impressive speed of 80 words per minute. Growing up, I was always amazed at how my mother would crank the fascinating apparatus and command its keyboard with so much dexterity with her fingers. Her writing was also impressive. I often wonder if I learned to type at 19—despite being teased by my college friends—because I couldn't imagine living without the skill that had lifted my mother and aunt out of poverty. It wasn't just the skill of typing that I inherited from my mother—it was the sense of responsibility and duty to those I loved. This deep-rooted desire to support and uplift my family would shape my life in ways I didn't fully understand at the time.

Understanding My Savior Complex

For as long as I can remember, I have felt completely justified in giving much time and relentless effort to support my family. I helped my father tirelessly before and after his devastating bankruptcy. I cheered my mother through chronic periods of excessive anxiety and depressed moods, as well as provided ongoing assistance and comfort to close family members who struggled with stress, anxiety, depression, traumas, and addictions. More importantly, in my adult life as a father, I provided around-the-clock emotional assistance to my youngest son while he was learning to face and ultimately accept challenging medical and psychological conditions. I will elaborate on this journey later in the book.

A few years ago, under the guidance of a skilled psychotherapist who integrated psychedelics into her practice, I had a profound realization: since childhood, I had unknowingly developed a pattern known as *the Savior Complex*. Carl Jung, a Swiss psychiatrist and psychoanalyst, introduced the complex concept as an unconscious cluster of related thoughts and feelings (1). According to Jung, complexes can manifest through various emotional responses, often leading to difficulties in certain situations or with specific people. For instance, someone with an "inferiority complex" might consistently feel inadequate and struggle with low self-esteem, even when there is no objective reason to do so.

Over time, my Savior Complex increasingly influenced the direction and meaning of my life, jeopardizing my career, relationships, and overall well-being. The Savior Complex, also called the *white knight syndrome*, originates from a deep-seated urge to save others at the expense of one's own needs. My therapist helped me recognize and accept that this pattern stemmed from compulsion rather than the needs of those I was helping. She opened the door of consciousness aided by the powerful effect of a psychedelic-induced state of *ego suspension*. Since I often refer to the *ego* in the book, let me offer a simple definition from Sigmund Freud that he first coined over 100 years ago (2).

An Austrian neurologist and the founder of psychoanalysis. Freud initially used the German word *Ich*, which translates to "I" in English, to describe the conscious, self-aware part of the human mind. It was later translated to ego in the English-speaking world. According to Freud, the ego is a critical component of the human psyche, functioning as the mediator between the primitive desires of the *id* and the realities of the external world. The id is the unconscious part of the psyche that operates on the pleasure principle, seeking immediate gratification of desires without consideration for consequences.

Meanwhile, the ego is the part of the mind that mediates between the conscious and the unconscious. It is responsible for testing reality and creating a sense of personal identity. In simpler terms, the ego is our sense of self, pivotal in managing our thoughts, feelings, and actions in or about the external world. Ultimately, it took me many years to understand my Savior Complex because it was active below the radar of my consciousness. As a result, I never understood until much later in my adult life why I was so anxious all the time, why my abuse of alcohol was conveniently rationalized as a need to soothe, and why recurring bouts of depressive episodes were making my life often so miserable. For the most part, I was hiding my suffering as much as I could, fragmenting my life skillfully to make sure no one around me, even my family members, would ever know about my emotional troubles. When I reflect on this now, it is ironic, if not embarrassing, that my background in psychology did not help me see sooner how much the complex had managed to control my life.

Surprisingly, I have been drawn to people I perceived as needing rescue or fixing since I was 12. I have felt a strong urge to help solve others' problems, even when they were not asking for my help. In the pursuit of helping others, however, I often needed to pay more attention to my needs and well-being. Thanks to my therapist, I eventually realized that my self-esteem and self-worth were closely tied to my ability to help others. In other words, I felt worthless if I could not help someone. Some of my most precious relationships were formed under the Savior Complex and, therefore, became unhealthy or unbalanced because they were based on the dynamic of one person needing my help and me providing it unquestioningly. This meant I needed help setting and maintaining healthy boundaries, as I would systematically prioritize others' needs over mine.

Over time, my desire to help others has not always been inherently negative or pathological. After all, I have had a successful career, holding executive positions in private and public companies. I also co-created the first full-service neuromarketing agency, training thousands of senior executives to achieve better marketing results by applying neuroscience. However, in my personal life, my addiction to helping others has been driven by a neurotic need for validation. As a result, it led me to unhealthy dynamics with friends and close family members. I was trapped in this pattern without knowing how to loosen its grip. It took me the last ten years to clear up the anxiety and stress associated with it, and once I fully accepted the diagnosis, another four years to tame the insidious role of the complex by relentlessly working on my *neuroplasticity.*

Neuroplasticity (I will often refer to this phenomenon as *plasticity*) is a remarkable property of our nervous system to reprogram itself. It can be influenced by genetic factors affecting neurotransmitter function, receptor sensitivity, and structural brain characteristics. Plasticity correlates to specific traits of our personality, especially the degree to which we are open to other people's ideas and suggestions. While our ability to reprogram our brain is innate, it can be impaired by many years of stress and anxiety. Using two of the best personality systems in the world, *The Big Five* and *the Enneagram*, which I explain later in the book, I relentlessly studied the underpinnings of my lack of openness. Once I accepted the brutal verdict of living with low plasticity, I started the process of increasing it, gradually allowing the Savior Complex to release its grip on my brain while adopting habits that would sustain a previously unimaginable level of openness, happiness, and flow.

Why Read OPEN?

OPEN is a neurospiritual healing guide to reprogram **S**tress, **A**nxiety, **D**epression, **A**ddictions, and **T**rauma out of your life. I refer to these five conditions using the acronym *SADAT. Comorbidity* levels between them are notably high due to shared risk factors and overlapping symptoms. Comorbidity is the simultaneous occurrence of multiple disorders within an individual, often arising from shared risk factors or overlapping symptoms. For example, Kessler and his colleagues (3) found that 58% of individuals with major depression also had an anxiety disorder. Meanwhile, Brady (4) reported that 40%-60% of people with PTSD also had substance use disorders. Additionally, Swendsen (5) highlighted the substantial overlap between anxiety, depression, and substance use, especially under chronic stress. Since this interconnectedness complicates diagnosis and treatment, it often requires integrated therapeutic approaches. In other words, treating one condition may lead to improvements across multiple mental health issues.

If you suffer from any of the SADAT conditions, then you can think of this book as your physical, mental, and spiritual wellness coach. It will help you navigate the complexity of healing SADAT conditions. My goal is simple: to help you find the home of your soul again, a place of internal peace where you can enjoy your life with purpose and bliss every day. With scientific rigor and considering the extensive teachings I have received during my long personal healing and spiritual journey, *OPEN* presents the scientific and spiritual benefits of psychedelics to treat SADAT as well as the power of well-established contemporary wellness practices, many of which are inspired by shamanic rituals used by Indigenous communities for millennia.

The *psychedelic renaissance* is real (6). This catchy term refers to the renewed scientific interest and exploration of psychedelics, like psilocybin and LSD, for their therapeutic potential, especially in treating mental health disorders such as depression, anxiety, PTSD, and addiction. This movement is gaining momentum around the world, challenging previous stigmas around psychedelics and highlighting their potential to revolutionize psychiatric care where traditional methods may fall short. Its importance lies in redefining mental health treatment, building upon neuroplasticity, and consciousness research. It represents a significant shift in how society views and utilizes psychedelics, moving away from the stigmatization of the past and toward a more informed and potentially beneficial understanding of these powerful substances. Despite the enthusiasm, the legalization of psychedelic-assisted therapy has been slow, and PAT programs are only available in a few countries, such as the USA, Canada, Australia, Brazil, the Netherlands, Switzerland, Israel, and Germany. Therefore, the delays in the approval process continue to impact millions of people seeking effective alternative treatments for debilitating SADAT conditions.

Meanwhile, ancient healing traditions are reclaiming their authority and influence since many Western views on mental health are finally beginning to evolve. Joe Tafur, M.D., a prominent supporter of integrative healing models, advocates the value of hybrid models combining PAT and ancient healing modalities. In his seminal book *The Fellowship of the River,* Dr Tafur states:

"I acknowledge the emotional and spiritual dimensions of human health. In these times, we are dealing with an epidemic of soul sickness, i.e., spiritual illness. We must expand our medical and cultural paradigms to adequately address the flood of spiritual maladies" (7).

Thus, facing an epidemic of soul sickness, the number of trained professionals in alternative treatments is rapidly expanding. As a result, the neurospiritual field is still developing, and the exact number of professionals thoroughly trained in *neuropsychotherapy* (8), spiritual, and psychedelic approaches are not well documented at this time.

If you are a mental health professional or practitioner, *OPEN* offers a compelling resource that complements clinical expertise with innovative, neurospiritual insights. Designed to address the deep roots of stress, anxiety, depression, addiction, and trauma (SADAT), *OPEN* provides a unique blend of evidence-based neuroplasticity practices and spiritual traditions, including insights from psychedelic-assisted therapies and integrative wellness practices.

Unsurprisingly, the pharmaceutical industry appears hesitant to promote alternative treatment options that could compete with a profitable catalog of conventional prescription drugs, which often offer only temporary relief from the worst symptoms of SADAT, rather than providing a long-term solution (9). However, this book is not a critique of mental health professionals or pharmaceutical companies involved in addressing SADAT's complexities. Instead, *OPEN* seeks to educate and inspire wellness advocates to support and advance alternative therapeutic protocols that deliver credible, lasting results.

Of course, many authoritative books have already been published on many topics discussed in *OPEN*. For instance, the historical and medicinal value of psychedelic plants is brilliantly presented in *The Plants of Gods (10)*.

Here are other recommended reads to complement *OPEN:*

- On the power of ancient healing traditions: (11)
- On the effectiveness of narrative therapy: (12)
- On the nature of consciousness: (13, 14)
- On the neuroscience of spirituality: (15, 16)

Also, I highly recommend publications from exceptional scientists and thought leaders like Deepak Chopra, Gabor Maté, and Dan Siegel (17-19). However, no other publication has tackled the neurospiritual dimension of SADAT as a unified cluster of conditions with shared psychological and neurobiological foundations. In today's information-rich environment, countless articles and social media posts discuss mental health, wellness, psychedelic medicine, psychotherapy, and psychology. Yet, navigating this overwhelming and often contradictory information can be stressful and confusing—especially for those already managing the demands of SADAT.

OPEN is designed to address these challenges comprehensively. To the best of my knowledge, it stands alone as the only book to holistically explore the complex, multi-dimensional nature of SADAT's maladaptive behaviors while offering a straightforward, three-step process to eliminate its root causes. *OPEN*'s protocols and healing modalities deliver remarkable outcomes grounded in rigorous research and verifiable evidence. Now, let's examine the key focus of each chapter.

Chapter ONE Summary: Living Fully OPEN

Chapter ONE sets the stage for an *OPEN* life in action. Without enough consciousness and plasticity, I experienced toxic levels of stress, anxiety, and traumas that started very early in my childhood. That emotional baggage ultimately morphed into patterns of depressed moods and addictions as I became an adult. It took me decades to understand and accept that *the Savior Complex* caused my suffering. As a result, I could not consider new paradigms of healing. I was in a complete state of denial that my stress and anxiety were self-inflicted. The breakthrough happened once I started to engage with medicinal plant-based treatment modalities that raised and sustained new levels of awareness about my condition.

Throughout my healing journey, my chronic lack of openness not only hindered my progress but also impacted many of my family members. In this reflection, which blends biographical insights and current experiences, I explore why neuroplasticity has become central to my wellness practices. These include rituals and personal work that challenge me to explore consciousness fully. A few years ago, spending an entire month in Peru would have been unimaginable. However, as you follow my daily account of the program I recently attended, you will see how I embrace new experiences with a single goal: to live a life grounded in high consciousness, mental flexibility, and optimal health. Staying open is now my guiding principle.

Following the day-by-day account of my retreat in Peru, you will also learn how ancient traditions can provide relief and reset many SADAT conditions. Not only do I share information about my experience during the shamanic initiation, but I comment on the progress of the participants who attended this program with me. This way, you can fully appreciate how the Shipibo approach can impact so many lives simultaneously, even though the SADAT challenges are all completely different in scope and severity. Through my eyes and theirs, I hope you will understand that these ancient healing modalities are truly extraordinary. Two stories create the narrative structure of this chapter.

The first story is about my battle with *the Savior Complex* and the related toxic stress, anxiety, addictions, and depression I endured for decades. I made the difficult but necessary choice to describe personal struggles that have impacted me and many of my family members. I believe that sharing these personal experiences makes unpacking the science behind the central topic of brain plasticity less intimidating and more tangible. By anchoring your attention on the emotional nature of my healing journey, I trust it will help you appreciate the damage that constant worries and feelings of despair create in our lives. I am not ashamed to share intimate life details if they can help you and many others heal deep wounds. My personal story and the many people I have met on similar journeys prove that there is a path to relief that revolutionizes our understanding of SADAT.

The second story (*italicized in the text*) provides a day-to-day account of the 26-day shamanic retreat I completed in Peru in November 2023. I chose this challenging program to deepen my connection with the medicinal and spiritual traditions of the *Shipibo*, an indigenous tribe of the Amazon. I also wanted to ensure my research for this new book would not be filtered through my lenses as a psychologist and neuroscientist but also include the remarkable teachings of ancient plant-based approaches to healing the body and the soul. I share notes I wrote after attending ten Ayahuasca ceremonies to give you a real-life version of the effect psychedelics can have on our consciousness. This ritual has been performed in the Amazon for thousands of years, one that has gained popularity among Westerners in the last decade but needs to be better understood.

The parallel processing of these two stories provides context as to why this book is so important to me personally and, I hope, to the millions of people affected by toxic levels of SADAT. Having studied the effect of stories on the brain for over two decades, I know how powerful narratives are to excite neurons and stimulate emotions. Stories will likely improve your retention of *OPEN's* teachings since we retain stories with far less cognitive effort than facts!

Additionally, I openly discuss my medicinal use of psychedelics and their therapeutic benefits. The field of *psychedelic-assisted therapy* (PAT) is still in its infancy and remains hotly debated. However, psychedelic-assisted therapy has expanded rapidly worldwide (20). The movement is critical in advancing therapeutic options that complement or replace conventional solutions to address SADAT. While many people continue to see psychedelics as harmful, some psychedelic-assisted therapy options have been authorized by the FDA already, for instance, the use of *ketamine* for treatment-resistant depression, chronic pain management, and PTSD.

An *MDMA* protocol was expected to be approved in 2024, but the FDA rejected the application, claiming that more studies are needed to demonstrate the safety and efficacy of the treatment (21). Psilocybin is also under FDA consideration, but the timetable is unclear. Current clinical trials, particularly those targeting *major depressive disorder* and treatment-resistant depression, have shown promising results. For example, studies involving a novel psilocybin analog, CYB003, have demonstrated robust and sustained improvements in depressive symptoms, with a significant percentage of participants achieving remission. The ongoing positive outcomes from these trials suggest that psilocybin could become a viable treatment option within the next few years if phase 3 trials continue to confirm its efficacy and safety. Undeniably, psychedelic-assisted therapy is gaining traction, excitement, and support among growing members of the medical and psychological community, even though the legalization process is slow.

Chapter TWO Summary: The Self-Healing Power of Consciousness and Neuroplasticity

The two stories shared in Chapter ONE provide the context for a comprehensive discussion on the critical importance of consciousness and plasticity. Both topics are seldom explored together because of many competing views from evolving scientific and spiritual perspectives. So, in Chapter TWO, I share practical definitions of the most critical building blocks behind our instinctual, emotional, cognitive, and spiritual states. Then, I present the latest research on the neuroscience of SADAT and its relationship to consciousness and neuroplasticity.

Diving into some of the scientific concepts highlighted in the book can sometimes be challenging. However, the general framework of this chapter does not require prior knowledge of the brain and its complex intricacies. Meanwhile, as a frequent speaker on consumer neuroscience, media psychology, persuasion science, personality psychology, and mental health, I know too well how important it is to reduce cognitive load to keep my audience engaged. Thus, I made Chapter TWO "easy on your brain." I included many tables, graphs, and illustrations to help you pause, reflect, and digest the content with limited effort. I am convinced you will appreciate gaining a solid scientific foundation upon which you can learn why the neurospiritual model of *OPEN* works.

Chapter THREE Summary: OPEN Your Mind

In this chapter, I present the role of consciousness and its relationship to raising plasticity. A simple definition of consciousness is our ability to observe our experiences. Consciousness research is complex because of the many fields involved in investigating it and the equally plentiful definitions used to describe it. As established in Chapter TWO, consciousness and plasticity can explain and predict many maladaptive behaviors underlying SADAT. For instance, SADAT may keep you at sub-optimal levels of awareness most of the time, compromising your ability to self-diagnose or rewire toxic patterns. Therefore, I make the construct of consciousness as practical as possible since it is central to understanding how mind states and perspectives shape our reactions and attitudes toward our mental health.

Meanwhile, this chapter aims to describe some of the most potent ways to achieve higher levels of consciousness (meta and hyper levels), states through which you can learn to observe your thoughts and feelings rather than be entrapped in them. However, this process can be challenging if you are unaware of the impact of numerous *barriers to consciousness*. This chapter will help you quickly realize the degree to which you may fall into the deep holes of low consciousness because of *ego resistance*, *projections*, *self-limiting beliefs*, *cognitive biases*, *self-doubt, and distractions*. Finally, the role of psychedelics and natural healing traditions is discussed extensively to demonstrate how these modalities can achieve impressive gains in consciousness at record speed.

Chapter FOUR Summary: OPEN Your Self

Our mind is full of conflicting narratives that are difficult to process when energy is spent attending to demands that exceed our resources (stress, trauma, addiction), when predictions hold too much uncertainty (anxiety), or when moods that color our life are sadness or despair (depression). Once you have raised your consciousness, however, you can clear your past emotional marking, limiting beliefs, negative self-talk, and other *monkey mind* patterns that are known to feed or worsen SADAT. Only then can you leverage your innate capacity to reclaim your story. Uncovering its positive thread will free mental energy to reprogram old *coping mechanisms* that no longer serve you. To help you in this critical process, I will discuss two personality models that offer remarkable insights to identify and rewire maladaptive traits you may have inherited or adopted over a long period. The field of personality neuroscience has evolved considerably in the last decade, and it confirms that you can reprogram any patterns of behavior or traits that tend to lower your consciousness and worsen the symptoms of SADAT (22). Finally, I will discuss the power of manifestation and the placebo effect. Both topics demonstrate remarkable healing properties that *materialistic scientists* have unfortunately dismissed for decades.

Chapter FIVE Summary: OPEN Your Life

Chapter Five focuses on adopting *OPEN* rituals that are critically important to protecting your mind from relapsing into SADAT conditions. This is the process and practice through which you can reprogram your routines to solidify new pathways of consciousness, plasticity, and flow.

My review of pertinent literature reveals that *OPEN* rituals can suspend our ego to free a large amount of brain energy blocked by your **Default Mode Network (*DMN*)** while reinforcing the encoding of new patterns that serve you. The default mode network is active when you focus on your inside world—like daydreaming, thinking about yourself, remembering the past, or planning for the future. It is like the brain's "idle mode" that kicks in when you are not doing a task that requires your outward attention. Finally, this chapter also explains how working with *entheogens* may benefit meta-consciousness and plasticity.

Flow Into Your OPEN Future

The conclusion of *OPEN* emphasizes the transformative power of neuroplasticity and consciousness in achieving mental well-being, resilience, and flow. Drawing from scientific insights and personal narratives, I illustrate how everyday practices—like breathing exercises, creativity, and spiritual rituals—can cultivate adaptive changes in the brain. These *OPEN rituals* enhance cognitive flexibility and emotional regulation, helping to alleviate stress, anxiety, and trauma. By consistently engaging with these rituals, you can nurture neuroplasticity to strengthen mental health and deepen self-awareness.

Finally, I discuss the holistic nature of healing, showing how integrating activities like meditation, stargazing, and music can reinforce positive mental patterns and create balance. Through a lifelong commitment to self-discovery and growth, I suggest that you can achieve flow—a state of deep immersion that promotes optimal brain function and fosters joy. Consciousness, raised through mindfulness and therapeutic experiences, is framed as essential for reframing personal narratives and building empathy. Ultimately, the journey toward mental resilience and fulfillment lies in purposeful, daily engagement with these neuroplastic practices, empowering each person to shape their life dynamically.

CHAPTER ONE

Living Fully OPEN

"The wound is the place where the Light enters you."

--Rumi, Mystic and Poet

Even though the air conditioner blasted cold air in my face all night, I slept well on my first night in Peru. I feel rested from the long trip. Since I still have a decent internet connection, I quickly checked my emails and caught the latest news, especially updates on Ukraine and the Middle East wars. I realize how much I feel the need to understand why these terrifying events are happening and why there is so much hate between the leaders and the populations involved. How many lives must be lost before we begin to comprehend the human madness in these horrors? I feel helpless in front of such suffering, which is very unpleasant. I sense the urge to do something that can change human consciousness.

Fortunately, over the last ten years, I have found ways to accept what I cannot control, which means witnessing or " holding" terrifying events without absorbing the pain and sorrow of innocent victims in the process. Most importantly, my work with plant medicine has sharply increased my emotional resilience and enabled me to return to neutral at will. Therefore, I see the next step of my life journey as a relentless pursuit of more awareness to eliminate what is left of my old unconscious pattern: the Savior Complex.

Today, we are going to the Belén market. This is my second time visiting this iconic place.

Figure 5: Motorcar in Iquitos, Peru

We hailed a motorcar right out of our hotel. Motorcars are not just a mode of transport in Iquitos; they are a cultural emblem, a testament to the Peruvian spirit of ingenuity and resilience. In the city, they weave seamlessly through the congested roads, their nimble form allowing them to slip through gaps where larger vehicles dare not venture. The drivers, often local entrepreneurs, adorn their rides with personal touches – stickers of favorite football teams, religious icons, or even family photos—turning each motorcar into a moving tapestry of individual stories.

Motorcars are the connective tissue linking remote communities. They traverse roads less traveled, bringing goods, news, and sometimes just a friendly chat to the more isolated corners of the city. Their affordability and practicality make them accessible to a broad spectrum of society, fostering a sense of independence and mobility. Riding a motorcar is an adventure.

The open-air experience, the wind in your hair, the unfiltered sounds and sights of Peru passing by–an immersion into the daily rhythm of life here. The motorcar embodies the Peruvian approach to life: vibrant, resourceful, and always in motion, a dance of survival and celebration amidst the diverse landscapes of this enchanting country. After a short and fun ride to the market, we see the banner of the entrance at a distance and the crowds infiltrating this fascinating cluster of stands and shops. As I hop off my ride, with the engine's echo fading into the distance, I realize these motorcars are not just vehicles; they are the pulsing heartbeat of Peru.

As I meander through the labyrinthine aisles of Belén Market with my friends, I am engulfed by a sensory mélange that encapsulates the essence of the Amazon. The market brims with life, its stalls bursting with exotic fruits, chicken, fish from the river, and curious bottles of traditional medicines. The sounds of haggling mixed with the rich aromas of Amazonian cuisine create a vibrant atmosphere with the pulse of local culture. This is the market's charm, a raw and authentic immersion into the heart of the Amazon, offering a treasure trove of unique finds and gastronomic adventures.

However, visiting the market can be challenging. The bustling crowds are quickly overwhelming, and navigating the narrow passageways tests your patience. Also, dealing with vendors can be intense for the uninitiated, not to mention the risk of being robbed by skilled pickpockets. In 2015, I was alerted by local police to avoid venturing alone into some of the corners of the market. The worst part of navigating the market is the hygiene standards, especially around the food stalls and fish vendors. To put it bluntly, they are a stark reminder of the market's raw, unrefined nature. Despite these drawbacks, the Belén market is an unforgettable spectacle, fully displaying Iquitos's vibrant spirit and resilience.

Figure 6: The Floating Market, Iquitos, Peru

I am excited to return to a place that etched many incredible memories in my brain. Noticing how much I need a goal whenever I go anywhere, I decide to find handcrafted pipes for rapé ceremonies. I discovered the sacred tobacco several years ago and have used it occasionally as part of my plant medicine rituals. Following our guide, I enter a small smoke shop full of local craft. There, I quickly find a wide variety of wood pipes and an impressive array of rapé blends. Enjoying the ambiance of this treasure hunt, I pick a blend made of bobinsana, commonly called "Maestros in the Amazon," which is considered essential for healers. The ancient tradition of using rapé is sacred among the indigenous tribes. Shamans prepare the ceremonial blend, a mystical concoction of Nicotiana Rustica, and other medicinal herbs to imbue the plant's spirit.

With each application, blown swiftly but gently in your nostrils through a carved pipe or Tepí, you might feel an immediate rush, a forceful yet grounding connection to the earth beneath and the nature around you. After buying the bobinsana snuff, we explore the floating village next to the market. When I visited Peru eight years ago in March, the town was perched on wooden stilts over the river. However, in the dry season, the river is low, and all the houses are no longer surrounded by water, so we take another motorcar to reach the river. These colorful, three-wheeled vehicles buzzed everywhere like a swarm of industrious bees. In every hue imaginable, from bright yellows to deep blues, they adorn the streets and dirt roads, each a burst of life amidst the traffic chaos.

Amidst this picturesque scene, the stark realities of life on the water are evident. Still, the harsh living conditions are also a study of human ingenuity. Houses, often rudimentary and constructed from local materials, bore the marks of the river's whims – some tilted slightly, others showing water stains from seasonal floods. Basic amenities are never taken for granted here. They are luxuries since access to clean water and proper sanitation are perennial challenges. The community's tenacity in facing these hardships is humbling and inspiring. Their lives, deeply intertwined with the river, now critically polluted, are a testament to the enduring spirit of humanity and its ability to adapt and thrive in even the most challenging environments. In the floating market of Belén, life is not just about survival; it is about living in harmony with the unpredictable rhythms of the Amazon.

Wrapping up our busy day of exploring the Belén market, I must regroup and ground myself for what comes next. I am grateful for the visit. As we reach our hotel, surviving our fourth motorcar ride in just a few hours, I invite our friends to share this spiritual voyage of the soul to join us for a rapé ceremony. I tell them it is a good way to strip away the layers of mundane worries and open the pathways to deeper consciousness. We go to my room and find comfortable sitting positions. The ceremony begins.

As soon as you inhale the curious tobacco, intense physical reactions, though momentarily overwhelming, provide cathartic releases, purifying the *body and soul. Rapé brings me back to clarity and focus in these moments, sharpening my senses and heightening my awareness of the intricate tapestry of life surrounding me. Also, the communal aspect of the rapé ceremony fosters a profound sense of unity and empathy. Sharing this potent experience with our friends, sitting in a circle enveloped in the sacred aura of the Amazon, our spirits become woven together in a silent, unspoken bond. The rapé snuff, I realize, is more than a plant medicine; it is a key to unlocking a more profound spiritual connection with nature, community, and my inner self.*

Each rapé ceremony has contributed to my spiritual journey of introspection and discovery to achieve my plasticity. It is a reminder of the delicate balance between the physical and spiritual realms and a testament to ancient wisdom that continues to thrive in the heart of the Amazon. A journey with rapé is not just an encounter with a traditional practice but a transformative pilgrimage into the depths of spiritual awakening and the ongoing pursuit of more openness.

Living With Brain Deficits

My father was forced to quit high school at a very young age to run the family's small automobile repair shop. As part of a respected French program called *Les Compagnons du Devoir et du Tour de France* (also referred to as "Les Companions") offered to young apprentices of his trade, he traveled throughout France, primarily by foot and hitchhiking, to earn his credentials as a master body shop craftsman. This expertise eventually made him one of his generation's most brilliant automobile inventors.

I was born as the third child in the family after my parents had a girl and a boy who were already 6 and 5, respectively. My older sister became a nurse, my brother a social worker, and my younger sister a professional storyteller. I was the only child interested in working for the impressive business my father created and ran for over 30 years. He grew it fast despite limited resources. Moreover, it all started when he inherited the tiny body shop from his father. Sadly, his dad died in his early fifties of cancer. His mother also passed away in her early fifties. I never got to know either of them. Assuming the responsibility of their small business alone before he turned 18, my father quickly transformed it into the world's largest manufacturer of mobile retail trucks and airport ground equipment. He called it *SOVAM, Societe des Véhicules Andre Morin.*

Driven by a passion for disrupting the automobile industry, my father boldly attempted to make affordable sports cars, also called SOVAM, molded from high-density resin in the mid-sixties. At the time, no other large automakers had used this technology. Alain Prost, a famous F1 world champion, performed some of his early rallies driving a SOVAM. While the stylish automobile was remarkable and earned distinctions in auto shows, the business venture was a flop. To survive, my father had to swiftly diversify his product line to manufacture airport equipment for transporting passengers and luggage. Despite developing and building innovative products, his company went bankrupt in the early eighties, running out of cash to finance the manufacturing and marketing of too many brands. Nonetheless, it still operates successfully today.

Figure 7: SOVAM (1965)

I remember my father was always busy working on his latest invention until I left for college. As a result, I spent most of my childhood alone since my older sister and brother were sent to boarding school before I was 8. Fortunately for my parents, I was the kid who always did his homework without help. I knew I would not cause any trouble if I got good grades. I was especially cautious not to disturb my father. He was the dominant person in the family, even though my mother, I learned later, had been instrumental in starting and running the business, especially taking care of accounting. Since my father was a risk-taker with no formal education, my mother was constantly terrified by his bold moves. Sadly, her fears and her worst nightmare would ultimately materialize.

In 1962, he had a terrible accident that propelled him out of his car through the windshield. He would have never survived if he had been wearing a seat belt. I have often wondered if that explains why I resist wearing my seat belt. It was truly miraculous that he survived the accident. However, he stayed in a coma for nearly two months while my mother was expecting my younger sister. As a result of the traumatic brain injury he suffered, my father struggled with deficits in his short-term memory and had temper issues for the rest of his adult life. I suspect he damaged some nerve fibers in his left frontal lobe, the part of his brain that hit the windshield. This area is critical for processing short-term memory and controlling impulsivity. By now, you may have guessed that my passion for neuroscience was partially fueled by the need and desire to decode my father's psychological and emotional conditions! Much more on that later.

33

DAY 2: Iquitos PERU, November 10, 2023

After a 2-hour bus ride, we arrive at the retreat center in the jungle. We met the group participating in the program yesterday: a joyful mix of young and mature folks like me, all seeking the same level of healing and connection with spirit, all determined to learn and be guided by ancient traditions from the Amazon. I chatted with many people, even though I am usually less social. The irony of my core personality is that I am highly introverted, yet I have spent most of my life speaking in front of groups and crowds!

Today is different, however. I feel like I am traveling with old friends. The collective excitement and fear are palpable in our conversations during the ride. On our way to the center of Iquitos, we quickly stopped to enjoy some coconut juice freshly prepared by locals. Finally, after 2 ½ hours of hectic driving, the bus pulls off to the side of the road as if it were out of gas. This is the end of our journey. We must walk 30 minutes to the Shipibo village, where Aya Healing Retreats (AHR) is located. As we hike along the dirt road, I breathe in the earthy scent of the jungle and take in the lush greenery of the forest, yet I cannot help but wonder if I will be truly safe in the wilds of the Amazon. I look forward to meeting the owner of AHR, Elio, who welcomes me with a big smile. Back in September 2019, I attended a challenging Aya retreat Elio organized in Ibiza. Since then, Elio started AHR in partnership with a highly respected Shipibo shaman, a fantastic accomplishment in just a few years, during which he also survived the severe business blow of the pandemic.

Born in Italy, this 40-year-old man is the poster child of the entrepreneur of the future. Intelligent, humble, and incredibly sensitive, Elio is

Figure 8: My Tambo at AHR

the conductor of a beautiful orchestra of locals who trust him to bring committed and respectful visitors to the jungle. AHR is not a retreat center for casual Ayahuasca tourists. This expression describes the many people who believe that the ancient brew can solve all their problems in one weekend. The facility also bears the beautiful name of El Canto Chuyachaki. AHR offers rustic but effective lodging options for all participants to enjoy privacy in small huts called tambos. After all, what more than a bed with a mosquito net, a desk, a chair, a toilet and a shower do we need here?

Each tambo has space for a hammock to rest and relax between workshops, plant treatments, and ceremonies. The construction is typical of a Peruvian Shipibo village, with many structures occupied by the landowner's family, the maestro who will guide our Aya sessions: Don Miguel Lopez.

Don Miguel also welcomes us with a big smile on his face. He introduces us to his team one by one. I immediately feel immense gratitude for this opportunity to share space with Don Miguel Lopez, his family, Elio, and the AHR crew. We are in good hands. On this voyage, I am accompanied by three dear friends who have done extensive transformative work with us over the last two years. I know they would never have made the bold decision to attend a 26-day retreat in the Amazonian Jungle without psychological preparation but also trust and faith in my judgment of Elio.

DAY 3: Iquitos PERU, November 11, 2023

The agenda of day 1 calls for a light day in activities and teachings. First, we receive a general introduction to the program that helps us manage our expectations. Elio is firm sometimes, insisting that aspects of the retreat are not pleasant. "You will find that you want to quit at some point, possibly after a few days. However, during your stay, I ask you to trust the maestro (Don Miguel), especially Ayahuasca. We are here to help you heal, to help you strengthen your mental clarity and connect with the divine nature of the Jungle."

Today is the first day of my total commitment to a life cut off from processed food, coffee, constant news feeds, emails, and text messages. I am still wondering to which extent I can continue to check my emails and news. It is incredible how much we can sometimes deny our reality. "Repeat after me," Christophe says the little voice in my head: "It will not be possible to check your emails, send text messages, or check the news here in the thick of the Amazonian Jungle!" What seems like a punishment is a gift in disguise.

I settled in my tambo yesterday. There is also additional space to practice yoga. When I first moved to Hawaii, my wife and I had a daily yoga routine, which we kept for over 5 years but eventually abandoned. We miss the mindful practice and are both looking forward to resuming it. The tradition of a plant dieta insists on isolating all participants, including couples. Fortunately, we are still next to each other. Sleeping apart is enough to meet the bar! Don Miguel reminded us that we need to continue to abstain from specific foods and sex or masturbation during the retreat but also two weeks after the dieta officially closes.

After lunch, Elio invites the group to walk to a beach on a river and take a refreshing dip. The air temperature typically fluctuates from pleasant in the early morning to hot and humid until 4 p.m., when you can finally enjoy a refreshing breeze. It is challenging to handle the humidity, but living in Hawaii is a plus, as we are used to it. Other participants who are coming from cold conditions are less lucky!

When we return to the village, I meet with Don Miguel to discuss my personalized plant treatment and open our master plant dieta. This tradition of dieta has always fascinated me but also made me nervous before coming. Don Miguel offers four master plants to choose from. Since I had bought the bobinsana tobacco a day earlier at the Iquitos market, I requested to diet bobinsana, and he agreed by formerly opening the dieta. Don Miguel asks me to start communicating and invoking the plant's spirit by drinking brews made from its extracts several times daily.

Bobinsana grows between 4 and 6 feet and has been used by Peruvian shamans for thousands of years to heal wounds of the heart while helping those doing the diet to feel more love and empathy. It can also strengthen boundaries against the energies of others. Finally, it is known to boost creativity. I could not think of a better match, given my addiction to the Savior Complex. We close the day by visiting a Noya Rao tree after sunset. The energy around this rare and majestic creation of nature is mesmerizing, mainly because the leaves of Noya Rao exhibit fluorescence due to the presence of specific compounds that react to ultraviolet light. This makes the ancient tree even more mystical. This phenomenon is a fascinating example of how plants can develop unique chemical and physical properties, possibly due to evolutionary pressures or specific ecological roles.

DAY 4: Iquitos PERU, November 12, 2023

This is the day things are starting to take shape. First, we started with a vomitivo session. As the name suggests, we were all invited to drink a plant mix that activates intense and rapid vomiting if we chug four liters of water in just a few minutes. I was familiar with the ritual, having done dozens of such sessions with a mapacho concoction prepared to create the same effect. I knew detoxifying the body and preparing us for our first Aya ceremonies was standard practice. I admit that being in the presence of primarily strangers for this ritual can be awkward, if not wholly embarrassing. However, most participants had already done it in one form or another. We completed the routine in less than 20 minutes, laughing and happy to pass the first step!

Then, we were reminded to fast the entire day. I have never done this in my whole life, so I am nervous. Alternating drinks of bobinsana and water makes it easy, however. Starting a plant dieta is a very formal process in the Shipibo tradition. It is the only way to receive the spirits of the plant and requires following a strict eating routine while continuing to abstain from alcohol and sex. It is also essential to maintain a meketi attitude. The meketi attitude in Shipibo culture refers to a principle or way of being that emphasizes living in harmony with oneself, others, and the environment. It involves living in a way that is in balance with the natural world and the community.

This includes a deep respect for the environment, understanding its cycles, and taking only what is needed. Meketi also refers to an inner balance and peace. It suggests a way of being that is centered and grounded, allowing one to navigate life's challenges with calmness and wisdom. This attitude extends to interactions with others, promoting cooperation, mutual respect, and community well-being. The Shipibo people place a strong emphasis on communal living and shared responsibilities. Meketi is also connected to the spiritual beliefs of the Shipibo, who have a rich tradition of shamanism and plant medicine. Living in a way that is aligned with Shipibo's spiritual practices and beliefs is an integral part of Meketi.

The resistance to many Shipibo practices was evident initially from many participants, including me. However, I have one firm intention: being open at all costs. So, I find the courage to comply with and follow the treatment plan. I am also prescribed to drink a plant that grows everywhere on my property in Hawaii called tanti rao or sensitiva. It is made of tiny leaves that retract when you touch them as if they were shy or eager to interact with you. The plant offers multiple benefits for emotional stability and blood circulation. I told the Maestro during our one-on-one session that I was having debilitating leg cramps at night, and so this is one of the many plants he prescribed to treat the condition. I rested until we were called into the maloca, the sacred place for all our Ayahuasca ceremonies.

I am not as nervous about our first ceremony as I had expected. I already feel the protection of the mapacho and the bobinsana mixture. Also, I have complete trust in Elio, his crew, and Don Miguel.

Aya Session #1

As I prepare my space in the maloca, my anxiety suddenly spikes. I am back in Peru, where I had very challenging Aya sessions nearly a decade ago. The retreat center where I stayed back then could have been better. We were all packed into a tiny house for all the ceremonies and sleeping arrangements, with *no option to walk more than a few steps outside in the jungle without putting ourselves at risk from highly poisonous snakes and other wild creatures. We had limited food and no access to a functioning shower. While an exceptional curandero delivered the Aya sessions then, I still had a bad taste of the Peruvian jungle ceremonial experience.*

Washing the memory of that episode from my brain lasted a few minutes. I took a few deep breaths to reset my nervous system. After all, I have been through 50 Aya sessions since 2015, and this retreat center is light-years better, offering individual tambos, a beautiful maloca, a convenient kitchen, a medicine house, and plenty of space to walk around medicinal plants. It is more than I ever expected. Also, I have complete trust in Elio, and I can tell we are in the hands of an exceptional maestro.

*Besides, going through a vomitivo, fasting for a day, and starting our plant dieta has primed my body and soul to receive the brew. Also, my dieta provides another unique opportunity to create a spiritual relationship with Aya. So, as I wait for the maestro to call me for my dose, I hold on to my clarity and peace. I am protected by the energy of bobinsana, of course, but also **mapacho**, a strong tobacco used in many ceremonies to provide guidance and protection.*

I ask for a small cup of Aya, assuming the effect will be mild. However, unlike other sessions during which the effect was often overwhelming and scary, the ramp-up is slow and easy to adjust physically and emotionally. Within about 20 minutes, however, I started getting the first visuals. The beginning of my hallucinations with Aya is often messy and challenging to make sense of. However, this time, I asked her for guidance in answering essential questions.

A surprising creation story

Figure 9: A Creation Story (AI)

The first narrative is unexpected. Aya transports me quickly to stunning scenes of the cosmos, in which massive structures appear to me. I see that millions of planets formed the shapes of the structures. However, once I look closer, I realize this cosmic display is meant to set the stage for something big. Aya wants me to understand Earth's creation story. She helps me see that our world was created by star people billions of years ago as an experiment.

As the vision unfolds, I encounter alien creatures belonging to different fractions that do not agree on how our planet should evolve naturally and socially. Some want to rule using evil forces; others prefer peace and harmony. I have never paid attention to narratives that place aliens at the center of the creation of our planet. So, I received this first story with skepticism. However, by raising my plasticity with the help of Aya, I was able to release judgment until the end.

A family loss

The second narrative starts with no transition. It is about the dynamics of my family of origin, especially my parents. Again, I appreciate that my father was such a dominant force, but I realize now that I rarely questioned his authority. While I genuinely believe he was an exceptional man who did the best he could, I can feel today that many of his decisions devastated my mother and affected all of us in profound ways.

Figure 10: A Family Loss (AI)

For instance, my father's priority for most of his life was growing his business, not raising children. Since my older sister and brother were somewhat of a distraction to his goal, he decided, presumably with my mother's blessing, that they both be placed in religious boarding schools at a very young age. Neither my father nor my mother was raised religiously and did not attend church. I can see now that this choice was most likely motivated by the desire to "outsource" my siblings' education and ensure they would be under tight control. As this narrative unfolds, I realize I was cut off from my older sister and older brother for most of my teenage years. I feel profound sadness upon this unexpected epiphany.

Additionally, I have a new understanding that I was forced to play the oldest sibling's role without any desire to do so. While this reframe is challenging to integrate, it helps me grieve my unconscious loss and resistance to becoming the "second" first child of the family.

Meanwhile, with no warning, the narrative shifts to a different topic. This time, Aya tells me not to be afraid to terminate activities that compromise my ability to find more spaciousness. Next, she works on my right knee, which has caused me so much pain over the last few years, and confirms that I am healthy overall. She insists that I can work and make money for many years to come because of the value of what I do. Finally, she tells me that my sons are doing fine and require very little guidance and support from me at this point in their lives.

As challenging as the ceremony was, the closing was enchanting; listening to Elio and his assistant Anna's delightful singing brought the atmosphere back to joy and peaceful closure. I feel rested and happy with my first session, even going through difficult emotions and some expected purging, a process often regarded by first-time drinkers as an adverse effect of Aya when it is, in fact, one of the central benefits of its healing power. Resetting the body and mind before integrating the teachings is essential.

Saving the Day

A few years after his dramatic car accident, my father was driving a new model of his sports car (a SOVAM) and missed a turn, most likely because he was speeding. This time, however, thanks to the ingenious technology he designed to make the car's body practically indestructible, his SOVAM sectioned an electrical wood pole with hardly any scratch. He walked out of this one without injury. I do not remember this event since I was only a few years old, but later, I do recall how none of us were very excited when my father drove us anywhere. As I entered my teenage years, I would often remind myself that he was the survivor of two accidents that could have killed him, one leaving noticeable cognitive and emotional scars. Still, I did not fully understand the consequences of life they left him with until I became a neuroscientist. The accidents clearly explained his struggles with memory and temper.

So, he learned to live with severe cognitive deficits without complaining. He always carried a small notepad to write down to-dos and ideas in his shirt pocket. He would also wake up several times at night to note critical insights that would become central to his inventions. We often joked about his behavior during family gatherings. However, while he was calm and composed most of the time, we all feared he might explode in the scariest outburst of anger with limited warning, a byproduct of his traumatic brain injuries. This reaction would happen if he felt disrespected or we woke him up from his nap. I remember one day vividly; I must have been 15. I brought him his coffee while he took his sacred 30-minute nap at lunchtime. I accidentally bumped his leg doing so and received a punch in my face. The incident devastated him, and he was stunned when it happened. It devastated me, however. I felt awful; disturbing his precious moment of rest was not my intention.

In another dramatic anger episode, he chased my older sister after she defied my mother during a family dinner. I tried to intervene as he destroyed the door to the bathroom, which my sister had taken refuge in. He also pushed my mother towards the stairway as she was trying to calm him down. Fortunately, I managed to prevent her from falling down several steps. I was only 12 years old when that happened. While these events were few and far between, they caused uneasy tension for many years. For instance, I would never invite friends to my house because I was unsure if my father would be working, resting, or if he was in a good mood or upset. Disturbing him or making noise was the ultimate threat to his emotional balance. One day, which is etched in my memory, he grabbed a rotary phone ringing non-stop and threw it out of the house in fury, smashing a window with the force of his throw.

Nevertheless, beyond his emotional instability, my father was also in agony when he needed to write more than one paragraph. This was not a good time to bother him, either. He would lock himself in his office for hours, struggling with the composition of a letter or a talk in front of all his employees. All this explains why my mother was so essential to the success of his career. She did all the writing and typing for him but also provided the cognitive legwork he could not perform. By doing so, she enabled him to shine in front of his workers, a workforce reaching almost 800 employees by the mid-70s. In 1979, my father received one of the most prestigious distinctions a French citizen could ever receive (*L'Ordre National du Mérite*), an award that sadly did not bear my mother's name. I felt she deserved it as much as he did in many ways.

Unlike my siblings, I loved spending time in the factory at a young age, especially watching my father in action. He had this incredible energy to command the attention and respect of all his workers. He was one of them, after all. He, too, learned the essential skills that make the magic of transforming steel into unique shapes. So, while wandering throughout the factory, he could grab a tool and show anyone how to use it. Most feared him because he could see the tiniest finishing flaws. I saw him repeatedly point out problems, defects, and opportunities no one else could see. He was firm and often compulsively obsessive but would command admiration and respect at all levels in his company.

DAY 5: Iquitos PERU, November 13, 2023

Meanwhile, for the last four days, I have been suffering from severe headaches, and I had hoped that by now, they would have gone away. I decided to see the bone doctor (Teo) for help. In the Shipibo tradition, a "bone doctor" is often referred to as a huesero, a traditional healer or shaman who specializes in treating physical ailments related to the bones, joints, and muscles. The huesero combines massage, bone manipulation, and spiritual practices to treat conditions like fractures, dislocations, arthritis, and other musculoskeletal issues. This practice is more than physical; it also involves spiritual elements, as the Shipibo believe physical ailments have spiritual causes. The role of the huesero is highly respected in the community, as they possess knowledge passed down through generations, blending practical skills with spiritual wisdom to restore health and balance to the body and spirit.

When I arrived for my appointment, Teo offered to perform a deep tissue massage on my neck. The first minutes were painful, but I felt a significant release. My next Aya session would eventually pulverize whatever stress was left. In past sessions, whenever I had headaches, I would often spin into frustration and fear, which would make them worse. Part of this trip was about noticing how I face discomfort or sickness and to which extent I resist those experiences. I can see today how surrendering to what happens leads to better outcomes than trying to block fear.

After the massage, I drank my master plant brew (Bobinsana) and the tanti rao several times today. Unlike Aya, both are easy to swallow. I also bought a pipe made of Bobinsana wood from one of the Shipibo villagers. I plan to use it for all ceremonies and rituals from now on. In the afternoon, we were invited to take a bath infused with our master plant. This tradition is another opportunity to connect to its spirit. It is refreshing but also enhances our sensory experience of the sacred plant.

I closed the day feeling rested, grounded, and ready for what comes next. Journaling helps me integrate my experience. Moving from one activity to the next feels so good without considering how much time has passed. After years and years of intense work and travel schedule, I finally taste what it means to be unattached to goals or projects. In just four days, the protocols prescribed by the Shipibo tradition have taken me out of this continuous loop from which comes many worries and stress. Now, I can start reprogramming old patterns that do not serve me.

DAY 6: Iquitos PERU, November 14, 2023

I woke up at 5 am when the light was already intense. I love starting my days early, regardless of how much sleep I have had. The night was busy with dreams and resetting my capricious bedding, especially the mosquito net. We had decided against taking malaria pills because the area we are in has no known cases. Usually, I get easily worried about not getting a good sleep score from my Garmin watch. I know too well how poor sleep can affect memory as well as put me at risk for cognitive decline. Conveniently, my sleeping scores and a long list of other biometrics I tend to review daily are not computed on my watch unless I am connected to the internet. I will have to wait another 20 days to do so!

I feel rested despite getting less than my usual 7 hours per night. Today is a day of restoration, as it will be after each Aya ceremony. I embrace the routine and welcome each opportunity to diet my master and treatment plants. I also visited the beautiful medicinal garden early this morning, a special time to smell and listen to the jungle awakening. Finally, I decided to write and sing an **icaro** to honor Bobinsana, as suggested by our Maestro. What was once an interesting, if not curious, practice became a creative and somatic experience I had previously underestimated. This is worth unpacking further. What Is an Icaro?

The term icaro derives etymologically from the Quechua verb "ikaray", connoting the act of "dispensing curative properties through the smoke." In a broad ontological sense, Icaros are songs imbued to mitigate physiological, affective, and spiritual afflictions. Icaros are designed to invoke the spirits of plants, engendering healing, safeguarding against evil influences, and, intriguingly, even inciting the attraction of romantic affinities. The intricate and multifaceted therapeutic potential of icaros has fascinated researchers, scholars, and healing practitioners for hundreds of years. The creation and use of icaros are inextricably intertwined with Amazonian shamanism.

Icaros played a pivotal role in my plant dieta. Creating and singing an icaro helped me cultivate a heightened sense of interconnectedness and communicative resonance with Bobinsana. Research inquiries into the tradition of icaros have predominantly gravitated toward spiritual protocols and their resultant impacts on the outcomes of shamanic rites. Their vocalization appears to transcend the exclusive purview of spiritual or ceremonial properties.

Here is an example of an Icaro sung by Don Miguel, the Shipibo shaman dedicated to our health and transformation during the retreat. You can access the video of this performance at https://tinyurl.com/zu5ksph2.

Below are the translated lyrics invoking ancestors and spirits so that they can protect and heal the people present in an Ayahuasca ceremony:

NOCON SHAWAN // Welcome / calling the ancestors
NOCON SHAWAN CAIBOBO // Welcome family from afar
NONRA MATO JOWE ACAI // Now the spirits welcome you
NESKA NESKA SHAMANKIN // I call you to this healing place
MATO JOWE ACAI // I call the spirits to heal you
MATO JOWE AYUNSHON // With this prayer, I will heal you
MATO ISHON BANONKIN // Listen to this healing prayer
NESKA NESKA RANIKE // My ancestors heal the sick
NON SHAWAN ANIBO // I call my sickly family
NOMABO BEASHKIN // Spirits of the plants shall come
CHITI CHITI SHAMANI // They all dance among us now
RARO RARO SHAMANI // With their joyous dance, they heal
NON MATO ISHONON // To you all, I give this prayer

When I wrote The Serenity Code (23) to explore how self-love habits could reset our autonomic nervous system and increase our neuroplasticity, I found impressive evidence that music can increase motivation and goal-directed behavior. For example, listening to music while exercising can make a workout more enjoyable and help us to push ourselves harder. Music can also create a more positive and motivating atmosphere in the workplace, a hospital, a retail store, a school, and, of course, at home. Music can also activate the mesocorticolimbic system, a brain circuit involved in reward and pleasure.

Listening to music that we enjoy can cause the release of dopamine, norepinephrine, endorphins, and oxytocin. These neurotransmitters produce feelings of pleasure, euphoria, and well-being. Music can reduce stress and raise arousal levels. It can slow the heart rate, lower blood pressure, and minimize muscle tension. Music can also be used to distract us from negative thoughts and feelings. Finally, music can improve our immune system function by producing white blood cells that help fight off infection and reduce stress hormone levels.

In addition to these top health benefits, music has also been shown to be beneficial for a variety of other conditions, such as pain management, anxiety and depression, sleep disorders, autism spectrum disorder, Alzheimer's disease, and Parkinson's disease. My mother just turned 90, and as you would expect, her cognitive functions show some deficits. However, as a teenager, she sang as a duo with her twin sister, and both toured professionally under the name "The Jerusalem Sisters." Whenever I visit her and ask her to sing a song from her youth, her face lights up, and not only does the melody come back, but the associated lyrics appear to roll out of her tongue with no cognitive effort whatsoever!

Since the musical component of the practice of an icaro is critical to its overall impact, there are many ways in which Icaros may promote healing. Unfortunately, separating the direct effect of using icaros under the influence of Aya during any ceremony is difficult. Research does show that participants in Ayahuasca ceremonies with icaros (such as the ones I am attending in this retreat) have reported that the songs helped them heal from specific physical conditions, such as chronic pain, autoimmune diseases, and even cancer. For instance, a research team investigating the benefits of icaros (24) found that Ayahuasca drinkers who participated in ceremonies with icaros reported significant improvements in their physical health, including reduced pain, improved sleep quality, and increased energy levels.

Meanwhile, icaros are often sung in a specific language, such as Quechua, but more importantly, they contain unique healing sound frequencies. The frequencies of icaros are not well-documented, but they typically fall in the 432 Hz to 528 Hz range. These frequencies are believed to be particularly resonant with the human body and can be used to promote healing and well-being. Research from Nakajima and his colleagues (25) demonstrated that sounds directly affect our autonomic nervous system. Specifically, high-frequency sound waves play a more significant role in stress relief than low-frequency waves. When icaros are sung, the healing frequencies can directly affect the body, where they can work to heal imbalances. As a result, icaros can also be used to clear negative energy from the body and to promote spiritual growth.

In addition to their specific frequencies, icaros integrate the healing energy of plants with powerful medicinal properties, such as Aya and master plants. The plant spirits are believed to be transferred into songs to reduce stress, induce altered states of consciousness, relax, facilitate self-reflection, generate insights, and provide support and guidance. As noted earlier, the benefits of listening to icaros appear to transcend the physical realm. For instance, icaros contribute to the critical goal of creating a sacred space where people can feel comfortable exploring their inner selves and facing their most profound challenges. Those who sing icaros use their voices to create a sense of harmony and balance, which can help people feel more grounded and centered. I have had direct experience with this benefit in every Aya session I have participated in over the last 15 years.

Additionally, icaros can evoke deep emotions, such as sadness, anger, and fear. By allowing people to experience these emotions in a safe and supportive environment, icaros can help people process their emotions and move toward healing. Finally, icaros connect people with their spiritual selves and help them see themselves as part of a more extensive interconnected web of life or cosmos. This sense of interconnectedness helps one feel more loved, supported, and accepted.

Thus, icaros are instrumental in providing emotional healing during ceremonies. For example, Researchers have found that people who participated in Ayahuasca ceremonies with icaros reported significant improvements in their emotional well-being, including reduced anxiety, depression, and anger (26). In addition, some people reported that icaros helped heal them from deep emotional traumas, such as childhood abuse and neglect. Another participant claimed that icaros helped her process being raped and reclaim her self-power.

Here are the lyrics for my Bobinsana icaro. Since I wrote my icaro in French, I included both French and English versions below. Curiously, the rhyming component is not lost in the translation.

Bobinsana, Bobinsana, j'aime ta brillance
Bobinsana, Bobinsana, Je cherche ton essence
Bobinsana, Bobinsana, enlève ma résistance
Bobinsana, Bobinsana, donne-moi la délivrance
Bobinsana, Bobinsana, guide moi
Bobinsana, Bobinsana, aide moi
Bobinsana, Bobinsana, inspire moi
Bobinsana, Bobinsana, aime moi

English Translation

Bobinsana, Bobinsana, I love your brillance
Bobinsana, Bobinsana, I look for your essence
Bobinsana, Bobinsana, remove my resistance
Bobinsana, Bobinsana, give me full release.
Bobinsana, Bobinsana, guide me
Bobinsana, Bobinsana, help me
Bobinsana, Bobinsana, inspire me
Bobinsana, Bobinsana, love me

Meanwhile, some academics believe that icaros are an "inter-species communication" in which the apprentice intercepts and interprets the phytochemical signals inherent in plant communicative processes (27). Also, other researchers suggest that kené designs, a traditional Shipibo art form often used in icaros, can be seen as a visual representation of the healing power of icaros (28).

Figure 11: Kené Design

At 1 o'clock today, we were invited to take a special sauna treatment. The ritual was prepared following the Shipibo tradition. It involved sitting up on a wooden structure that immediately reminded me of the French Guillotine. Once I was perched on it, Don Miguel's shamanic apprentices put a vinyl tarp to cover me entirely. Then, two attendants placed a large steel pot containing a wood-boiled concoction of medicinal plants just one foot under my feet. They removed the pot's cover to release the steam. The scented vapors escaped quickly, and the heat wave engulfed me in seconds, a sensation like a traditional sauna. However, the ritual was designed to provide additional therapeutic benefits. I enjoyed it immensely. Day after day, this program continues to amaze me.

We ended the afternoon schedule with a lecture from Don Miguel on Shipibo traditions and belief systems. I admit that I may not have grasped the depth of his stories. Like many maestros in the community, Don Miguel, a former schoolteacher, was called to serve as a healer or "Onaya" at a young age. He was also extensively trained and supported by his parents, especially his grandparents. With humor and humility, he shared some revealing accounts about what it meant to grow up as a Shipibo and ultimately surrender to the call of the responsibility of becoming a maestro: long periods of plant dietas, which means ongoing sacrifices, including abstinence, food restrictions, and isolation. To become an Onaya, a shaman needs to study the plants and connect with the spirits of the plants. An Onaya often holds a respected position within the community, serving as a guide and advisor in health and spiritual well-being matters.

Onayas play a crucial role in maintaining the balance and harmony of the community. They are believed to connect with the spiritual world and the cosmos deeply. They often act as intermediaries between the physical and spiritual realms, using their knowledge and skills to navigate multiple dimensions for healing and guidance. The concept of Onaya is a fundamental part of Shipibo culture, reflecting their holistic understanding of health, where physical, mental, spiritual, and communal well-being are intimately connected. The training to become an Onaya is like earning a Ph.D. in botany, combined with years of meditation, diets, and Aya ceremonies. Don Miguel ended his lecture by insisting on keeping our bodies clean, our minds clear, and our hearts open to love and kindness.

Aya Session #2

As I enter the maloca this evening, I notice the palpable energy of the room is different. A big storm started right as we all finished settling in. In the past, I would discount, if not entirely dismiss, the importance of preparing my space ahead of the ceremony and taking time to place all I needed to feel safe and spiritually protected. However, I have come to appreciate whatever help I can get to embrace the benefits of Aya. So, I brought my mapacho pipe, some mapacho tobacco for protection, a bottle of Agua De Florida, a red flashlight to ensure I could go the bathroom without disturbing other participants, a nice pillow, and my journal. Agua De Florida, or Florida Water, is primarily used for spiritual cleansing and shields the ceremony's place and the participants.

The floral and citrus scent attracts positive energies and spirits while repelling negative ones. Additionally, it is often used in rituals to mark the beginning and end of ceremonies and bless sacred objects and participants. Some also use it for physical cleansing before the ceremony to prepare the body and mind for the experience. I also brought my handpan tonight as Elio invited all participants to sing a song or play an instrument at the end of the ceremony. My wife bought me this beautiful instrument a few months ago for my 63rd birthday, and I have not found time to practice seriously. I immediately fell in love with it, however. I knew this trip would finally give me the time and space to connect with the magical grooves that can quickly enchant and calm your brain.

Tonight, the room is filled with the energy of thunder. Flashes of light enter the screened section of the maloca. Hollywood could not create a better atmosphere to make us all shiver in wonder. I remained calm and composed. I am determined more than ever to use my newly expanded mental agility to accept, trust, and surrender to the unfolding.

I took the same amount of Aya as I did for the first ceremony. As I lay down, I wonder how long it will take to feel the effect. It is difficult, if not impossible, to predict the speed at which the brew starts to work. Within minutes, though, I can detect bodily sensations typical of Aya's signature. Tonight, the brew traveled fast through my body to activate the release of DMT.

*This potent psychedelic compound acts primarily on **serotonin**, a critical neurotransmitter that regulates mood, cognition, and perception.*

Healers believe that the effect of DMT is a function of how clean your body and mind are, so I took it as a testament to the benefit of my dieta, but also the many years of clearing massive amounts of anxiety and stress out of my nervous system. Not to mention that I finally quit decades of heavy drinking as soon as we moved out of Honolulu to Kona on the Big Island. This bold move radically transformed my life. I feel so grateful each time I think about it.

The power of meta-consciousness

Figure 12: Meta-consciousness (AI)

I am only 20 minutes into the session when the visuals start. Mild and tolerable at first, they quickly become brighter, faster, and erratic. I already know this session might be far more challenging than the first one. Fortunately, I am ready. In the past, I would have resisted the unfolding, asked assistance from the shaman, and possibly begged for ways to lower the effect, a move that would trigger the same response from the curandero: "You are under the effect, Christophe, just ride it, it will stop after a few hours. Stay grounded and invoke the spirits of the plants to protect you, accept what is, trust and surrender". Tonight, though, I do not need help or guidance. I understand exactly what to do: trust and surrender. I patiently wait, playing the observer role the whole time, accessing a state of "meta consciousness" (meta means above in Greek) and refusing to let fear control me. And it works. The first purge immediately clears the excessive flow of disturbing images.

Now, the teachings can come through. I recognized long ago that Aya is not just about the visuals or the fireworks, as many Aya facilitators call them. Therefore, I always enter each session with no expectations for visuals. I understand at a deep level that there is never a "bad trip" with Aya since the medicine heals me regardless. However, I must admit it is hard not to cherish the magnificence of Aya's cinematic productions.

As I like to do during all my sessions, I asked Aya how she chose today's programming. Humor is often an excellent way to disarm the worrying machine. This time, Aya seemed to interpret my internal dialogue as a request to change the direction of the narrative. I was happy about that!

Manifest the life you want.

Figure 13: A Beautiful Barge (AI)

First, I see a crisp, bright scene featuring a beautiful, long barge decorated with magnificent bouquets of exotic flowers, sides covered with petals, and stunning draperies. I notice a crowd of children, elderly, and sick people slowly moving towards the back of the barge. I am so intrigued by the story. I cannot wait to see who that lady at the front of the barge is. I imagine she is someone important.

A beautiful, graceful woman dressed in gorgeous royal clothing greets the visitors. She smiles with an unforgettable expression of grace, love, and compassion. I finally recognize her face: it is my wife. I am so happy to see her in this extraordinary setting. The scene is stunningly accurate and relevant to what she does daily as a soul-centered psychologist: tirelessly giving her time and expertise to those needing guidance and comfort to ease their suffering.

The story fades, and the icaros of Don Miguel soothe my soul again, carrying me into a space where sound becomes medicine, and melody dissolves the weight of my thoughts. Each note feels like a gentle thread weaving through my spirit, unraveling tension and reconnecting me to something ancient, something beyond words. His voice, infused with wisdom and reverence, rides the rhythms of the jungle—flowing like the river, whispering like the wind, steady like the heartbeat of the earth itself.

As the icaro unfolds, I feel its resonance in my chest, in the space between my ribs where old wounds linger. The melodies are not just heard; they are felt, vibrating in the core of my being. It is as if Don Miguel is guiding my consciousness, his song is like a bridge between worlds, calling in spirits of healing, protection, and transformation. The sound is both anchor and wings— keeping me grounded while lifting me into a realm where the sacred and the personal blur into one.

With each repetition of the melody, my body softens, and my breath deepens. It is not just music; it is a transmission of energy, an invitation to surrender, to let go, to trust. In this moment, I am no longer just a listener; I am a participant in a timeless ritual, where song and silence dance together, cleansing, soothing, and restoring me to myself.

As we all regained access to "ordinary consciousness", I took the bold step of playing my handpan during the closing part of the ceremony, a difficult task since I only had a few hours of training. However, in the spirit of the ritual, I was happy to share my instrument with the beautiful and brave people in attendance tonight. I did not feel I was playing it, but rather that the instrument was playing me.

DAY 7: Iquitos PERU, November 15, 2023

This morning, we had a sharing circle after each ceremony. Hearing participants' testimonials is never easy, but it is critical to each person's healing process. By now, I must have heard over 1,000 such stories. Also, I have hardly ever seen a participant regret the experience or dispute that they learned something profound from it. Even if the session appears only to bring agony and pain, there is always a therapeutic thread people can pull out of any ceremony. I have witnessed people struggling with multi-generational trauma, rape, aggression, addiction, war, gun violence, sexual abuse, emotional abuse, divorce, and so many more debilitating physical, emotional, and psychological conditions.

There seems to be no limit to the wall of suffering that can block us from enjoying life. There is no doubt in my mind now that Aya is on a mission to destroy that wall. "Whatever it takes" seems to be her motto. Whether it is through the purging process, often perceived as a punishment by people who start working with her, or the stomach pains, the exhaustion, the headaches, the dizziness, the feeling of being overwhelmed by the swirling and sometimes scary visuals, she is always prepared to deliver incredibly soothing moments of joy, rest, bliss, love, wonder and peace. If you work with Aya for a long time, you will discover the nature of her healing spectrum. You will gradually understand that there is little you can do to control her moves, to predict if she will start your session on one end of her pain-pleasure scale or the other. However, you will eventually understand that it does not matter where you begin or where you end; you will receive healing each time.

I played my handpan today for at least two hours. What a feast! I am already learning a new groove. I love the time and spaciousness this program creates for me. I also had my second sauna and did at least three Bobinsana baths. I am in the flow of the program by now, delighted by the blend of teaching moments and the cleansing effect of the master plant dieta.

The Bankruptcy

By age 10, I knew I wanted to be part of SOVAM. I aspired to pursue a business education and eventually run the company, continuing my father's legacy. Unfortunately, that dream never materialized. A series of poor decisions, rooted in my father's struggle to manage the growth of a large, complex industrial operation, led to the company's bankruptcy (see Figure 14 as an example of products engineered by SOVAM in the transport of passengers in airports). SOVAM's downfall was caused mainly by my father's reluctance to make layoffs that would adjust the workforce to more realistic sales forecasts. As the second-largest employer in Parthenay, thousands relied on his management decisions. Many of these workers had become his family, so much so that he lost sight of his own.

As it was common practice in France at the time under a socialist government, bureaucrats from Paris started to intervene to save the company. They orchestrated a complex financial restructuring procedure, re-investing a balloon of massive severance payments owed to his lifelong, loyal workforce. Doing that brought fresh capital into the business and allowed the people who reinvested their severance money to own part of the company. Strangely, my father indirectly, but not willingly, donated his business ownership to his employees through this process. Without external investors lining up to acquire SOVAM, this financial scheme was the only way to save the company and keep a large portion of the workforce. However, it also meant that my father lost control of SOVAM while bearing the entire financial liability of bankruptcy since he had signed multiple personal loan guarantees.

Going through the financial fiasco was a significant trauma for my entire family. We all expected it at some point because my father was financially reckless. I remember attributing his failure to his lack of formal education, a belief that probably fueled my obsession with getting graduate degrees. I was the closest to him, and I could see quite early that he lacked the basic skills to read financial statements or launch new products with any marketing and financial discipline. Before I turned 18, I had already decided to pursue a business degree with the vision of joining his team. After a tough prep year in Tours at the Descartes lycée, I was accepted at AUDENCIA in Nantes and completed my BA in Business in 1982 with honors.

At that point, I was supposed to serve my 2-year mandatory military service, but I managed to get out of it. I could not conceive of spending two years away while my father's business already showed troubling signs. This is how I got out of it: I brought x-rays of my horseshoe kidneys to my military service recruiting interview, and I claimed I was not fit for service. Having horseshoe kidneys is a curious but benign condition that can lead to chronic urinary infections – which I had only gotten once a few years earlier! With a grim look on his face, the recruiting officer ordered me to go to a military hospital for a complete physical evaluation. This was not a good day for me.

I had to wait another 30 agonizing days to drive to Bordeaux to fulfill my obligation. When I arrived at the military complex, I was consumed by anxiety and fear. Would they confirm my earlier diagnosis from years before? Would this failed attempt to trick the system backfire and compromise my opportunities to serve as an officer instead of a private? Would I be dispatched within days to a base in a remote village far away from Parthenay? After checking in, I was asked to wait for the radiologist so I could get new X-rays. I did not expect that and felt consumed by dread. Finally, after at least an hour of anxious waiting, I entered a tiny room and met a young doctor serving his military service at the army hospital. He smiled at me, carrying the X-rays. He was not fooled a bit by my move to use the excuse of my horseshoe kidneys to be exempt from service. He asked, "You don't want to do your military service, do you?". I said "no" with a scared facial expression and told him I needed to save my father's business. Miraculously, he agreed to dismiss me from my obligation to serve. The nightmare was over! This was one of the best days of my life!

Figure 14: Aérobus (1978)

Since my father's company quickly expanded internationally with his airport equipment line, I pursued a fast-track MBA in the USA. Fortunately, my good grades from AUDENCIA granted me a scholarship, making the financial burden of the university tuition tolerable for my parents to cover. I excelled in the program. I studied hard and returned to France after just 13 months. Unfortunately, by the time I graduated in the fall of 1983 with a 4.0 GPA, it was too late to save SOVAM. I was devastated. Instead of joining my father's business, I stayed a few years in France working as a consultant and helped my parents rebound from their bankruptcy's long and grueling consequences.

DAY 8: *Iquitos PERU, November 16, 2023*

I woke myself up this morning as I found my dreams annoying. Two narratives emerged, which have returned regularly in the last few months. Both were related to my addiction to my phone and my fear of disappointing a customer. The first one, however, brought back the painful memory of losing my best childhood friend.

The Loss of My Best Childhood Friend

I met Gerard when I was 15 years old. We were inseparable until I left for college. He lived on a small, humble farm twenty miles from my house, a distance that was difficult to reach at the time, other than by bus or car. We shared a passion for sci-fi books and would debate the meaning of complex plots for hours. Gerard was smart and direct. He often pointed out my arrogance and sense of superiority, which he attributed to the social status of my parents. We loved identical girls and the same movies; we cried and laughed together more than I could ever recount. When I left France in 1988, I maintained regular contact with him and cherished our catch-up calls. Sadly, though, he cut all ties with me nearly ten years ago over a troubling misunderstanding.

His daughter had traveled to San Francisco at the worst possible time for me to welcome her to our home. I had just returned from a brief trip to Hawaii, during which I had suffered a severe ankle sprain and finished my vacation on crutches, exhausted and in excruciating pain. Also, I was furiously working on my PhD dissertation between intense client engagements and business trips, as was my wife. Neither of us felt we had the physical or emotional space to host and entertain his daughter and two traveling companions for a few days. Yet, with little warning, Gerard expected me to do so. Several years earlier, he had graciously welcomed my youngest son to stay in his home while spending a semester at a French high school near Parthenay. However, instead of hosting his daughter, I paid for a hotel in San Francisco where she could stay with her friends. I justified the move by arguing that, since we lived about 30 miles from San Francisco in a suburban area with few things for young people to do, not to mention they did not have a car, it would make more sense for them to stay in the city instead.

Gerard did not see this move with the same eyes. He was furious and severed our lifelong friendship because of what he perceived as an act of selfishness and betrayal. I tried multiple times to reach out and justify my rational behavior. I apologized for disappointing him and misreading the situation. Nothing could convince him. So, I eventually had to accept his decision and move on. Sadly, I understand the underlying causes of this outcome now: my complete lack of consciousness and plasticity. I was locked in my ego and failed to open my heart to understand his hurt and disappointment. Many years later, I am still pained by this loss. However, I finally know why.

In my dream last night, I had stolen Gerard's phone by mistake and could not find a way to reach him to give it back. I felt ashamed. I wonder if that had anything to do with the backstory I just shared unless somehow the dream highlighted my emotional attachment to a working phone, which is not an option in the retreat center. The dream quickly shifted to a variation of the same story for which my phone was, yet again, a central character. I was supposed to reach my elderly mother and could not. The phone interface was going wild. I could not figure out how to navigate it. So, in both cases, dialing back either Gerard or my mother was impossible and frustrating in the dream. Nathalie, my younger sister, and my oldest son were also trying to help me, but nothing worked.

In the second dream, I was late showing up for a workshop and entirely ill-prepared to run it, a nightmare I have had hundreds of times during the last couple of decades. I could not find where it was being held and did not have my handouts. I remained confident I could deliver the content without slides, but I was unhappy being so unprofessional. I forced myself to wake up when I realized it was a dream and quickly acknowledged the value and teachings of the disturbing narratives. I am resting now, preparing for the third ceremony tonight. I just had my bath in Bobinsana water. What a beautiful and refreshing routine.

Icaro #1

This morning, we learned to sing our first icaros with Don Miguel. It was special to learn a small amount of the Shipibo language in the process. Most words have multiple meanings depending on the context. I am truly fascinated by the coherence and resonance of the messages carried by icaros. It is an essential ceremonial practice to guide each Aya ceremony and deliver important messages presumed to be delivered by the plants. In the past, I have noticed how a shaman can manipulate the energy of an entire session by singing icaros.
Here are the lyrics of the one we learned today, translated into English.

RAO NIWEBO ↑
Sacred medicine, rise.

NOYA RAO RIOSBO ↓
Moon medicine, descend.
JOYO, JOYO NIWEBO --3 ↓ ↓ ↓
Blessings, blessings of the sacred medicine, descend deeply.

NIWE JOKON ROMNA –2 ↑ ↑
Sacred energy, rise up.
KANO MAYONTANANRA
With you, we are connected.
JONI SOWA SOWAWA –3 ↓ ↓ ↓
The essence of life flows down deeply.

PANPA RONINKA ↑
Let the spirit of healing rise.
KANO PANPO RONINKA ↑
With your healing spirit, rise.
NETE BAWA JOYONI
Great world, bless us.
KANO IRA MANKE ↓
With guidance, lead us.
IKI I-SHAMANI –3 ↓ ↓ ↓
This is the path of the healer, descend deeply.

NAI WAPORONINBI –2 ↑ ↑ //
From the stars, come closer.
NETE ANI RAKATA –3 = ↓ ↓
Let the world be renewed, flowing deeply.

RONO, RONO PAKESHON –2 ↑ ↑
Spirit, spirit, come forth with strength.
MATON YORA TIBI
Protect us, guide us.
NANE AKE TANSHONRA
Light up the path of healing.
NETE ANI KEPENI –2 ↓ ↓
World of spirits, descend.

RIOS SEN SHINAMAN -2 ↑↑
Pure light of the rivers, rise up.
NETE TORRI CAMPANA
World, echo like a bell.
TSARA IMAYONTANAN
Let us walk together.
MATON YORA ABANO –3 ↓↓↓
Protect us deeply, guide us deeply.

This translation is an interpretation based on the structure and presumed intent of the original phrases, which often involve invoking nature, spiritual protection, guidance, and connection with the elements. The arrows add a layer of rhythm, signaling how the chant should flow or the tonal emphasis during recitation.

Aya Session #3

I enter this third ceremony feeling grounded and ready. I am proud of myself. In the past, my heart would accelerate before receiving Aya, and my mind would race, wondering what would happen. Not today, and hopefully not ever again. The medicine acted quickly, probably within 15 minutes. I am excited that I will soon learn to prepare Aya. I am especially eager to figure out if there is a way to cook the brew to make it react faster and even lower the purgative effect of the medicine, even though I recognize its role in the healing process. The preparation of Aya is an essential skill for Don Miguel and his crew to master. I have so much admiration for our medicine man. He is so humble and passionate about his work that it can only bring admiration and a big smile to your face.

The visuals are coming fast and furious. I do not recognize Aya's typical signature. The images are immediately beautiful and crisp, featuring landscapes from what appeared to be another planet. The colors are stunning, and the characters are intriguing, but I cannot find a specific narrative that enables me to stitch it all together. I admit that I am obsessed with discovering the thread of a story.

Plasticity as a core meta-trait

The buffet of sensations is non-stop for about 20 minutes. Then, Aya decides to "work on me." I purge a few times, and quickly after that, I run to the bathroom. Diarrhea starts to do its unpleasant but expected job. Unfortunately, I feel it will be a central theme of the evening. I must have walked the short distance between my mat and the bathroom a dozen times, feeling dizzy and confused by the visual distortions.

However, I keep thanking Aya for clearing stuff that does not serve me. The last two years have been so intense renovating our retreat center that I packed much stress along the way. Aya was determined to get it all out, in whatever form it had been stored in my emotional body.

The worst of the session seems to be over. So, I engage her in a dialogue on decoding neuroplasticity --the self-healing property of the brain, otherwise known as our capacity to rewire thoughts and traits – also commonly referred to as "plasticity." She helps me clarify how to benchmark plasticity using a revised version of

Figure 15: Plasticity as a Meta-trait (AI)

my current Serenity Code assessment. I currently use a simple algorithm to compute a plasticity score based on The Big Five (BF) model. The Big Five is a remarkable personality assessment tool based on the five most predictive traits of people's behavior. To remember them, use the acronym OCEAN: O stands for OPENNESS, C for CONSCIENTIOUSNESS, E for EXTRAVERSION, A for AGREEABLENESS, and N for NEUROTICISM.

Studies suggest that plasticity is highly correlated to openness and extraversion. While I believe the Big Five model is the best personality model today, I am aware that research has challenged its validity (29). Still, it provides a good cartography of the "blueprint" of our personality, essentially what we inherit from our parents and ancestors.

Aya insists I continue using the Big Five and consider Enneagram scores to calculate plasticity. The Enneagram is an ancient personality model based on nine typologies formed over time by our coping mechanisms. There is a good correlation between the Enneagram typologies and the Big Five traits. However, while many Enneagram experts do not share this view, I consider the model an excellent cartography of the "imprint" of personality. After all, to survive, we must adapt, and as a result, we may alter some of the programming of our DNA.

I am excited to get more clarity on this critical piece for my book. As I did for my prior work, I like to provide ways for people to benchmark their personality profiles as it can point to the root causes of SADAT.

Narrative transportation

Figure 16: Narrative Transportation (AI)

Our dialogue quickly shifts to another fascinating topic: the effect of stories on our brains. I feel immense excitement as Aya confirms the importance of narrative transportation in increasing plasticity. Narrative transportation is the process of becoming fully absorbed in a story, where your emotions and beliefs are influenced by the narrative, creating a vivid, immersive experience. As a Media Psychologist, I have spent many years decoding the power of stories, especially commercials, customer testimonials, and corporate videos. Therefore, I have rich data showing how stories affect our brains. Good narratives move the brain's activity from the bottom up, from the "PRIMAL brain" to the "RATIONAL brain." The PRIMAL Brain (also called the Reptilian Brain or Survival Brain) is the older, instinct-driven part of the brain, responsible for essential survival functions like fight-or-flight responses, emotional reactivity, and instinctual decision-making. The RATIONAL Brain (also known as the Neocortex) → This is the more evolved, logical, and analytical part of the brain, responsible for higher cognitive functions like reasoning, planning, decision-making, and language. It allows us to process complex narratives, evaluate long-term consequences, and override impulsive reactions.

Therefore, plasticity can increase to the extent that we allow new or revised narratives to achieve the bottom-up effect. I realized that I needed to unpack the relevance of this phenomenon in the book, and I felt good that it would be a critical piece of my book.

After receiving these insights, I focused on my bodily sensations to manage several more trips to the bathroom. The ceremony was intense but beautiful. Don Miguel's icaros bathed all of us in love, protection, and wisdom.

To conclude the ceremony, Elio and Anna delighted us again by singing beautiful songs. As Elio invited us all to participate, I found the courage to grab my handpan to play the groove I had rehearsed over the last few days. I was still under the influence of Aya, plus I could not see the notes in pitch dark, but I delivered a nice piece of music.

DAY 9: Iquitos PERU, November 17, 2023

Today is our day of rest. Little did we know that we would be induced to experience more rest than ever. Before breakfast, Don Miguel invited us to drink another medicinal concoction. It is called "Chiric Sanango" and is an excellent remedy against rheumatism and arthritis. I have had chronic pain in both of my knees from running thousands of miles since I was 16. I was happy to indulge. After the necessary warnings about the possible side effects of the brew, we all drank our shots with a big smile. "Expect to feel tingling in your arms, numbing of your tongue and lips for most of the day," confirmed Don Miguel, laughing. I had none of these symptoms. However, I felt tired all day and could hardly move around the center, which was also the whole point of the brew.

In the afternoon, Elio invited us to a presentation on a humanitarian project run by his partner. A Peruvian woman with the grace, presence, and intelligence of a goddess gave us a fascinating account of an effort she and a few of her friends were making to aid and support an isolated indigenous village of the Amazon. Pictures and videos of the project reminded us how fragile life is for thousands of indigenous tribes across Peru. Sadly, the Peruvian government is poor and has little interest in supporting them, according to Estephania. The village she was helping had just suffered a catastrophic storm that destroyed critical education and health infrastructures. She shared images of the damage, ranging from ripped solar panels to flattened houses. The construction of these structures is so primitive that these kinds of storms can wipe out an entire village in just a few hours. Global warming increases the frequency and strength of such climate events. Despite the dire consequences of the storm, I could not help but notice the smiling faces of the children gathered for a picture opportunity. How can we ignore the predicament of these people?

She closed her presentation by inviting us to support a fund-raising project to bring more school supplies for the kids and organize a Christmas party where they could all share a chocolate cake. She also insisted that the community was not begging for support without contributing. She presented a catalog of beautiful dresses made by local artists that would soon be offered online to bring income opportunities to the community. The presentation moved me. We bought some honey from Estephania and confirmed our interest in supporting her cause.

This morning, we were invited to sniff another medicine known as "maruca". Don Miguel explained that this would sting a bit, but that was very good for our sinuses. We were all obliged and had fun doing so. After that, one of the facilitators taught us how to participate in a "despacho" ceremony, a ritual performed by many communities of the Andes to process feelings of spite, resentment, or heartbreak, often related to romantic relationships. Thus, a despacho ceremony might refer to informal, personal, or community practices to express or process these emotions, particularly in love or romantic disappointments. Such ceremonies can take various forms, depending on the cultural background, personal preferences, and the specific nature of heartbreak. They might involve activities like gathering with friends to share feelings, listening to music expressing heartbreak or empowerment, burning or discarding items associated with a past relationship, or other ritualistic actions that symbolize letting go of pain and moving forward. This ceremony involved preparing a living prayer of flowers, leaves, nuts, spices, rice, and many natural or artificial décor items as we wanted. Once the arrangement was done, we closed it in large leaves, wrapped in paper, and buried it near a tree. The day went fast, and I felt ready for my fourth Aya session.

Aya Session #4

Since it was hot all day, the atmosphere of the maloca became heavy this evening. I am excited about this session, though. So far, I have requested ½ cup for each session, which has given me a tolerable dose of purging and insights. So, I am asking for my usual dose.

I can already feel its effect only 15 minutes after drinking the brew. However, this evening, the onset is very different. I sense Aya is both gentle and generous. As I have mentioned, the visuals are not the most essential part of the ceremony. It is nice not to focus my attention on decrypting them, either. Instead, I expect another fascinating dialogue with Aya about purging, plasticity, and self-love, all intimately related to my new book.

Purging and plasticity

I have been purging during every Aya session, which is unusual. I do not resist it, as I know it has healing properties. However, I wanted to understand this process's significance and purpose better. Most people who participate in Aya sessions tend to dread or resist the physical discomfort of that reaction. Purging can happen in many ways. Vomiting and diarrhea are the most common forms of

Figure 17: Clearing Toxicity (AI)

purging. Both can occur with or without warning and sometimes simultaneously. However, laughter, burps, gas, sweat, heat, and tremors are all considered forms of release from the body, providing multiple options through which the process unfolds.

I asked Aya about the relationship between purging and plasticity. I had not thought about the topic until I felt the effect. She was quick to oblige. She explained that purging was critical to clear emotions, thoughts, or body aches that do not serve us. However, there is more. Releasing anger, sadness, terror, regrets, shame, and self-judgment can prepare our body and brain to make space for new emotions and thoughts. She told me we must complete this process to rewrite new stories. Essentially, purging prepares the brain's plasticity to kick in full force. Without it, we may continue to carry more toxicity. Aya concluded that she always chooses healing over comfort.

Achieving optimum plasticity

What is plasticity if not our ability to invite others' ideas or our own without judgment? If so, how can we rewire the neuropathways formed by our stories unless we clear the clutter of our minds?

Aya tells me that keeping the highest level of plasticity is central to our mental health. It maintains a child's view of all our experiences, good or bad. Upon opening our minds to the possible, to the new, we can begin reconstructing our narratives. Aya asks me to imagine the state of my brain and body just after birth.

Assuming I am not carrying inter-generational traumas, Aya insists that, at birth, my brain is like a blank sheet of paper. My PRIMAL brain is pre-programmed to help me breathe, swallow, digest, distribute blood and air, power up my muscles, and help me survive. I noted earlier that this part of our programming also plays a significant role in the blueprint of our personality. However, we need much more cognitive capacities to thrive. Aya confirms that we rely on our experiences to acquire valuable

Figure 18: Achieving Plasticity

knowledge, skills, beliefs, and coping strategies to encode and activate the most evolved part of our brain: The RATIONAL brain. I know these functional systems (PRIMAL and RATIONAL) very well since they have been central to my research and writing on decoding persuasion and serenity for over twenty years. However, this insight into our baseline plasticity at birth was an important reminder. Aya pointed me in the right direction by confirming the building blocks of consciousness and plasticity. I was thrilled with this first teaching of my session.

However, there is more, Aya continues. Once you reach a state of emptiness, you can finally rid your body and mind of negative emotions and thoughts. In my past studies of the effect of stories on the brain, I have been able to measure the strength of the activity of neuronal pathways during the visual and emotional processing of specific scenes. For instance, I am very familiar

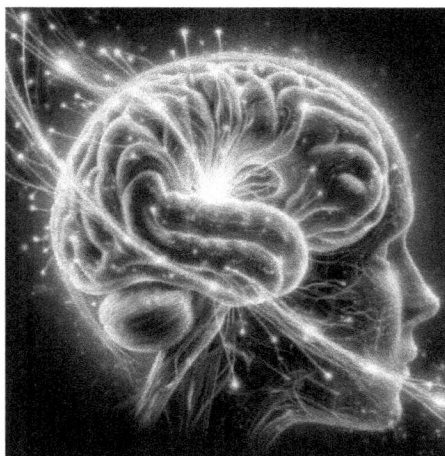

Figure 19: Self-love (AI)

with the power of negative emotions and thoughts on our memory system and their impact on our facial expressions. I have always assumed that this "negative" bias was related to marking our brains with emotional and cognitive events that could threaten survival. However, I had not given as much thought about the impact of these neurophysiological patterns on our plasticity. Now, it was so clear. Aya could enable and accelerate the rewriting process of our stories to clear past traumas and struggles.

Self-love

We are now transitioning to the topic of self-love. Self-love is the entire focus of The Serenity Code, which, for the most part, was "downloaded" to me by Aya several years ago. However, when I wrote it during the pandemic, I did not examine the relationship between self-love and plasticity. Now Aya invites me to do just that. After purging, according to her, we must soothe and move from cleansing to healing. That process can only happen when we fully accept and surrender to our divine gifts.

In essence, true self-love does not just heal the past but sets the direction of our future. Self-compassion, positive thinking, acceptance, gratitude, and the release of self-judgment and self-doubt are often considered essential mental states to enable self-love. I understand now that it is much more than that. It is a practice that can help us rewrite the programs that guide our future attitudes, thoughts, and beliefs. Aya stresses that without a self-love practice and, better, a ritualized version of that practice, the benefits of the purging process are short-lived. She told me that I should continue to engage in daily habits that reinforced the importance of self-love, so I decided to make sure it would remain a central theme in my new book.

Tapping in pure divine consciousness

Meanwhile, somewhat unrelated to our discussion, Aya takes another turn to insist that we are not born as terminals of our creators; we receive their full power. Essentially, they grant us a holographic copy of their consciousness to maximize our ability to survive and thrive.

The point of this surprising exchange was clearly to note that we are not as fragile as we often see ourselves under the weight of our fears. We may be divine, too. If so, we have the full power to manifest

Figure 20: Divine Consciousness (AI)

our future and the innate ability to take from all our experiences whatever we need to live extraordinary lives.

So, I asked Aya what powers the gods have to influence our human destiny. According to her, the gods or our creators can manipulate many climate events, like storms and tornadoes, but they cannot change the course of climate change due to human activity. As I questioned Aya on this aspect, she clarified that while she was not considering herself divine per se, she was speaking for the gods, invoking their spirits to guide my spiritual journey.

I was in complete awe of what had just happened. I figured it might take several days to unpack the gifts I received. I was grateful for this powerful session and excited to integrate the teachings into the writing of OPEN. Also, I realized that I needed to add a section on the neuroscience of purging. This would also help me gain a better understanding of eating disorders, even alcohol abuse, and their psychological relationship with the act of purging. In the traditional Western view, we tend to judge these acts of release as symptoms of mental disorders. What if the people who fight these addictions are longing for the benefits of the emotional and spiritual release provided by purging? What if we could reframe the conversation about these sensitive topics by reconsidering the nature and emotional basis of these maladaptive responses?

The Return to France

When I returned to France in 1983 after earning my MBA, I thought I could still save my father's company. However, it was way too late. All I could do was help him sort out the mess he had gotten into. The business was lost. Worse, he had signed personal guarantees that stripped him of everything he and my mother owned. Most of the consequences of his negligence and risk-taking addiction had been anticipated by all my family members, especially my mother, who was desperately trying to hang on to as much of the silverware as she could. For her, more than my father, going through the devastating process of bankruptcy was excruciating. Determined to help ease the blow, I spent countless hours driving back and forth from Paris to Parthenay, a grueling distance of 250 miles each way. I worked as a marketing consultant in the big city since working at SOVAM was no longer an option. These incessant trips made my birth city more depressing than I even remembered. Within months, my parents lost their home and most of their precious belongings. Thirty years of hard work had gone up in smoke. Worse, their marriage quickly started to fall apart. It lasted a year, but their sacred vow miraculously survived the dramatic unfolding.

Meanwhile, my mother took a job selling jewelry one hour away from Parthenay, swallowing her pride and showing us all how strong she was. Fortunately, my father negotiated a consulting contract with the restructured entity that emerged from the SOVAM bankruptcy. I helped him start a consulting business based in Parthenay called DIRECT EXPANSION. We sold a few consulting local gigs together, but mostly, I was trying to keep him engaged with projects so that he would preserve his self-esteem and secure a minimum income in the process. I also helped him work on repairing his marriage, even though I had no psychological credentials at the time to do so. My father had never shared much with me up to this point. During that difficult time, he described the emotional toll he was experiencing, particularly from the stress of losing his business. Afterward, he focused on finishing an invention he had kept in a drawer but never had the chance to protect. In a short year, he completed all the technical drawings, built a prototype, and filed another remarkable patent.

DAY 11: Iquitos PERU, November 19, 2023

Today was a rest day before our next ceremony. As we did a couple of days ago, we got a shot of Chiric Sanango to continue healing our bodies from the effects of rheumatism. Then, after breakfast, we were invited to participate in the sharing circle. This would be an excellent opportunity to hear more about how Aya might help people purge and rewire, two of the fundamental building blocks of a process I am determined to investigate in OPEN. Once again, I was amazed at the consistency of the narrative across many different personal situations. I feel lucky to be among such a remarkable group of people willing to step in and do some hard work to resolve their suffering. We are all used to burying so much of our wounds that it is easy to assume that our case is unique.

I recognize that this retreat center has attracted people who might be facing more challenges than the general population. However, after hearing the stories and witnessing the warrior-like attitude of the participants, it is evident that sexual abuse, childhood traumas, eating disorders, abandonment issues, and other debilitating psychological troubles are more widespread than I had ever imagined. Meanwhile, our society does not typically encourage or reward people who open up about these issues, but rather those who pretend everything is okay.

We received our last sauna today. What a fantastic treatment to steam in a bouquet of medicinal plants. Afterward, Teo applied another round of medicinal plasters on my knees to lower inflammation. I feel so pampered here, going from one treatment to another. The crew is so attentive and dedicated to their healing work. Considering the conditions in which we stay, it is not your usual spa, but I am convinced the long-term effect will far exceed any wellness retreat I have ever attended.

Meanwhile, a big storm broke several days of dry and hot weather. Thunder rocked the sky, and the jungle danced to the wind's strength and the rain's weight. The atmosphere of the center could not be more shamanic. I looked forward to tomorrow. Our fifth session would close the first part of the program. What comes next is a five-day period of silence during which we will eat only twice a day instead of three times. The healing journey is far from over.

DAY 12: Iquitos PERU, November 20, 2023

I broke my media dieta last night. I got tired from drinking another shot of Maruca in the morning, taking a medicinal sauna, and receiving two plasters for my knee later in the afternoon. All these treatments were intense and, as a result, impacted my energy. So, I wanted to watch a movie that would uplift me a bit. I know how stories can quickly stimulate the production of neurotransmitters that affect our moods. I found one recent movie that had been automatically downloaded for me

Figure 21: Balloon Dream (AI)

before I left for Peru: NYAD. The story was about a famous swimmer who broke the longest record for distance swimming by going from Cuba to the Florida Keys in 59 hours, covering 110 miles in water with dangerous jellyfish and sharks. The film was outstanding. I had not seen Annett Bening for many years and was surprised to see how beautiful and graceful she has remained. The character she played was my age when she attempted the impossible. This extraordinary athlete was trying to prove she could break a record she had set over 30 years earlier. Jodie Foster played as her coach. What a dream team of actors it was.

I was moved by the story and loved how it ended with such a powerful message of hope. As Nyad succeeded on her fifth try, this is what she said in front of the crowd cheering her up:" I want to say three things: One, never, ever give up; two, you are never too old to chase your dreams, and three it looks like a solitary sport, but it takes a team." I burst into tears watching the end. I *thought it was the essential theme of my journey: pursue big goals, never stop chasing my dreams, and, most importantly, nurture and protect my team, especially the most precious team members of my life: my wife, my children, my extended family, and all my soul brothers and sisters.*

Meanwhile, I woke up from a disturbing dream about an alien invasion. In it, I lived alone on the top floor of a skyscraper and noticed dark clouds moving over a large city. They morphed into threatening dark shadows that appeared sentient. I tried to exit the building using elevators, but they were all used by workers fixing them. Somehow, I managed to get down. I was looking for my youngest son, who was supposed to stay with me, while my daughter (imaginary, as I do not have a daughter) was staying with her mother. I was apprehensive about my son. As I walked through the city, large crowds gathered, and red balloons dropped from the sky. I was confused about the scene as I did not see how balloons could harm me. They were exploding on the ground, releasing deadly gas. I also heard loudspeakers with a voice claiming that the army was in control and had already killed many of the attackers. I did not feel fear throughout the unfolding, however. I walked along the busy venue quite peacefully, confident that the situation would resolve itself and that I would find my son. I woke up before I could.

We learned another icaro from Don Miguel today. I love the Shipibo language. It seems to work like the Hawaiian language: few words, multiple meanings. All convey evocative sounds and images, the plants' spirits, the healers' power, and the promise of direct connection to the divine.

Icaro #2

JONI SOWA SOWAWA −5 ↑↑↑↑↑
Essence of life, flow upward in abundance.
NETE ANI MEARA-a −-4 ↑ ↓ ↑ ↓
World's spirit, weave through us.
KANO IRA MANKE-jee −6
With guiding light, lead the way.

NETE JAKON RAKAya-a ↑
World's pure energy, rise.
NOA JONI VIRIVI
You, source of all life.
NETA ANI RAKAya-a ↑
Let the spirit of the earth ascend.

RIOS JEMA KEPENI-i
Rivers of light, bless us.
TORRIN EWA JOYATA-jee ↓
With sound, resonate deeply.
KANO IRA MANKE –5 ↓↓↓↓↓↓
With guiding wisdom, descend to us.

ININ KEWE NIWERA-a –4 ↑↓↑↓
Sacred breath, weave within.
RAO NOMA RONKENI-jee ↓
Sacred medicine, protect us.
KORI CHIKI JOYATA –6 ↓↓↓↓↓↓
Golden spirit, resonate deeply.

KANO BAIN TA-aNA-a ↑
With boundless love, rise.
NETE KANO RIOSBO ↑
Great world, and river, rise.
ISA YOYO IRAI-i–6 ↓↓↓↓↓↓
Let all spirits descend with grace.
NETE KANO NIWEBO-o ↑
World's sacred essence, rise.
MANIRA BEAKI-i --7 ↑ ↓↓↓↓↓↓
Pure heart, join us and descend.

  Today, my oldest son turned 35 years old. I am so proud of him. He has become such a fantastic young man. I tried several times to text him today after walking back to the road from which we hiked, but the jungle did not cooperate. We had a massive storm that started in the morning and ended late at night. I am used to heavy rainfall living in Hawaii. However, the experience I witnessed was even more powerful than I anticipated. Observing the raw might of Amazonian nature in full force was mesmerizing. Within moments, small streams transformed into rivers, and puddles swelled into ponds. Several hours after the storm began, the solar-powered electricity system failed as the prolonged absence of sunlight depleted its energy reserves. This vividly demonstrated the overwhelming dominance of nature's forces, as the relentless storm deprived even modern systems of their ability to function.

  Tonight is our fifth Aya ceremony. I look forward to the unfolding. This program has delivered more than I could have ever hoped for. It clarified my resolve to continue becoming a healer, gave me the courage to write and share my story, and gave me dreams and insights to help me communicate critical teachings. I felt that I could inspire many people who struggle with SADAT and the Savior Complex to navigate the ups and downs of these conditions with *grace and ease.*

Aya Session #5

The maloca's atmosphere tonight is chilly and soothing. The storm calmed down, but the power was still off, giving the room more energy. I am ready to have another dialogue with Aya. I took my usual dose and laid down. Within minutes, I start to fall asleep. I remember waking up when Don Miguel was already singing his icaros. It is so good to surrender and let his beautiful voice rock my soul as I emerge from a deep rest.

The first part of the ceremony did not have disturbing visuals, though. Was Aya willing to give me a pass tonight? Not so fast. The creation story returns, but this time with a very different spin. Surprisingly, this theme is central to many of my sessions. I cannot say I have been very interested in this narrative in the past. I struggle with the fantasy-like aspect of it all. I think that Aya is testing my new level of plasticity. What did I have to lose by letting her open my mind further?

Could extraterrestrials be our creators?

Aya starts by telling me that aliens decided long ago to live on our planet because the Earth offers ideal conditions they could not find elsewhere in our galaxy. However, we cannot see them because they live at different quantum frequencies. They inhabit a parallel world that does not intersect with ours. As she unpacks this surprising story, I instantly think about the growing number of people who report being able to channel or even see extra-terrestrial beings, possibly aliens that live behind the veil of our

Figure 22: Alien Story (AI)

human consciousness. What Aya is sharing with me could easily explain the countless sightings of UFOs, ghosts, spirits, and all the mystical creatures in many ancient stories. I finally let this revelation come through my brain without filters, judgment, or denial. Though a purge is coming. I hold my bucket with each purging sensation, waiting for the release. Once that ended, I spent a good chunk of time in the bathroom with intense diarrhea. I came out with the

70

cleanest intestinal track ever. Meanwhile, Aya has more plans for me tonight.

Bobinsana spirit

Now, I feel the presence of Bobinsana for the first time since I started the plant dieta. I receive a vision of the spirit of the master plant, its aura emanating a blueish color. It is beautiful and eerie at the same time. During the storm earlier today, massive rainfall poured on the retreat center. It filled the pond at the entrance of the property in record time. As I

Figure 23: Bobinsana

walked around the center once the storm weakened, I saw a gorgeous bush of Bobinsana emerging from the pond. I took a picture and zoomed it, only to find that it revealed an alien-like form that made me immediately think of the plant's spirit.

Passing to a higher frequency

Figure 24: Higher Frequency (AI)

Toward the end of the session, the vision faded, only coming back later with a stream of familiar, random, and disturbing visuals. I ignored them and remained composed and unimpressed by the dark energy behind the scenes. It went away quickly. I had swept them out of my consciousness with ease and speed.

In the final sequence, I saw either my father (or myself) pass to the other side. I saw a body resting on a bed and a beautiful blueish smoke leaving through the crown of the head.

I was not afraid of this vision. It was quite the contrary. Seeing consciousness transfer from the body back to its source felt peaceful. I immediately noticed that the color was the same one I saw radiating from the Bobinsana visual I received during that session.

DAY 13: Iquitos RU, November 21, 2023

I slept only a few hours, but it felt enough. Today is the first day of our silent 5-day retreat. I am taking this experience one hour at a time. After doing a mindful yoga session, I went straight to my hammock and let my mind wander for 45 minutes, a very unusual routine for someone like me. During that time, though, I started to think about consciousness and energy fields that power up the human brain, animal brains, and plants. Suppose we define consciousness as the ability to engage with the world around us. In that case, even rocks and water meet the definition of holding a primitive yet fundamental level of consciousness. If that is the case, we all are likely connected to a primary frequency of consciousness that helps us communicate without words and effort.

I went to my plant bath and returned to write my notes from last night's session. Sharing will happen in just a few minutes. I took Chiric Sanango again today. This time, I felt its presence throughout my body. The fatigue was soothing rather than crushing. I could sense the healing flowing in my veins and muscles.

We shared what happened with our last Aya session today, even though it was our first day of silence. Surprisingly, most people were struggling. Elio reminded us that this is typically the most challenging time of the retreat. He shared the archetype of the Heroes' Journey, as explained by Joseph Campbell. Going to the underworld is the first part of the journey, and coming out of it is the second and final part. I realized I was not at the bottom as he shared the familiar U-shape curve illustrating the hero's emotional trajectory. Instead, I had managed to maintain a healthy balance between minor ups and downs. I was proud to be able to stay neutral. That was a testament to the work that I have done for nearly 13 years.

It was evident to all the participants that we had both learned and applied many of the teachings from our past journeys, not just with Aya but other entheogens. I felt the Chiric Sanango most of the day until its effect diminished in the early afternoon. I relaxed for several hours in the hammock, letting my mind wander again. I was learning to be still rather than worry about what to do. It felt so good. As the day closed, I practiced my handpan beside a gorgeous bobinsana plant. I was playing the tunes I learned this week for my master plant. It is so rewarding to feel that I can spend time learning and practicing this fantastic instrument. I have played in all the ceremonies to the apparent delight of the participants. This evening, I also practiced the new Icaro Don Miguel shared with us yesterday. Ready for day #2 of our silence retreat.

The ClipCar Adventure

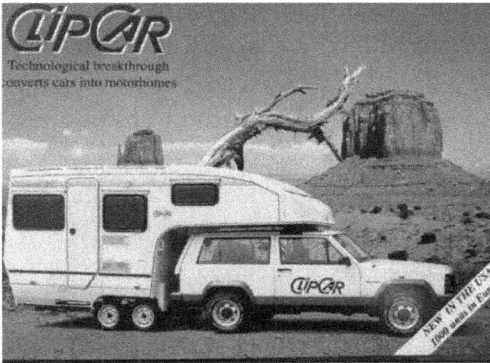

Figure 25: ClipCar

Meet CLIPCAR, a new generation of trailers that can safely and quickly attach over the roof of any car and effortlessly pull it with a standard hitch. This genius product enabled any mid-sized vehicle to tow a comfortable RV as if it were a simple extension of the car. The concept makes both vehicles perform and look like one RV, simplifying all driving and parking operations. It was a remarkable achievement. Only my father could have created this technical miracle alone. But he received plenty of help from me and the rest of my family to market it. I also created a new company that would eventually manufacture CLIPCAR, aided by government subsidies.

Unfortunately, the company was short-lived. Setting up a small shop with the ability to build CLIPCAR was very difficult. We decided to close it and license a local manufacturer to continue the manufacturing. Moreover, while this venture did not bring my father a fortune, it boosted his confidence and rallied our family's energy, admiration, and love around him. Since the CLIPCAR royalties could not support my parents until retirement, as he was only 55 then, my father started looking for another job. Amazingly, considering his age and lack of formal education, he quickly got a remarkable position as Head of Production for a large vehicle transportation company, a job he kept until he retired in 1991. Against all odds, my father had successfully turned his life around, proving he was a great inventor and an exceptional leader. More importantly, he ensured that SOVAM's bankruptcy would not define his legacy.

During this time, I was thriving in my market research consulting job in Paris with FIGESMA. Ironically, I met the owner of FIGESMA, Jean-Louis Lasnier, because of bankruptcy. The French government had hired him to assess the viability of SOVAM and its markets at the peak of the crisis. Despite his ambiguous role with the government, my father liked Jean Louis and urged me to discuss job opportunities with his firm. This was a life-changing move for me.

Jean-Louis was a sharp, joyful man who immediately captivated me. I joined his company as a consultant for a modest compensation package. I wanted to leave Parthenay quickly and learn to become a stellar marketing consultant under his mentorship. I worked with him for four years and was quickly promoted to leadership positions. Jean-Louis treated me very well. I loved working with him on fascinating market research projects for large and mid-sized companies. It gave me opportunities to save companies from disastrous strategic pitfalls by providing them with critical clarity and guidance. He also gave me a love for BMWs and fast driving. It would take us only two hours to drive from Paris to Lille, where his clientele was plentiful. That is an average speed of 110 miles per hour!

Meanwhile, even though my father's life had resumed a semblance of normalcy, the loss of SOVAM left scars on many of my family members, especially on my mother and me. I had exhausted myself for years supporting him. Against my better judgment, I even contracted loans and spent thousands of hours helping the launch of the CLIPCAR. All of that, of course, was driven by my compulsion to save him and my mother. However, I also convincingly dragged the rest of the family members into the process: my brother, my two sisters, their boyfriends, and my wife at the time, who did not always appreciate my obsession with doing so. This pathological need to help my parents often interfered with our plans. In retrospect, I cannot blame her for resenting the situation. I brought her from the United States to live in Paris with me (her dream more than mine), and I spent most of my weekends in Parthenay helping my family.

DAY 14: Iquitos PERU, November 22, 2023

I had a fair night, finally. However, I woke up, yet again, after a strange dream. All I could remember was hosting a few friends in an apartment, presumably my home. It was a couple made of one man and one woman. They cuddled with me in my bed, which made me very uncomfortable. I suggested to the woman that I had an extra room. She said she preferred to go home. As I was checking the room I intended to share with my "forced guests," I saw one of my cats lying on some disgusting green gel that he appeared to have eaten already. He was not dead but seemed rather drunk. I was concerned about his health and woke up. Trying to make sense of this dream was challenging. I understand that many messages we receive from dreams are hidden unconsciously and can rarely be interpreted literally. This one was difficult to decode, however! I played with a few interpretations.

Hosting friends in my home could symbolize my sense of responsibility or my role in supporting or caring for others (Savior Complex!). Welcoming someone in my house could also represent openness or vulnerability as I shared my personal space. The part of the dream where a couple was cuddling with me could suggest an invasion of my boundaries. My discomfort might indicate a real-life situation (the retreat) where I feared others would infringe upon my personal space or boundaries. Finally, the woman's decision to leave instead of staying in the extra room I offered could signify a part of me that wishes to retreat from uncomfortable situations rather than confront them directly. Meanwhile, cats in dreams often represent independence, curiosity, and intuition. The cat that ate something harmful (green gel is close to the appearance of Aya!) and appearing drunk could symbolize a concern about someone or something close to me being affected by a negative influence. The fact that the cat was not dead but seemed unwell could reflect a fear of something or someone being hurt. This might relate to my concerns about how specific actions or behaviors (either my own or others) affect my well-being or that of others I care about. Finally, waking up worried about my cats' health could mean that this dream was indeed connected to my real-life concerns, possibly about my pets (I left my five cats for a month!), my environment or my home being compromised somehow.

Today is day 2 of our silence. It is challenging to cross people and not greet them with a quick gesture or facial expression. This reminds me of our innate social nature and how much we all need to connect with other human beings, especially those we know and appreciate. Meanwhile, I am making significant progress writing OPEN, even though I woke up in the middle of the night thinking it was a complete waste of time. My monkey mind was messing with me like Aya sometimes takes me to darkness. However, after finishing The Fellowship of the River from Dr. Joe Tafur, I was reenergized again. So many of his key points align with my thinking, though we come from very different perspectives. I finished a course with Joe just before coming to Peru and enjoyed his authenticity, humility, and passion for traditional healing practices. It is refreshing to see a medical doctor embrace the neurospiritual approach, as Dr. Gabor Mate, Dr. Dan Siegel, Dr. Deepak Chopra, and many other doctors have already.

We met in the maloca today for a lecture on master plants. Don Miguel insisted on the sacred nature of the plant dieta. He told us stories of completing multiple dietas before becoming a Maestro. He claimed that while the process was long and complex, it was gratifying once you had acquired the power to communicate with the spirit of the plants. Each master plant has its personality and different ways to connect with a dieter. It can take months, if not years, to engage fully with the spirits of a master plant. Here are some key differences between the plants, according to the Shipibo people.

Noya Rao

This tree is very sacred among the Shipibo people. The energy of Noya Rao is revered as well. It teaches us to live in balance and harmony while helping us find the truth within. It is believed to bring greater capacity to think, intuit, and reveal the unconscious, allowing us to face our limiting beliefs. It brings air energy, enabling shamans to shapeshift into birds or crocodiles, especially if they are in danger. Don Miguel explained that a shaman (Onaya) may use his powers for evil purposes instead of healing and protecting. Such an individual would be called a "brujo" in a negative context, indicating a sorcerer or witch who engages in harmful magical practices. Interestingly, "the plants do not discriminate between healers and sorcerers," Don Miguel pointed out.

Chuyachaki

This plant is known to help shamans connect with animals. It is common for people dieting Chuyachaki to have visions of an older man with big shoes. The plant's spirit can also appear as an insect or an animal. The tree itself is considered the king of the forest. It treats pain and anxiety and provides clear thinking while bringing the body into profound union with nature.

Chaikuni

This master plant can help you find plants you can use to heal others. Dieting this plant can make you relatively weak, but you feel transformed at the end of your diet. Don Miguel also reminded us that many master plants may test us to see how serious we are about the dieta. It is commonly expected that one might have visions of a beautiful woman or delicious food to tempt the dieter.

Marusa

Marusa is excellent at healing problems with the female reproductive system and balancing the nervous system by stimulating the parasympathetic branch of the Autonomic Nervous System. It can also increase self-confidence. Don Miguel did not elaborate on Marusa since it is not a plant we could diet at his center. However, he did mention that this is a very jealous plant.

Bobinsana

The last master plant Don Miguel discussed tonight was the one I am currently dieting: Bobinsana. He said it can take quite a while before you feel its spirit, but it is kind and generous. It can also clear headaches. The plant loves water and may appear in our visions as a mermaid. It is known as a gentle spirit that heals and opens the heart. It can also stimulate colorful and lucid dreams.

That might have explained how intense and vivid my dreams have been since I arrived.

We also learned a new Icaro today. I remember the melody well; this is one of my favorites, and Don Miguel and other shamans often sing during Aya ceremonies.

Icaro #3

NESKA NESKA SHAMAKIN −2 ↑ ↑
Calling, calling the healer, rise.
YORA PARI ABANO −2 = ↓
Spirit of protection, descend.
PANA YON PARIBANO −2 ↓ ↓
Embrace us with deep healing.

PONTE YOUPARIBANO −2 ↑ ↑
Bridge us to higher healing, rise.
KANOBO PONTE BANO --2
With the bridge, we are connected.
YOI YOI SHAMANKIN −2 ↓ ↓
Hear, hear the call of the healer, descend.

SOWA SOWA SHAMANKIN −2 ↑ ↑
Flow, flow with healing spirit, rise.
NETE PARI ABANO ↓
World's spirit of protection, descend.
NESKA NESKA SHAMANKIN ↓
Calling, calling the healer, come down.
NANTA NANTA SHAMANKIN −2 ↓ ↓
Guide us, guide us, healer, descend.

YORABO ABANO −2 ↑ ↑
Protect us, spirit, rise up.
PAyaN YONPARIBANO ↓
Let love surround us, descend.
OWIN YONPARIBANO ↓
Let calm embrace us, descend.
MANA YONPARIBANO −2 ↓ ↓
Let peace envelop us deeply.

Today is day #3 of our silence. I woke up at three, having only slept for about five hours. It is my average here, but I am not worried about it. I have a very slow routine anyway. We eat only twice daily until Saturday, when we resume our regular schedule. I had some significant insights from my book today while reading The Transcendent Brain by Alan Lightman (16). The first part was disappointing, as he tried hard to prove that the soul is only material. However, the second part was excellent, discussing our current understanding of consciousness from a brain perspective. It reminded me that I had discussed this topic extensively in my doctoral dissertation and now believe it will be a critical section of OPEN.

Meanwhile, we learned to prepare tinctures made from our master plant today. It is an exciting process that involves grating the bark of the trees or mashing their leaves, then adding sugar cane alcohol to extract the medicinal elements. We buried several jars underground to let the concoction steep for about ten days before we could filter the bark to extract the concentrate. We closed the evening by going to the Noya Rao tree. It was a beautiful evening and a lovely opportunity to reconnect with all the participants amid our requisite period of silence.

Christina, the savvy and seasoned Spanish woman who has spent much time in the jungle with master teachers and who has engaged in multiple plant dietas, the one who shared her knowledge about making tinctures, motivated us to learn more about "spagyrics", a field related to alchemy that investigates plants' psychic and spiritual powers. Spagyrics is a holistic approach to health that originated in the Middle Ages and was developed by alchemists like Paracelsus. The word "spagyric" comes from the Greek words 'spao" (to draw out) and "ageiro" (to gather), reflecting the process of separating and then recombining elements of plants. Spagyrics involves fermenting, distilling, and extracting medicinal properties from plants. The process aims to purify and concentrate these properties. The practice focuses on isolating three critical components of plants – the salt (body), the mercury (spirit), and the sulfur (soul). After purification, these components are recombined.

Spagyric remedies are intended to treat the body, mind, and spirit. The belief is that physical ailments are connected to spiritual and emotional states. Alchemy in spagyrics is not about the literal transformation of substances into gold, but rather the transmutation of natural substances for healing. It emphasizes the interconnection between the macrocosm (the universe) and the microcosm (human beings). Spagyrics also incorporate the concept of vital energy, suggesting that the remedies contain the life force of the plants, which *can help balance the life force of the person taking them. It is important to note that spagyrics is considered a form of alternative medicine, and its practices and efficacy are not typically supported by conventional scientific research.*

DAY 16: Iquitos PERU, November 23, 2023

I had another decent night, finally. Just six hours seems enough to go through the day without feeling fatigued. I was energized to return to my desk around 4 AM and spend long hours on a very uncomfortable stool, despite all my attempts to mitigate the discomfort using my bed pillow, which falls off every 10 minutes. This is turning out to be a writing retreat, and I love it. I believe all the medicine I have ingested so far, and especially dieting Bobinsana, fuels my eagerness to write. I have already typed more than 35,000 words to prove it!

DAY 17: Iquitos PERU, November 24, 2023

I woke up again at 3:30 this morning, ready to resume writing. I have done this every day since I arrived in Iquitos 17 days ago. I am making good progress on the structure of the book. I have an excellent outline to discuss the relevance of focusing on plasticity to heal many psychological disorders. Meanwhile, I am not as worried about the limited food we receive daily. My body has adapted well to the intermittent fasting.

I managed to practice my tunes on the handpan and continued to enjoy reading Alan Lightman's book, The Transcendent Brain. I was seduced by his belief that the brain can produce all states of consciousness, including spiritual states, states of awe and wonder, and many others that have puzzled scholars, scientists, philosophers, and religious leaders for thousands of years. However, while I appreciate his brilliant intellectual exercise, I need more convincing. Even if I accept the fact that the brain is indeed able to create emergent states that transcend the more basic functions of regulating our body, producing emotions, and performing complex computations, Lightman does not share any evidence that higher states of consciousness, especially those made by feelings of oneness are not enhanced by energetic fields outside the brain.

Also, since Lightman mentions only one account of a friend who shared a remarkable LSD experience, it suggests that he has not worked with psychedelics himself. If that is the case, how could he understand what journeying under the guidance of plant teachers does for consciousness? In other words, how can anyone believe that our brain is incapable of tuning to higher frequencies of consciousness without direct experience of being in a non-ordinary state of consciousness (NOSC), such as those induced by psychedelics?

After decades of denial, I now believe hallucinations and spiritual insights produced by NOSC far exceed the potentiality of our biological wiring.

Tonight, we were invited to another engaging workshop. Christina taught us all how to prepare a tincture made with camalonga. This is a strong medicine that shamans use to provide spiritual protection. It comprises ½ male and ½ female seeds found in a seed about the size of a walnut. Camalonga is always purchased in pairs in this spiritual tradition, so she recommended we mix a jar with five pairs to make the medicine strong. Once you crack the seeds open, they must be cut into small pieces and placed at the bottom of a jar, filling a quarter of its volume. After that, add a cut of male garlic and 4 or 5 cuts of a small white onion to fill another quarter of the jar. Then, add two small blocks of camphor plus sugar cane, and that is it. Three huayruro beads are also included to add flavor to the concoction. The tincture can be dropped on the tongue or sprayed on the face. Christina insisted that we do not underestimate the power of this medicine. Don Miguel uses it for each ceremony to protect himself and often mixes it with Aya.

In closing, Christina shared many fascinating Indigenous creation stories in the last part of our workshop. She recommended we look up IMIKA from the Tubu tribe on YouTube (30), where he narrates many of the creation stories. I smiled when I noticed the tribe's name reshuffled, which would make the word UTUB like YouTube. Christina works with many ancient tribes all over South America. She said that elders feel it is urgent to communicate essential messages to as many people as possible to awaken us all to the imminent danger of climate change and the devasting impact of consumerism. In the past, I have often received these messages with a lot of skepticism and judgment. However, I am consciously working on my plasticity to ensure this does not happen again during the retreat. It is not always easy!

I look forward to closing the day so we can resume our routine tomorrow evening with the first of the last set of five Aya sessions.

Synchronicity at Play

A few years after the SOVAM bankruptcy, and once my father's new career opportunity flourished, I managed to focus back on my career. Indeed, after spending four exciting years at FIGESMA, I wanted to test my marketing skills on a bigger fish. I joined a knitting yarn franchising company called PHILDAR, based in Roubaix, at the border of France and Belgium. At the time, PHILDAR was the largest producer and retailer of knitting yarn worldwide and one of the largest franchisors in Europe, with over 3000 retail stores.

If Parthenay was dark and depressing, Roubaix was ten times worse, yet I decided to live there for a year to pursue this exciting opportunity. I did not want to move permanently from Paris and welcomed the emotional space between my wife and me. Just a few months after our wedding, we had ongoing bouts of friction, and I knew intuitively that the marriage was in trouble. PHILDAR agreed to rent me a studio for a year and pay for a one-year pass to take the fast train TGV every Friday to come home. One year was all I could take anyway. I enjoyed the position of Head of Global Market Research. I was proud to be the youngest executive on the management team of a large company at the age of 27, but living in Roubaix was not a long-term option. I have fond memories of traveling extensively throughout Europe to research women's motivation to buy knitting yarn. Fewer and fewer women were knitting; therefore, the erosion of the popular tradition was threatening the company's future.

I remember feeling strange coming in and out of knitting yarn stores to visit franchisees since most men did not knit or buy yarn then. However, I was stimulated at the prospect of becoming a hero (yet again) as PHILDAR faced severe headwinds. Using my training at FIGESMA, I wanted to uncover patterns of attitudes and behaviors that could save a business employing thousands of workers and serving millions of customers. In just a few months, I created promising new retail concepts in the UK and France with complete freedom to exercise my research and creative skills. I loved the job and might have considered staying longer if the company had not been based in Northern France. However, my soul had other plans.

One day, walking through downtown Lille, a beautiful city a few miles south of Roubaix, I ran into Luc Doublet, a former client from FIGESMA, for which I had completed a few projects. He was excited to see me, as he was actively looking to hire a senior executive who could open and run his U.S. operations in San Francisco. He remembered I held an MBA and appreciated the consulting I had provided to some of his companies while at FIGESMA. Therefore, he offered me the job on the spot—synchronicity at its best again.

Within a few months, on February 18, 1988, I moved back to the USA to my favorite American city. In 1983, after graduating from Bowling Green State University in Ohio with an MBA, my college best friends and I embarked on a cross-country road trip. As we journeyed across the United States, attending rock concerts and embracing youth adventure, San Francisco was our final destination and also where our car died. I instantly fell in love with the city's irresistible charm and entrepreneurial energy. Even then, I knew I would return—indeed, I did.

*I woke up at 2:45 AM this morning and could not go back to sleep. I had weird sensations that reminded me of my **bufo** experience last year. This was one of the most potent psychedelic experiences of my life. I would not recommend it to many people, considering the intensity of the effect. Here are my notes from the session.*

Bufo: the god molecule (notes from prior session)

I made the difficult decision to receive bufo today. I have absolute confidence in my friend's ability to serve it, as he has extensive medical and shamanic training. Therefore, I am ready to surrender fully to this famous psychedelic medicine, often referred to as "the God molecule," due to its intensity and potential to initiate powerful mystical experiences where one can feel at one with the Universe. With bufo, I do not need to decide how much I should take, which is somewhat liberating.

Figure 26: Bufo Vision (AI)

I was asked to stand in front of my friend and prepare to fall back into the ready arms of facilitators once the medicine took effect. Within seconds of inhaling the toad medicine, I felt my body drop and was transported into what seemed to be a cosmic experience of eternal nothingness. I instantly began to experience an incredible feeling of lightness and bliss. I felt I became like an electron or maybe a distant planet.

I was the closest I have ever been to the God molecule, the essence of divine energy. I felt I was merging with a higher power, receiving profound insights into the nature of existence. My friend who served the bufo had informed me that he would play the gong close to my head about 20 seconds after he delivered the dose to be sure I had received the correct amount. I heard nothing. I felt disconnected from reality for what seemed to be a long time, though it turned out to be only about 20 minutes. I sensed intense vibrations throughout my body until I regained ordinary consciousness. During the session,

I felt my friend's presence and the team assisting him. I was never afraid and felt bathed in love and protection with ease and bliss.

The essence of who I am was so clear to me. I felt immense gratitude towards my caretakers, wife, family, and myself. I felt that I needed to take care of myself to enjoy my body until I died, presumably to ensure I would be ready for a new life after death. This is not a thought that ever entered my mind until I was under bufo. This session profoundly transformed me. My work with Aya for many years and the spiritual and medical guidance I received from my friend prepared me perfectly for it.

So, I woke up last night because I was vibrating in my torso for at least 10 minutes. I got concerned, but it eventually receded. I wonder if that was a case of reactivation, a well-documented phenomenon among those who have experienced bufo. The term "reactivation" describes the spontaneous recurrence of certain aspects of the psychedelic experience after the initial experience has ended. It can happen days, weeks, or even months after using 5-MeO-DMT (bufo). These recurrences can include sensory, emotional, or cognitive experiences similar to those felt during the actual psychedelic session. The intensity and duration of reactivation can vary. Some individuals might experience brief and mild effects, while others may have more intense and longer-lasting experiences. Various factors might trigger a reactivation, including stress, use of other drugs, particular states of consciousness like meditation or falling asleep, or even seemingly unprovoked spontaneous occurrences.

The neurological basis for reactivations is not fully understood, but it is hypothesized that 5-MeO-DMT may cause changes in brain function or structure that led to these spontaneous recurrences. Coincidentally, we had a stimulating discussion on bufo last night with all the retreat participants. Almost half of them have already participated in bufo ceremonies.

Meanwhile, I wrote between 3 AM and 7 AM, did yoga afterward, took my Bobinsana plant bath in the medicine house, and performed essential rituals. We are encouraged to perform a precise sequence of spiritual steps before we drink our master plant. In my case, I sing my Bobinsana icaro and then blow mapacho smoke around myself and inside the glass holding the mix. Finally, I pray. When we moved from Honolulu to Kona, I gave up my daily run because there was no easy way to do that around our property. I used to combine my run with prayer, so my routine faltered after we moved. I am happy to bring back yoga and praying as part of my daily routine here and am very committed to continuing when I return to Kona. I should know better since I researched the power of praying for my last book, and the health benefits of carrying out the sacred ritual are awe-inspiring, such as lower stress, stronger resilience, improved immune functions, better sleep, emotional balance, and more.

Later in the morning, we learned another Icaro with Don Miguel. Before that, however, Don Miguel asked us how we were all doing. Most participants reported that they were doing very well. This was a sharp contrast with a few days ago. However, two of our three friends attending the retreat were struggling to integrate the first half of the retreat during our silence. The approach was working us hard. I asked Don Miguel if my vibrations could have been related to Bobinsana, and he said yes. He also suggested that after two weeks on the plant dieta, we should start feeling the plant's spirit. Maybe that was the first sign.

Here is the Icaro we learned today.

Icaro #4

NIWE TORRI CAMPANA...Je –3 ↑ ↑ =
Sacred bell of nature, rise and resonate.
NETE ANY KEPENE –2 ↓ ↓
World spirit, descend and bless.

NOYARAO RIOSBO...Je –2 ↑ ↑
Moon medicine, rise up.
JOYO JOYO NIWEBO...Je =
Blessings of the sacred, remain with us.
JONI SOWA SOWAWA –2 ↓ ↓
Essence of life, flow down.
EN KIYO KIYO BAN...Je ↑
Awaken, light of the stars.
MIN RAMI NEWEBO =
Your essence is pure and sacred.
EN KIYO BOKINRA...Je –2 ↓↓
Come down, starlight, and guide us.

SOWA SOWA BOKINRA ...Je ↑
Flow, flow through us, star spirit.
PISHA PISHA BOKINRA =
Shine, shine upon us, light.
NETE SEN(e)AINKO =
World spirit, hold us steady.

MAya AKETANASH –2 ↓
With love, encircle and protect us.

MA(a)NIRA BEKANAI ...Je ↑
Pure heart, rise up.
NOKI BEIRANIRA
Guide us with gentle strength.
WEI BERANTANA...Je
Calm us with your presence.
CHITI CHITI SHAMANI –2 ↓↓
Quietly, softly, healer, descend.

MATO KANOMAYONBAN...Je ↑
In sacred union, we rise together.
JA NETEN PIKOSHON ↑
Stand strong with the spirit of nature.
MANA MANA SHAMANBAN….Je
Blessed healer, come to us.
MIN YORA MEABO –2 ↓↓
With great love, stay by our side.

This is another beautiful Icaro. Don Miguel insisted that we put power in our voices and move as we sing. He said singing opens the heart. I am preparing now for the sixth Aya session of the retreat. I feel tired from not getting much sleep and have a light headache. However, I feel ready for the next ceremony.

Aya Session #6

My head started hurting about an hour before last night's Aya session. It got worse throughout the ceremony. Usually, I am patient and know that Aya will take care of most of my headaches. However, this one is different. I was in and out of consciousness for at least an hour, during which time I could not focus on any of the visuals. As I emerge from my mental fog, I notice Aya's signature is unusual this evening. The visuals are mostly very crisp at times, moving slowly, and do not include disturbing elements like snakes, insects, and deformed beings, which show up often for me.

Galactic war

The first series of images shows an army of soldiers marching to the top of a hill, pushing an enormous weapon. Initially, the soldiers reminded me of a battalion of warriors from the Middle Ages. Still, subsequent scenes show commanders whose faces are all covered by black masks, which immediately makes me think of Darth Vader and his soldiers from Star Wars. The next scene shifts the story to images from the chaos of war.

Figure 27: Galactic War (AI)

At that point, I think the battle I witness is part of a galactic war involving evil forces. A few seconds later, the story shifts to a different dramatic unfolding. I see spaceships killing dinosaurs on Earth. I begin to wonder then if these ancient creatures were all exterminated by aliens rather than by the giant asteroid that caused the big explosion, wiping about 75% of Earth's species 66 million years ago. Could aliens have played a role in this ancient event? Embracing my capacity for plasticity, I welcome these stories without judgment, considering their symbolic significance. The recurring alien theme in my sessions made me realize Aya might be urging me to pay closer attention to it.

The Hunger Game

The second story brings me back to the story of the presumed kingdom I discovered a few sessions ago. I see a group of women wearing traditional dresses enjoying a tea party near the banks of a beautiful river. They all have exaggerated makeup on their faces, like the psychopaths in the Hunger Games movie. Later, I witness several of their bodies being pulled out of the water. I cannot tell if they have drowned or been killed, but the atmosphere is eerie. I don't know what to make of this unless it relates to the earlier war story. Dark forces are at play in both stories so far, no doubt.

Between two worlds

Finally, we are approaching the end of the session, and more visuals are returning to me, which is rather unusual. This time, though, I see patterns of electrical frequencies running vertically on a dark background, with one band more noticeable than the others. Now, I feel like stretching by moving my arms above my head and joining my hands. When I do, I notice that the most robust electrical bands in these frequencies follow my hands. Though this effect is stunning, I assume it is a coincidence.

Figure 28: Between Worlds (AI)

However, the best way to find out is to test my hypothesis through a simple experiment. The mind of a scientist never stops! So, I start to move my hands up and down very slowly. Shockingly, the band follows me. Now, at this point, I am still closing my eyes, and we are nearly three hours past my medicine ingestion. This is astonishing!

Towards the end, I performed my groove on my handpan to the group's delight. I am still amazed at how quickly I have learned to play this magical instrument. What a gift playing music is!

DAY 19: Iquitos PERU, November 26, 2023

Today, we learned to cook Aya. Don Miguel asked us to smash pieces of the woody vine that had already been cut into 12-14-inch pieces. Then, the flattened strips of Aya were mixed and boiled with chacruna leaves. Chacruna contains significant amounts of dimethyltryptamine (DMT), a powerful psychedelic substance structurally like serotonin and found in many plants and animals. Aya supplies the MOI inhibitor, preventing the breakdown of DMT in the digestive system. Adding a significant amount of water takes about 15 kg of Aya and 3 kg of chacruna leaves to prepare two liters of the brew.

On numerous occasions, scholars studying ancient plant medicinal rituals have documented tribal elders' claims that the spirits of the plants guided them in identifying the ingredients for the psychedelic brew. According to some Indigenous narratives, the knowledge of combining these plants was revealed through spiritual or mystical insights rather than through trial and error. The renowned ethnobotanist Dr. Richard Evans Schultes dedicated his life to researching what he termed the 'plants of the gods,' also the title of his seminal book on the ritual use of psychoactive plants (10). Schultes confirmed that, throughout history and across the world, Indigenous peoples have used psychedelics primarily in sacred ceremonies, aiming to facilitate healing spiritual experiences. This purpose contrasts sharply with the historical recreational use of psychedelic drugs in Western society.

So, the ratio of how much Aya is required compared to the quantity of Chacruna leaves is 5 to 1, but that can be changed to create different neurophysiological responses. More Aya in the mix may trigger more purging; less Aya may enhance the visuals from the effect of DMT in the chacruna.

Once Aya and chacruna are placed together in the pot, the batch must boil for about 13 hours; then, the first round of filtering occurs because only a tiny portion of Chacruna is mixed with Aya. Once the Aya and chacruna leaves have boiled long enough, the matter left in the pot is removed. After that, more chacruna is added and cooked for another hour. The leaves are filtered out of the mix yet again, and the remains are boiled until they thicken into a soup-like texture.

Many rituals, such as singing and praying, are associated with the process; it was beautiful and inspiring to see the respect the Shipibo people show for the different steps involved. Meanwhile, Christina encouraged us to learn more about Chaliponga, another potent DMT brew used by the Huambisa people, who have a deep cultural and spiritual connection with that medicine. Chaliponga keeps purging to more tolerable levels.

We closed the day with the sharing circle to discuss our last Aya session. It was poignant. All the participants reported considerable progress in addressing their pains, traumas, addictions, anger, or whatever condition they came here to get rid of in their lives. None of the stories compare on specifics, except that most participants here appear to be at the end of their rope when they first arrived. No one wanted to give up, however. Thanks to this healing tradition, they have been tapping into new hopes and dreams that they thought were once unattainable. Anyone invited to witness the sharing today as a trusted observer would be able to see the remarkable healing offered by this Shipibo community. It is not easy to comprehend, yet the benefits are unquestionable.

DAY 20: Iquitos PERU, November 27, 2023

I woke up at 3 AM again this morning, unable to fall back asleep. I started my writing routine. Today is our session #7 with Aya. I feel more vital than ever. This morning, we gathered for another vomitivo ceremony. This time, we used a different purgative for purifying the body. The process is never enjoyable, but the sensations you get afterward are compelling. We did our last sauna as well today. I will remember those treatments fondly and intend to build a medicine house with the same sauna set up for our retreat center in Kona. I did much writing today, as well as essential readings. I love this schedule. I want to create more space for writing activities and playing my handpan when I return home. These bring me so much joy while also stimulating my mind.

Aya Session #7

I decided to ask for two-thirds of a cup last night. The medicine was slow to kick in, but when it did, it was intense. I had many visuals for nearly three hours. What surprised me most was the emergence of three central themes that seemed to dominate the session: surgery, human harvesting, and the interconnectedness of nature. These themes felt deeply intentional as if medicine was guiding me to confront profound aspects of existence and my relationship with the world.

Surgery

I see Aya scanning my body for possible problems with my organs. It is a fascinating process. I notice a massive dashboard with three-dimensional transparent beings representing my body's internal physiology. Liquids flow in all directions, moving in and out through tubes. It looks like I am receiving a whole blood transfusion, or my blood is being filtered and purified. My spinal fluid and all the water in my body are also filtered and purified.

Figure 29: Surgery (AI)

It is a beautiful symphony of pulsating liquids nourishing my being.
Aya confirms, yet again, that I am in top health, and she tells me I will enjoy at least three more decades of vital life, a welcome notion since I am in my early sixties. She insists that she cannot predict when I will pass, but if I take good care of my body, it will last for a long time. Finally, I ask about my knees,

which have troubled me as an avid runner for decades. She reminds me that there is not much she can do for bones that are damaged by wear and tear and rheumatism.

Human harvesting

Following this remarkable story, the alien narrative reemerges, which has been a running thread in my sessions here. This time, I see a gigantic broom-like spaceship scrubbing the ground to scoop up people running for their lives. Initially, I thought the victims were killed in the process, but now I have the impression that they are harvesting humans.

Figure 30: Human Harvesting (AI)

The scene reminds me, yet again, of a Star Wars movie. Seeing lots of horrific details and grasping the drama and horror of the human targets is disconcerting.

I purged several times and spent a lot of time in the bathroom. Aya helped me resolve some constipation issues that had started when we were following an intermittent diet during our silence several days ago. When I asked her if this blockage suggested that I was stuck in my life, she merely suggested that it was a physiological issue she wanted to address. I like Aya's ability to get straight to the point!

Interconnectedness of Nature

Walking back toward the maloca, I am still tipsy and moving like a drunk. I decide to lean on the railing to look at the surroundings. The full moon provides enough light to witness an incredibly detailed net-light grid connecting all the plants, bushes, and trees. In the past, I have mostly seen geometric patterns that resemble a honeycomb. However, this resembles delicate strings, like fishing nets, ascending to the sky. It is stunningly beautiful. I could not ask for a more inspiring visual to remind me of the interconnected quantum nature of the universe.

I return to the maloca for more, but at this point, the medicine's effect is receding. However, I feel calm, grounded, and happy. The icaros gently rock my *soul until it is time for the open session, during which the group is invited to offer their songs. I am in awe of the quality of each song offered tonight. There is so much talent in our group.*

Figure 31: Interconnectedness (AI)

Besides me, one of the participants, who had not yet found the courage to contribute her voice to the group, surprised us all by singing two beautiful songs. Just moments before she began singing, she was still on her mat, purging violently and going back and forth to the bathroom. Knowing the depth of trauma from a prior sharing session, I could appreciate the severity of her condition, and I assumed that the purging was helping her go through a critical physical and emotional clearing process.

Later that same evening, she shared with us that she believed Aya had indeed removed a big part of her suffering in that way. Tonight, according to her, the despair and anger finally released their grip, enabling the young woman to showcase her victory by singing these uplifting songs. It was a magnificent moment and, yet again, an example of the type of soul surgery Aya can perform for people consumed by fear and anxiety.

I managed to pull myself out of my drunk stage to perform an improv on my handpan. I love this instrument so much. I used a headlamp while I performed. My wife told me later that it made my playing even more mesmerizing because the only parts of me that people could see were my hands and thumbs furiously bouncing up and down on the instrument.

Towards the end of the ceremony, I had a few more visuals of my passing. In the vision, my wife was with me and wanted to ensure we went together. We often imagined we would die together after a final embrace. This last scene was poignant and beautiful. The ceremony closed with a symphony of laughs and celebratory shouts. What a remarkable evening.

DAY 21: Iquitos PERU, November 28, 2023

This is our day of rest before our eighth ceremony tomorrow evening. We were offered Chiric Sanango again before breakfast. I continue to like how this medicine relaxes me so I can feel more flow. I am eager to see how the group is doing. As the sharing unfolds, the ceremony becomes a fantastic celebration of progress and hope. Each participant reports massive shifts thanks to Aya's teachings and, as a result, feels more robust and optimistic than ever before. Listening to other's stories and insights is part of the magic of this process.

Generally, we all think our suffering is unique and randomly created by our life circumstances. However, these moments of sharing demonstrate that we have more in common than we believe. Raising our consciousness is a process of accepting and reframing the nature of our suffering and trusting that if others are starting to heal, there is more chance we can all follow the same path.

I am biased toward providing individual therapy to people who work with plant medicine so that they can receive more support, guidance, and integration in the transformational process we are in. Unfortunately, it is not part of the Shipibo tradition.

I am ready for our eighth Aya session tomorrow. I hope she will continue to be kind and generous to me.

Becoming a father

I began my life as a career-driven CEO at 28 in the USA, filled with hopes and dreams of growing a vibrant business while building a family in the land of all opportunities. I ran the US subsidiary of DOUBLET passionately and fervently for seven years. I quickly inhabited the persona of a young executive with an avid desire to impress while continuing to save the people I

Figure 32: My Office in San Francisco (1988)

loved along the way. I will not elaborate on more personal details of my first marriage since it eventually ended in a challenging divorce process that lasted three years. I know today that I failed to recognize the extent to which I lacked consciousness and maturity to address my relationship issues.

I had tried to hold on to the idea of a perfect marriage for several years. The truth is that I should have paid attention to how much I had questioned my readiness and commitment from the beginning. I was so blinded by what I had to do to save my family that I missed essential signs pointing to the inevitable:

the soul contract with my first wife was going to be short-lived. Once again, I assumed I could save or fix whatever was not clicking. I deeply regret causing her and my two boys so much pain in the process. I am grateful for the two sons that she gave me. I will be forever praying for her happiness and well-being.

My first son was born in 1988 in San Francisco. What a fantastic day it was. He was born happy. Nothing seems to affect his optimism and resolve. My second son was born four years later, a time during which my marriage was unfortunately already deeply fractured. On the other hand, my second son needed more attention from a very young age. He was diagnosed with a brain tumor when he was only six months old. While the tumor was eventually found to be benign, I remember feeling terribly distressed and completely helpless to save him.

That feeling persisted when he was later diagnosed with *Tourette's syndrome* (TS) at the age of 6. TS diagnosis typically requires the presence of multiple motor tics and at least one vocal tic for over a year, with onset before the age of 18. A comprehensive medical history review helps rule out other conditions with similar symptoms, such as seizures or medication side effects. During the diagnostic process, a neurologist or psychiatrist observes the tics directly and confirms they have been consistent without a break lasting more than three consecutive months. There are no specific tests for Tourette syndrome; diagnosis relies solely on symptom history and clinical observations.

Many individuals with TS experience a peak in tic severity during early adolescence, with a noticeable improvement in late adolescence and into adulthood. For some, tics can become less frequent and intense over time, and in some instances, complete remission may occur after adolescence. However, about one-third of individuals with TS continue to have significant tics into adulthood. Meanwhile, long-term outcomes also depend on the presence of common co-occurring conditions. These can include attention deficit hyperactivity disorder (*ADHD*), obsessive-compulsive disorder (*OCD*), anxiety, and depression.

When I received the diagnosis, I believed there would be a quick fix to cure or control the symptoms of the rare neurological disorder. I was utterly wrong. Fast forward ten years to meeting him at the most challenging time of this condition: adolescence. Being a teenager with TS was more than our relationship could handle, especially in the middle of a divorce with his mother.

DAY 22: Iquitos PERU, November 29, 2023

Last night, I woke up multiple times with very annoying dreams. It was the first night I had slept nearly eight hours, but I felt exhausted from dragging myself out of bed. One of the central themes of my dreams was my involvement in a murder that involved members of my family. I was trying to cover it up with their help. It was not clear what the circumstances of the crime might have been, but I was clearly in trouble.

I felt guilty and ashamed in the dream, and the strength of those emotions triggered my brain to call for a break in the dream. I have had lucid dreams for many years. A dream is considered lucid if you can recognize within the dream that you are dreaming. As a result, you can interact with the dream to redirect the narrative or to stop it. I realized this morning that I often try to do this when working with Aya. While I trust the medicine, I cannot resist attempting to influence its plan. For the most part, I have mastered this over the past many years, but at what cost? I may not receive all the messages I need. My ego may be so strong that I am still filtering critical insights I need to integrate. I vow to try to maintain more plasticity in the ceremony tonight.

Later in the morning, we visit the medicine garden. I am excited about continuing my education about the plants we have been dieting and experiencing in treatments. Don Miguel walks us through his "green pharmacy" and explains the properties of multiple plants. As the remarkable teacher that he is, he passionately explains and demonstrates how each plant can deliver benefits ranging from relieving pain to stopping seizures, which is fascinating. While he does that, I wonder how indigenous people received all this knowledge over thousands of years. Shamans often claim that the plants themselves provided the knowledge. That would be the most credible explanation of all. Some plants can be served as a brew made of leaves or bark; others are prepared to be sniffed, crushed into a liquid for eye drops, mashed into a greenish paste to make an emplasto or poultice, to be placed on an ache or wound, or mixed into either cold or boiling water for baths and saunas. I recorded Don Miguel's lecture so that I could transcribe it later.

Finally, we are invited to partake in a "love bath" in the medicine house. The "bath" is a beautiful cocktail of plants that clears bad energy among family members and friends. The potion is meant to restore harmony and attract positive relationships. Don Miguel tells us that Shipibo people frequently spray their homes with this bath water to reap its benefits.

Aya Session #8

We have the privilege of drinking the Aya we cooked a few days ago tonight. I am excited to see where it will take us. I take half a cup, as this dosage has consistently provided me with the right balance of visuals and insights. The brew takes almost 40 minutes to kick in. I can feel the potency; a few minutes into vivid visuals with bright colors and impressive clarity, Aya's intensity *floods my consciousness. A few years ago, I might have asked the facilitators to do whatever they could to lower the effect. Eating honey or drinking electrolytes were two ways I believed, with little evidence, could mitigate the reaction. However, this time, as I have done during all the sessions so far, I stay at a healthy level of meta-consciousness. I can observe my state rather than fully immerse myself in it. That distance gives me freedom and peace of mind to appreciate the intoxication rather than resist or, worse, sink into a state of fear.*

Plasticity habits

As soon as the intensity of the visuals decreases, I ask Aya to continue giving suggestions on improving my draft of OPEN. I have over 40,000 words already, a therapeutic formula based on three steps: raising consciousness, reclaiming personal stories, and reprogramming plasticity habits into your life. I feel I still need the most help with plasticity habits. Aya guides me to keep my ego in suspension mode. By doing so, a large quantity of energy can be freed to heal and remove resistance, self-judgment, and self-pity. I can surrender to the teachings of any challenging experience. She also encourages me to reflect on the seven self-love habits I had presented in The Serenity Code and to identify other ways to catalyze the most plasticity. This is a productive dialogue that will fundamentally change the direction of the last step of the book. I am thrilled and immensely grateful.

Letting go of money worries

Figure 33: Letting Go of Money Worries (AI)

After the dialogue with Aya on plasticity ends, I purge a few times and go back and forth to the bathroom for at least 30 minutes. In that process, I stay grounded and maintain my ability to observe my state rather than dwelling on the inconvenience of what is happening to me. By now, approaching 60 sessions with Aya, I have long surrendered to the cleansing and purging routine. Better, I welcome it, as I fully recognize that Aya does not do this to punish me but rather remove toxicity that has no place in my body or my brain!

Surprisingly, this clearing phase opens a new conversation about my money worries. I had not planned to ask Aya for guidance on this issue. I managed to clear most of my SADAT symptoms with her relentless help a few years ago. However, I have somewhat forgotten how persistent my financial stress and worries have remained. We are talking about decades of fear of facing the same trauma my parents experienced when they lost everything. This recurrence of anxiety inspires a review of my relationship with money throughout my adult life, beginning with my first job in Paris after completing my MBA.

Aya prompted me to reflect on critical stages of my life that required significant investments: I lived a few years in Paris; then I moved to San Francisco, raised two boys, and put them through a private French school system. Then, I bought my first home in California, started my own consulting company, eventually moved to Hawaii, survived the economic turmoil caused by COVID-19, and finally acquired property in Kona, Hawaii, and converted it into a retreat center.

Doing this review of investments with Aya makes me immediately realize that, except for a few bumps during the financial crisis in 2008 and, of course, the first year of the COVID crisis, I have never failed to provide for my family. So, the first step in Aya's process was to remind me that this historical record did not justify my anxiety. More importantly, she suggested that rather than making so many irrational predictions about running out of money, I could spend energy focusing on my gifts: continue to speak my truth, write, and conduct breakthrough research. By doing that without the fear of generating enough income, I would continue to attract abundance.

Asking for help

Finally, Aya moved the focus from financial worries to personal worries for my sons. She helps me address an issue that has been bothering me for a while. Even though both of my boys are doing well, I wonder if long periods of silence from them without a phone call signal that they are not hiding their struggles from me. The Savior Complex is still partially active! Aya offers a simple suggestion: trust they will ask for my help when needed. This way, I can stop the self-doubt and remind them that asking for help is on them, not me.

Tonight, the virtual grid of light that awed me during the last session is in full swing again, both inside and outside the maloca. When Don Miguel sings in front of me, I can see swirling energy dancing to the rhythm of his icaros. I have so much gratitude for this man. I notice that he has been purging a lot tonight, but for the first time, I get the distinct impression that he is doing it on behalf of all of us. Outside, I watch the complex patterns of interconnectedness between plants, trees, and the sky with delight and awe. What a session!

DAY 23: Iquitos PERU, November 30, 2023

Last night was a short night. We finished the ceremony around 2 A.M. After just 23 days, I am making fantastic progress in my book. Aya has provided me with critical suggestions on how to finish my outline.

Once again, our sharing circle this afternoon was powerful. Most people reported massive progress in fighting their demons or gaining consciousness, enabling them to understand the why behind their suffering. I am ahead of many by having discovered how to stay in the state of meta-consciousness to maximize plasticity. By doing so, I can avoid getting sucked in the vortex of my emotions, especially fear which can quickly arise when Aya unleashes her flood of disturbing images. Also, some participants have less experience managing non-ordinary states of consciousness (NOSC) than I do. Remarkably, I can see how it can be taught, so I am excited about sharing this knowledge in my new book.

I look forward to sleeping better tonight after battling insects in my tambo for several hours last night. The rainy season started a few days ago here, and we had one storm after the next, which brought more insects to the jungle. Life in the jungle is difficult, but it reminds us that we are merely guests here.

DAY 24: Iquitos PERU, December 1, 2023

I slept longer last night. One of my dreams brought me back to when I was looking for a new job during the worst of the COVID crisis. At the time, I seemed adrift without much of a plan. I was desperate to find a position that could replace the sudden and massive losses from the cancellation of all my speaking engagements when the global shutdown began. One employment opportunity presented in the dream state was with a biotech startup that did not have any money to develop a cure for Alzheimer's. In the dream, I told the CEO I was very qualified to work for them since I had held prior jobs as CEO and CMO, but he said they did not have funding yet to hire me. In the last scene, I recall returning to a house with a large garden that had not been groomed for months. The narrative was confusing, but I had vivid visuals that reminded me of my Aya journeys. I guess I am being primed for tonight's session!

I decided to stick with the new brew tonight. It took a long time to kick in, and I quickly felt uncomfortable when it did. The visuals took a lot of work to decode for the first part of the session. By then, I could tell that Aya was taking me to the underworld, to the belly of my soul, where she had found some deep fears that had not yet awakened.

Addiction

Aya decides to focus on addiction first. I have struggled with compulsivity my entire life, mostly with alcohol, anxiety, and work. Millions of people suffer from uncontrollable or unconscious cravings. I knew I had no other option but to go along with her guidance on a deep exploration of what addiction is and what damage it could do to my health and my family.
She insisted that

Figure 34: Addiction (AI)

addictions are best defined as maladaptive behaviors or habits that reduce our cognitive control. They work like shortcuts from our PRIMAL brain, seeking rewards without much rational processing effort. Rather than think about a particular substance, Aya tells me to focus on the behavior associated with addiction, essentially a toxic habit that offers instant pleasure. Once I can decode that, creating space between cravings and action is critical. The more I do that, the more chances my frontal lobes can kick in to resist impulsivity.
Aya assures me that it may take a while to master the technique, but it will eventually change the relationship between the target of the craving and me. I feel weak and dizzy. I think I am going to faint. I must use all the techniques I know to return to my senses, especially my breath. Once the mental fog dissipates, I can finally enjoy gorgeous scenes in the forest and the sky. The virtual net connects all elements again inside and outside the maloca. Only a few days ago, I learned from the books I brought to Peru that Mahayana Buddhism illustrates the concept of interdependent origination and the

interconnectedness of all things in the universe by showing it as a large net created by the deity known as Indra, which stretches infinitely in all directions.

At each junction of the Indra net, there is a jewel, and each jewel is obvious and reflects all the other jewels in the net, representing the idea that each element in the universe contains the entirety of the universe. I realize I am seeing the Indra Net tonight for the second time.

Meanwhile, Aya brought me down tonight so that I could rise stronger. Making me face the issue of addiction was a tough one, but I needed the reminder. I had to

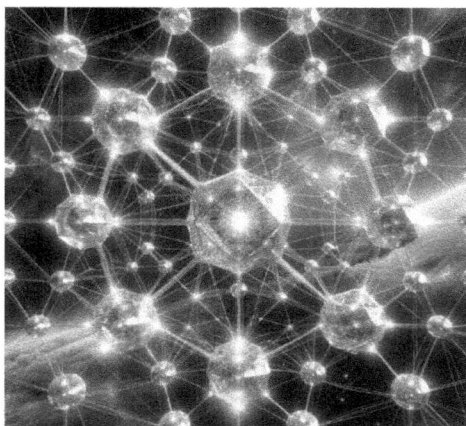

Figure 35: Indra's Net (AI)

remember that giving up heavy drinking over two years ago was a big, bold step towards wholeness, and I needed to nurture and strengthen my commitment.

I managed to play my handpan again tonight. I created a new tune that I was eager to share. It is so special to perform in front of the group and bring joy and relief while people are still under the effects of the medicine. One more session, and we are done!

Trauma and Resilience

I would stay with the Doublet family's residence twice yearly to report on my financial results and discuss strategy. I loved their husky since I had always had dogs when I grew up in Parthenay. One day, while I was visiting, Mr. Doublet adopted another abandoned husky from the dog pound. That same evening, the new dog was resting near all of us while we had dinner in the family house. I did not engage with him since he seemed subdued. The following day, however, I was the first to get up and started to go down the long stairway of the house to reach the ground floor. There, a husky was waiting at the bottom of the stairway. I mistakenly thought it was the husky I knew so well since it looked similar in shape and color to the rescue dog he had adopted the day before. Assuming he was happy to see me, I extended my hand to pet him.

Unfortunately, the dog interpreted my movement as a threat and quickly jumped at my throat. I was so shocked that it took a few moments to realize what was happening. I miraculously managed to deviate from his trajectory using my leg. However, he persisted and started to bite my hands and legs furiously. Fortunately, my boss heard the commotion and rescued me within a minute since I was now screaming my heart out. He immediately drove me to the ER, where I got over 20 stitches on both hands and was treated for massive bruises on both of my legs. I returned to San Francisco with bandages that made me look like a partial mummy. I remember now, with a smile, that I could not even zip my pants without help for days!

It took me many weeks to heal the physical wounds and decades to no longer fear dogs. Since fear is contagious, I communicated my fear of all dogs to many of my family members during that time. As I recall this terrifying event today, I recognize that I could have processed this trauma long ago using many of the modalities that I share in this book, making my life and the lives of others living with me much easier whenever I was around dogs.

At the time, though, this traumatic event reinforced my feeling that I would have fewer and fewer opportunities to run the entire company back in France, a future the owner had always suggested could be mine. I faced a similar situation to what I had experienced with my father's company: denial and disappointment. Nonetheless, a few months after closing a multi-million-dollar contract to supply and install all the flags for the Atlanta 1996 Olympic Games, I quit my job as CEO of DOUBLET USA. The Olympics deal was unprecedented in the company's history and made the owner appear on national TV and on the covers of major French business magazines. It was the perfect time for me to go.

A few years earlier, I had joined a TEC group composed of 12-15 CEOs who would meet once a month to learn from top subject matter experts on various business and personal topics and spend an entire afternoon processing business issues. I loved my TEC friends, and the format helped me grow as a business executive and, probably more importantly, as a person. I learned to be vulnerable with people I could trust, sharing the struggles I had been experiencing with my wife and sons. I discovered the Enneagram, a personality system that profoundly and positively influenced my life.

Figure 36: Olympic Games Contract (1996)

The friendship I created with my TEC group and the mentorship I received from our group chair, the late Jim Kelley, changed my life trajectory.

Thanks to the encouragement and guidance of my TEC fellows, I knew I would be supported after I left DOUBLET. My first move was to create a leadership training company called *The Leadership Network* using peer coaching. I thought I could easily replicate the common aspects of the TEC magic. I was naïve, however. First, I knew nothing about starting and running a peer-coaching company. Second, I missed not having the resources of an established company. I ended this venture after 15 months and started looking for a corporate job. One of my TEC colleagues offered me an executive position as Head of Marketing for his large grocery remarketing company, *GROCERY OUTLET.*

Created by two brothers a few decades earlier to offer close-outs at attractive discounts, the business had grown into a retail network of over 130 stores in 8 western states. My job was to form a strong marketing team and deploy the most rigorous research methods to identify the best new store locations. I was also tasked with creating memorable display signage, rolling out effective merchandising tactics, and starting TV and radio advertising campaigns to accelerate the business's growth.

While the company's culture was foreign to me, I was given "carte blanche" to execute bold programs. I had fun doing so for two years. My results were so impressive that the owners offered me a sizable bonus and promoted me to join the management team after just a few months. However, the best gift I received from joining Grocery Outlet was meeting the woman who eventually became my second wife. She joined my team a few months after I started. We had a blast working together. We left Grocery Outlet within a few months of each other as we were both pursuing different career paths. By sheer luck—or, as I prefer to say now, divine synchronicity—we both found new jobs in the same business complex in San Ramon, California. This was the crazy time of the technology bubble. Companies were fundraising left and right on a few PowerPoint slides. I joined *ZAPME!* in January of 2,000 as Chief Marketing Officer.

ZAPME! had raised 40 million dollars to create the largest US private network of computer labs for high schools, all powered by satellite internet connections. When I joined, I was responsible for nearly half of the workforce, producing one of the most ambitious educational portals ever built. My staff included educators, web designers, video producers, coders, writers, and business developers. The pace was fast and fun, but the company needed a better strategy. Soon after joining, I realized ZAPME!'s business plan was flawed. Our cash burn was very high, and targeting teenagers with ads was unpopular among lawmakers and consumer advocates. I started to spend much time controlling our narrative on top media networks. It took a year before ZAPME!'s CEO realized the business was not getting enough revenue to support the model.

So, we started divesting from the educational vertical and pivoted our activities to serve private networks interested in our Intellectual Property. The company went through multiple force reductions after this strategic move. I remember seeing employees leave as fast as they would appear. It was disconcerting. I did my best to soften the blow each time I would have to let people go. My turn would eventually come after just two years on the job. However, I managed to close a significant deal for a private network before that happened, and the CEO rewarded me with a sizeable bonus and a one-year severance package. That was the most unexpected termination package I had ever received. I was in the middle of a divorce at the time, and taking time off had not been an option. This exit deal allowed me to think about what I wanted to do next for the first time in a long while. I felt ready and supported to start my own business again.

I regrouped, finalized my divorce, and settled in a new apartment in Marin County. I was sharing custody of my two children, who were 8 and 12 at the time. This period was very stressful. Unfortunately, as happens so frequently in divorce cases, my sons were in the crossfire of a messy process. Despite the turmoil, my budding romance with a new partner flourished, making my custody arrangement more complicated but bringing the emotional support I desperately needed.

Day 25: Iquitos PERU, December 2, 2023

I woke up to an unexpected text message from our property managers in Kona this morning. A big storm hit a few days ago, and a large branch of one of our very tall ironwood trees cracked and fell near our cottage. Fortunately, the branch had stopped right over the roof by our massive banyan tree. The banyan tree is a powerful symbol of protection, deeply revered across cultures, particularly in South Asia. Its sprawling canopy, supported by multiple trunks formed from aerial roots, provides ample shade and shelter, making it a natural gathering place in villages. This expansive structure lies next to one of our most popular cottages, offering refuge from the rain and wind and a

Figure 37: Banyan Tree at Manta Soul

place of spiritual connection. Beyond its physical shelter, the banyan tree holds spiritual significance as a guardian of wisdom and life in many cultures. In Hindu mythology, it is associated with the god Vishnu, who is often depicted resting beneath its branches, reinforcing its role as a divine protector. In Buddhist traditions, trees related to the banyan, such as the Bodhi tree, also symbolize spiritual protection, serving as sacred places where individuals can seek enlightenment and inner peace.

The tree's deep roots in the earth and its branches stretching toward the heavens create a connection between the material and spiritual worlds, emphasizing its role as a bridge that shields and nurtures both the body and soul. I feel blessed to receive the protection of our banyan tree today, thousands of miles away from home.

DAY 26: Iquitos PERU, December 3, 2023

I got my usual 5 hours of sleep last night. I had lots of dreams, but not many that I can remember. I am now counting the hours before the end of the program. I am ready to go back home to Kona. I miss my dashboard of computers that make me so efficient. There is so much I can do with AI to accelerate my research for the book. I watched a terrific documentary last night called AlphaGO to break the monotony of my evenings. AlphaGO is an artificial intelligence (AI) program developed by DeepMind (a subsidiary of Alphabet, Google's parent company) (31). The film

details the creation and development of the application by the DeepMind team, led by Demis Hassabis. It explores the challenges in designing an AI capable of mastering Go, a game known for its deep strategy and vast number of possible moves far exceeding that of chess. Much of the documentary revolves around the historic 2016 match between AlphaGo and Lee Sedol, one of the world's top Go players. The five-game match held in Seoul, South

Figure 38: Playing GO with AI (AI)

Korea, was highly publicized and watched by millions around the globe. The documentary delves into the broader implications of AI development, highlighting the clash between human intuition and machine intelligence. The match with Lee Sedol is portrayed not just as a competition but as a symbolic event, representing a milestone in the evolution of AI. I did not know much about this ancient game and its popularity in China, South Korea, and Japan. It is also considered one of the most sophisticated board games ever created. When the match was held in 2016, most people still considered AI to be in its infancy.

What I found most interesting about this story is how many of the moves made by AlphaGO during each round were surprising, often utterly shocking to the experts monitoring and commenting on the games. The algorithms created by the coders were based on generating probabilities from prior games but also using machine learning, which enables a program to self-improve based on millions of iterations. Unlike a human player who tends to follow a particular style of play, AlphaGo generated moves based on millions of calculations, choosing playing options that most players would never consider. The champion admitted that he could not understand the logic behind most AlphaGO moves. Watching the film helped me realize how combining AI and human intelligence can reveal creative ways to solve problems. I have been quite excited about AI for a long time. While I can see its potential danger, it can be used for good, especially to develop new technological solutions and accelerate scientific discoveries. It can also disseminate knowledge that raises awareness of humanity's most challenging issues.

Don Miguel gathered us in the maloca this morning to explain the post-dieta. We have been drinking the master plants and receiving many other plant-based treatments for 25 days. To honor the spirits of the plants and avoid unpleasant reactions, he urged us to follow a strict dieta for a minimum of 10 days starting tomorrow. During that time, he asked us to continue to abstain from sexual activity and to refrain from consuming any cannabis and psilocybin. We should also avoid pork but can resume eating salt, sugar, and dairy. He insisted that our stomachs may have shrunk from the diet, so we should go easy on what we eat and use moderation.

Tonight is the last Aya session. It will be a celebratory event that the Shipibo people call arcana. Arcana refers to spiritual or protective songs, rituals, and Ayahuasca practices. We will finish the session with a dance and songs specifically intended to provide spiritual protection for the participants. These songs are believed to create a protective shield around individuals, safeguarding them from negative energies or harmful spiritual influences.

Aya Session #10

Tonight is the last ceremony. I decided to set an intention, which I have not done yet during this trip: celebrate my life. I do not have expectations or attachments to the outcome of the session. I am ready to receive whatever healing or guidance Aya will give me. The session starts with a few visuals; none are disturbing or uncomfortable, and I feel no signs of intoxication. I do not feel the need to purge, either. It is a welcome break. I feel relaxed and soothed.

Celebrating wins

To my surprise and delight, Aya starts the session by reviewing my achievements over the 40 years of my career. Numbers summarizing my accomplishments fly over my head with short titles, such as three million miles in the air, 250,000 books sold, 125,000 executives trained, 10,000 workshops delivered, 600 clients served, 43 countries visited, five books, and five research awards.

Figure 39: Celebrating Wins (AI)

This celebration orchestrated by Aya is a straightforward ego trip, but I appreciate it without shame! Meanwhile, Aya reminds me that my most important accomplishments are related to my role as husband, father, brother, son, cousin, friend, and citizen. What a beautiful moment of self-love.

Aya tells me to celebrate my wins more often. She insists that I am ready to complete the pivot from my business activity to coaching and retreat services based on my wife's courses in soul-centered psychology and my books on serenity and neuroplasticity. The synchronization of this transition could not be more perfect, she points out. I aspire to write more about these topics and research them until that ultimate moment when I leave my body for good! I feel I have at least ten more books in me.

Surprisingly, I ask myself why I have been so lucky at this point in my life. She tells me I have done the work by showing up relentlessly to claim our happiness. I felt deep compassion for those who cannot do it, and I am also happy that I can put my energy toward helping people heal.

As expected, we closed the final ceremony with songs from all the participants. I decided to sing an old French song about an artist missing a brother he never had and imagining what life would be like with him in it, which

I learned when I was 11. As I shared earlier, my older brother left for boarding school when I was eight, so I did not have a chance to connect with him meaningfully until he was in his mid-twenties. I missed him during my entire childhood. Performing this song tonight was another way to grieve that loss. Don Miguel sang a last icaro to close the dieta.

We are officially done with the program!

When Success Hides Despair

Taking advantage of my severance package from ZAPME!, I started a sales consulting business with a close friend based on his idea of bringing brain science to sales training. We co-authored our first book, "Selling to the Old Brain," in 2002, a painful process considering our views of the business model did not always align. Also, the rapid success of SALESBRAIN was threatened by a costly lawsuit challenging the copyrights of our intellectual property. It took us several years to settle it, and hundreds of thousands of dollars vanished doing so.

Because of the stress and intensity of the litigation, I continued to miss critical cues on the severity of my youngest son's despair, as well as the breadth of the psychological turmoil in which he was living each day with his TS. I tried my best to help him by trusting the medical system. I took him from one psychiatrist to the next, but this did not help much. With time, it seemed all my efforts were making things worse, significantly complicating our relationship.

Meanwhile, with perseverance and probably some luck, my partners and I built a thriving, active, and vibrant business that still operates today. Our second book, "Neuromarketing," was published in 2007 by Penguin and became an international best-seller. We sold nearly ¼ million copies, and the title was translated into a dozen languages. Clients were coming fast and furious, and we could barely meet the demand. Inspired by my positive experience with TEC and leveraging key relationships created during my membership, my partner and I joined the speaker network of *TEC* (which became *Vistage* in 2005). Thanks to the originality of our content, we quickly gained public recognition and clients by delivering thousands of workshops to CEO groups all over the USA and abroad. From 2002 to 2022, I presented to over 1500 Vistage groups and added more than 2.5 million miles to my United Airlines mileage account.

Throughout this seemingly perfect success story, the physical and psychological ordeal my youngest son was enduring remained demanding and often terrifying for me to go through. Unbeknownst to me, a Savior Complex was in full action for this extensive period, which ultimately spanned roughly 27 years, from 1992 to 2019. I went back and forth between highs and lows, mitigating one crisis after the next. Fortunately, after reaching the bottom multiple times, he finally emerged from the straight jacket of his challenging neurological and psychological condition. I will return to the circumstances of his healing, which are extraordinary and relevant to the teachings I share in this book.

Eventually, my excessive levels of anxiety and stress would push me to abuse alcohol and normalize the behavior. Unconsciously, maybe I thought numbing myself would give me more time to figure a way out. Instead, I practiced self-deception for 25 years until, finally, I could no longer hide my pain. I had reached the bottom. Maintaining a semblance of calm and control for many years had become unbearable. It was at that point, and with the help of psychedelic-assisted therapy, that I realized I was on the verge of a psychological precipice. Pretending otherwise would not only make me crash, but I would drag my sons and the rest of my family with me.

I should have figured this out earlier: My youngest son was my life teacher. I learned so much from him, even in the worst moments of our father-son story. I had been a sponge of his chronic despair for so many years, especially after he left for college. I could never step out of it and be conscious of my pathological condition.

DAY 27: Iquitos PERU, December 4, 2023

We ended the arcana ceremony at 2 a.m. last night. It was a beautiful celebration, and my Aya session was perfect. I could not have wished for a better end. I got up excited and tired this morning but eager to eat a regular breakfast with salt!

When Stress and Anxiety Test Your Limits

From 2002 (when SALESBRAIN started) to 2011, the pace of my life was quite insane. I absorbed the stress like a sponge, totally unaware of my very narrow state of consciousness and mental rigidity, otherwise known as a state of low plasticity. I re-married in 2007. A few years later, my wife pursued a master's degree in Depth Psychology to follow a newfound calling for soul-centered practices. Instead of being happy for her, that move worried me, as I saw it as a threat to our relationship. After all, she was doing the personal work I so desperately needed but could not conceive of ever committing to. I was too wrapped up in growing my business and continuing to support my sons to recognize the importance of her decision. She ultimately shared her interest in working with Aya for healing and insight. That was a terrifying idea when she first shared it, confirming my fears of being left in the dust. My monkey mind was building and constructing all kinds of possible scenarios, including the end of our marriage if I did not get involved. I joined her to attend our first ceremony in October 2010.

These early sessions with Aya did not change my attitude towards continuing to try to save my son after he left for college. My stress continued to spike, and the situation related to his Tourette's remained highly complicated. After a few years, however, he finally opened up to me about his challenges, especially his severe anxiety and depression. When he started to ask for my help, instead of resenting it, everything between us began to change. It also cleared the way for me to start to address my stress and anxiety, beginning in 2015. In addition, it motivated me to focus on writing *The Serenity Code* as early as 2018, which I completed in 2020 during the COVID crisis. COVID had instantly stopped my constant travel schedule and, therefore, most of my immediate income, forcing me to raise my plasticity to reinvent my career the year I turned 60. My growing interest in self-love became the key to my survival.

While researching *The Serenity Code*, I regularly shared with my son the benefits of self-love and its importance in clearing the body and mind of destructive thought patterns. For instance, one of the most important recommendations I made to him was to adopt a cat. In preparing to write the book, I researched the remarkable mental health benefits of connecting with pets. Indeed, science confirms that the presence of animals we care for can stimulate the production and metabolization of oxytocin, a critical neurotransmitter that brings comfort, joy, and peace of mind.

The direct effect of pets on releasing oxytocin to boost self-love is also well documented. Sunny, the beautiful feline who has played a central part in my son's life for several years, became his best friend overnight. Taking care of Sunny shifted his priorities in profound and long-lasting ways. It gave him even more reason to accelerate his healing. Once his mind started to clear his anxiety and depression, he was on solid footing to reboot his life. Moreover, Sunny remains his most loyal friend and companion today.

At a critical point, he finally started to have the energy and strength to complete the rehauling of this psychological treatment plan. He was prescribed so many meds to mediate his tics and to treat his anxiety and depression that it was difficult for him to clear the fog of the side effects. With the guidance of a doctor, he gradually stopped all his medications. I remember feeling nervous but also trusting that he was doing exactly what he needed to do. It worked beyond my wildest dreams. Once his life was back to enjoying more energy, more self-confidence, and far less anxiety, he pursued his amazing career as a cannabis grower and a brilliant medicine man. It is so impressive that he found the call of his soul as a healer after so many years of battling SADAT conditions. I am inspired and humbled to march in his footsteps.

Closing The Retreat and Key Takeaways

Our final sharing ceremony was extraordinary. All twelve participants—three men and nine women, aged 25 to 79—reflected on transformative journeys, significantly improving their physical and psychological well-being. Each had found renewed strength and clarity, as if guided by a spiritual compass. The journey was not without challenges; together, we completed a 26-day plant dieta, endured two vomitivos, fasted intermittently, kept silent for five days (with a few slip-ups), and participated in 10 Ayahuasca ceremonies. These experiences, though intense, mirror the difficulties of everyday life under stress, anxiety, depression, addiction, and PTSD. Here are three powerful takeaways I gathered from today's stories:

1. ***Raising Your Consciousness Awakens You****: Participants described newfound awareness of toxic patterns they once overlooked, denied, or failed to see. Without conscious recognition of such deeply ingrained psychological conditioning, it's easy to feel imprisoned by it. However, true awareness requires releasing energy held by fear and suffering, both in body and mind. While the purging was brutal, it provided a predictable way to clear this energy and reset consciousness—like reformatting a hard drive to create space for healing.*

2. ***Rewiring Your Narrative Frees You****: With greater awareness of their psychological roots, participants found they could rewrite their narratives, liberating vast reserves of emotional energy. This freedom restored their self-efficacy, transforming intentions into actions. Aya and master plants offered powerful insights that enabled this process, revealing insights each could share with clarity. No matter how skeptical one might be regarding the retreat, the experiences allowed everyone to decode the toxic threads of their life stories to invite more happiness and flow.*

3. ***Reprogramming Your Habits into Rituals Nourishes You****: The final takeaway emphasized protecting and nurturing elevated states of consciousness by embracing acceptance, gratitude, compassion, and, crucially, self-love and self-care daily. Many attendees realized the need for substantial changes—not necessarily leaving their jobs or*

relationships but aligning with their needs and wants. Moving toward growth means adopting practices that bring joy and peace. Several shared desires for more spaciousness, selective engagement with media, and cultivating daily rituals that nourish their soul. Activities like yoga, creating art, and playing music were popular themes that emerged.

All participants expressed intentions to continue working with plant medicine. Some are also committed to sharing their healing experiences within their communities, extending their wisdom to family members who face similar struggles. Sitting in this final circle, anyone could sense the profound impact of the Shipibo healing tradition. We celebrated our journeys with a dance, marking the end of our time together. Twenty-seven days ago, these people were strangers. Now, they feel like family.

Each of them will forever hold a place in my heart.

CHAPTER TWO

The Self-healing Power of Consciousness and Neuroplasticity

"What we are today comes from our thoughts of yesterday, and our present thoughts build our life of tomorrow: Our life is the creation of our mind."

--Buddha

In this chapter, I explain the neuroscience of neuroplasticity and consciousness, and especially how stress, anxiety, depression, addictions, and trauma (SADAT) may impact physical, emotional, and cognitive functions. There is no question that our innate ability to rewire our brains is complex and a relatively new field of research. So is the topic of consciousness.

Surprisingly, I have found that many members of the medical community rarely have a comprehensive education on the neurobiological basis of both consciousness and plasticity and how they interplay with many psychiatric disorders. I learned the hard way when I was going from one psychiatrist to the next, hoping to find an effective treatment for my son's Tourette Syndrome (TS), as many individuals with TS experience co-morbid conditions such as attention deficit hyperactivity disorder (ADHD), obsessive-compulsive disorder (OCD), anxiety, or depression. The simultaneous presence of these complex disorders in my son's TS condition complicated both his diagnosis and treatment options. I felt incredibly frustrated that conventional approaches to the treatment of TS were limited to a cocktail of pharmaceuticals with massive side effects.

This situation was one of the many reasons that motivated me to pursue a Ph.D. in Media Psychology with a concentration in neuropsychology (without the clinical certification). My passion for understanding human behavior has driven me to integrate insights from varying disciplines to develop innovative strategies and tools. From advancing marketing research to decoding neuroplasticity or teaching the intersection of AI and consumer behavior, my work reflects a commitment to applying broad-based knowledge in practical, impactful ways, improving well-being and decision-making across industries. Because of my interests, education, and research in interdisciplinary approaches, I am somewhat of a *polymath*. The term is derived from the Greek word *polymathēs*, which means "having learned much." As such, I thrive on combining knowledge from related fields, making connections across them, and applying their diverse skill sets to solve complex problems.

Chapter TWO starts by describing the brain in general and especially the nature of systems, functions, and networks that drive most of our behaviors. After that, I tackle the "hard problem" of consciousness. David Chalmers coined the expression nearly 30 years ago because he noted that consciousness was a problematic construct that could not be explained with scientific rigor and certainty (32). After all, scientists have remained unable to explain how the brain can produce a first experience of being.

Most agree that consciousness is an emergent process, i.e., a phenomenon where larger behavior patterns arise through interactions among smaller biological structures that do not exhibit such properties. The concept of emergence is widely applicable in various fields, including biology, physics, and social sciences. Therefore, consciousness is an emergent process because each neuron follows relatively simple rules and interactions. Still, collectively, neurons produce highly complex and sophisticated behaviors and thoughts that are impossible to decode at the level of each synapse.

Given this puzzling phenomenon, I propose an operational definition of consciousness that considers the roles played by three core psychological drives: the *ME*, the *I*, and the *WE* drive. The terms represent different aspects of our psyche and how we perceive and interact with the world around us. For decades, psychology has explored these concepts to understand various facets of human identity and social behavior. Here is a brief overview of what each term represents:

- **ME drive:**
  The *ME drive* is the primal selfish energy we need to survive. It reflects the instinctual aspects of human nature that prioritize self-preservation, personal needs, and desires. This drive is deeply rooted in human evolutionary biology and is essential for ensuring an individual's survival.
- **I drive:**
  The *I drive* is associated with the self and encompasses self-awareness and personal agency. This *ego-centric* drive focuses on our thoughts, feelings, and actions as autonomous beings. It represents the subjective experience of being a separate entity and is crucial for personal identity and self-direction.
- **WE drive:**
  The *WE drive* encompasses our connection and sense of belonging to groups and the human species. This includes our understanding of ourselves as part of a family, community, nation, or any other collective. The *WE drive* is fundamental to our social identity and influences our behaviors in group settings. It fosters solidarity, empathy, *oneness with nature*, and loyalty to the human community.

Imagine you have recently moved to a big city for a new job. As you navigate your new life, the different levels of consciousness influence your actions and inner experience, showing how each drive contributes to your journey and growth. On your first day, the *ME drive* is highly active. Navigating a new city alone, you prioritize your immediate needs: finding your way to the office, ensuring your apartment is safe, and understanding the basics of city life. The *ME drive* heightens your awareness of potential dangers and helps you make cautious choices. This primal, survival-focused energy keeps you alert and ensures your safety and well-being, especially in unknown surroundings. After a few weeks, you have settled into your routine and feel more comfortable. Your *I drive* becomes more pronounced, helping define your identity in this new setting. You start exploring your interests, discovering local art spots, joining yoga classes, and cultivating your style. You are building a sense of autonomy and self-direction through your *I drive*, giving you the confidence to express your unique identity. Your actions reflect a sense of agency as your ego begins setting goals for your career and envisioning a future that aligns with your values and goals.

Months (or years) later, you feel connected and rooted in the new city. You meet new friends, join a community gardening group, and volunteer with a local nonprofit on weekends. At that point, your *WE drive* comes into play, strengthening your sense of belonging to a larger community. You notice how the connections you build improve your understanding of purpose and reduce feelings of isolation. When your friends talk about city-wide events or issues, you feel empathy and a sense of responsibility for the well-being of your community. Your *WE drive* nurtures a feeling of unity with those around you, expanding your consciousness to include collective goals and a shared human experience.

This progression shows how each drive plays an essential role at different stages of your adaptation. Your *ME drive* ensures survival, your *I drive* establishes your individuality, and your *WE drive* fosters a sense of belonging and purpose within a broader context. Together, these levels of consciousness provide a holistic view of your journey, integrating self-preservation, personal growth, and social connection. Thus, these three drives are integral to navigating our internal and external worlds, influencing everything from personal goals and motivations to interactions and relationships. Understanding the balance and interplay between these drives can provide insights into individual behavior and interpersonal dynamics, as well as the nature of consciousness. Indeed, many scientific and philosophical views see consciousness as ranging from essential awareness to higher levels of self-awareness and introspection.

Therefore, consciousness is *"a spectrum of energy that provides varying subjective perspectives on our experiences."* Some are considered ordinary, such as wakefulness and sleep, while others are deemed non-ordinary, such as altered states induced by meditation, prayers, psychedelics, or various neurological conditions. Describing consciousness as energy captures its dynamic and intangible nature. The subjectivity of consciousness is a cornerstone of many theories, emphasizing that our personal experiences, background, and cognitive processes shape how we perceive reality.

Meanwhile, observing and reflecting on our experiences is a fundamental aspect of consciousness, often called *metacognition*. This includes awareness of external environments and internal states (thoughts, feelings, etc.). Metacognition is the ability to think about one's thinking, a sophisticated level of consciousness often referred to as a unique property of the human brain. Indeed, the capacity to articulate and communicate our experiences is a significant aspect of human consciousness, closely linked to language and social interaction. Describing and sharing our inner experiences is essential for social bonding, cooperation, and cultural development.

Recent discoveries and scholarly opinions have sparked rich debates on consciousness over the last decade. Consciousness researchers are increasingly bringing fresh and unique perspectives from burgeoning fields like quantum physics, genetics, neuroscience, AI, and psychedelic research. A comprehensive discussion of each domain is beyond the book's scope. However, I will highlight valuable insights into the physical and psychological mechanisms linking consciousness and neuroplasticity.

Finally, I will describe what happens to our brain's plasticity when we experience SADAT from the *materialistic* and the *non-materialistic* perspectives. The materialistic view considers the brain alone to be the central processing actor of our human experience, and the non-materialistic view places both scientific and spiritual states at the center of the origin and resolution of maladaptive conditions such as stress, anxiety, depression, addiction, and trauma. Imagine consciousness research as a grand symphony where each scientific field—quantum physics, genetics, neuroscience, AI, and psychedelic research- plays its unique instrument. Over the past decade, this orchestra has grown richer, with discoveries acting like fresh musical notes, sparking complex harmonies and debates on the nature of consciousness. In this book, I will sample parts of this symphony, focusing on the movements that reveal how consciousness and neuroplasticity resonate.

Figure 40: The Brain as a Corporate High-rise (AI)

Think of the human brain as a bustling corporate high-rise. Like skyscrapers, our brain took millions of years to become what it is today. The vertical organization of the layers is central to our ability to understand the logic of brain evolution. First, the bottom of our brain is less conscious and more automated than our top layer. Throughout the book, I refer to neurological processes that are either bottom-up or top-down, indicating the extent to which we tap into the subconscious mechanisms of our PRIMAL brain (bottom-up). Likewise, I call a process RATIONAL if we engage the more evolved layer of our brain, the neocortex. I refer to it as a top-down process.

The intricate network of connections between these layered sections ensures the seamless operation of the brain, much like a well-organized corporate headquarters. This analogy underscores the complexity and sophistication of our brain and its functions. Still, more is needed to help you fully appreciate the most complex structure in the known universe, with over 86 billion neurons, 85 billion other cells, and 100 trillion synaptic connections.

The PRIMAL And RATIONAL Brains

While many RATIONAL brain areas have extensive connections to the PRIMAL brain, it has uniquely enabled our species to perform more complex computations to allow functions like learning and memory, spatial and sensory processing, language, predictions, and decisions. In his seminal book *Thinking Fast and Slow*, Daniel Kahneman, a Nobel laureate in Economics, confirmed that there are two central systems through which our nervous system processes information (33):

- **System 1** (Fast Thinking): This is the fast, automatic, intuitive, and emotional mode of thinking. It makes quick judgments and decisions, often based on limited information and heuristics (mental shortcuts). System 1 operates effortlessly and is usually influenced by biases and subjective experience.
- **System 2** (Slow Thinking): This system is slower, more deliberate, logical, and conscious. It requires effort and is used for complex computations and rational decision-making. System 2 is more reliable for making reasoned judgments and decisions, but it is also more labor-intensive and can be lazy, often defaulting to the quicker System 1 processing.

Additionally, Kahneman explores various cognitive biases and errors that arise from these two systems of thought. He explains how they affect everything from daily personal decisions to investment or political choices. The book also delves into the interplay and conflict between them, showing how the fast, intuitive thinking of System 1 often dominates, leading to errors that System 2 fails to correct. Key themes include overconfidence in human judgment, the illusion of control, and the impact of cognitive biases on everything from stock market trading to planning personal events. He also suggests that our happiness and well-being are influenced by how we process and remember experiences. Overall, the book provides a deep understanding of the strengths and limitations of human thought, offering insights into how we can make better decisions, both in our personal and professional lives.

Since Kahneman and his colleagues recognize that humans pretend to or even aspire to behave rationally, they also confirm that the nature of our mind makes it hardly possible to make predictions and choices guided by logic alone. The biggest stumbling block that gets in the way of allowing us to think and behave rationally is the impact of *primal and cognitive biases*. These biases represent mental shortcuts that aim to accelerate our decisions, ensure survival, and avoid cognitive effort. According to scholars who research this topic, we have nearly 200 biases affecting our day-to-day, if not minute-by-minute decisions, from loss avoidance to our innate preference to seek information that validates our opinion. We will discuss this topic again later in the book.

Kahneman describes System 1 as the most primitive part of the brain, which is why, in all my books, I call it simply the PRIMAL brain. In contrast, System 2 is the newest part of the brain, and therefore, I refer to it as the RATIONAL brain since it pertains to a set of brain structures and networks that enable logic and higher states of consciousness. Together, these two systems help us feel, think, and generate millions of decisions and choices we need to make to survive and adapt.

Unquestionably, the PRIMAL is the oldest evolutionary portion of your mind. It protects us from external threats and keeps our body in homeostasis. Therefore, the PRIMAL brain is implicated in our unconscious responses to events that cause many of SADAT's maladaptive responses. The RATIONAL brain is the most recent part of our brain, giving rise to our capacity to speak, think, compute, and predict. The RATIONAL brain is central to how we have learned to adapt over our lifetime, relying on the more sophisticated wiring of our cognition to resolve the maladaptive responses of our PRIMAL brain.

The following tables highlight the significant differences between the two brain systems and the role of each lobe. The brain's lobes are specialized regions responsible for different functions, working together to manage thought, behavior, and perception. Each plays a distinct role in processing and responding to information. The frontal lobe manages complex tasks like decision-making, social behavior, and voluntary movement through areas like the motor and

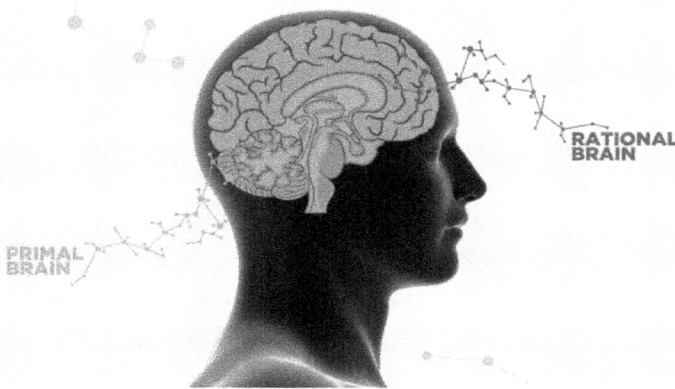

Figure 41: PRIMAL and RATIONAL Brains (SalesBrain)

prefrontal cortex, which are essential for planning and impulse control. The parietal lobe interprets sensory information, especially touch and spatial awareness, allowing us to navigate and interact with our surroundings. The temporal lobe is crucial for hearing, memory, and language, housing areas like the hippocampus for memory and auditory processing regions for understanding sound. Finally, the occipital lobe specializes in visual processing, enabling us to interpret visual cues like color, motion, and spatial relationships, which are vital for perception and environmental awareness. Together, these lobes support our cognitive, sensory, and motor functions, integrating diverse inputs into cohesive experiences.

PRIMAL STRUCTURES	RATIONAL STRUCTURES
The pons controls sleep and arousal	*The frontal lobes* control critical cognitive skills such as problem-solving, working memory, decision-making, concentration, emotional control, and predictions. They are often assimilated as a "personality control panel."
The Medulla Oblongata regulates critical survival functions like breathing and heart rate. *The reticular activating system (RAS)* controls sleep and mediates arousal.	*The parietal lobes* play a crucial role in integrating sensory information from various body parts, managing spatial navigation and orientation, and processing information related to object manipulation. They are also involved in various aspects of attention and coordinating complex cognitive functions.
The amygdala mediates fear and hijacks the body to confront or avoid situations. *The hippocampus* serves to organize and store long-term memories. *The hypothalamus* directs many responses to keep the body in a state of balance. *The thalamus* relays motor and sensory signals between the brainstem and the cortex. *The midbrain* allows rapid processing of responses to external stimuli.	*The occipital lobes* are primarily responsible for visual processing. They also help us understand and respond to the visual world. Visual perception is like an analytic process, with neurons sensitive to colors and others to contours, but an overall visual impression is created to form a coherent representation.

Table 1: PRIMAL and RATIONAL Structures

The next table outlines the evolution of the human brain from primal, survival-driven processing to more advanced cognitive functions. About 500 million years ago, our brain operated with fast, instinctual responses, prioritizing

FUNCTIONS	PRIMAL	RATIONAL
Evolutionary Age	500 million years	3-4 million years
Processing speed and options	Fast but limited (pre-verbal)	Slow but more sophisticated computational and language functions
Functional dominance	Survival instincts and emotional reactions	Cognition and logic
Time management	Present and short-term decisions	Past, Present, and Future
Level of consciousness	*ME only*--- low capacity to transcend selfish behavior	*ME and I (I in varying ways)* – capacity to deliberate and integrate moral rules and behaviors. *WE* may or may not be activated
Capacity to control	Very low	Moderate to high

Table 2: Functional Differences Between PRIMAL and RATIONAL

survival and emotional reactivity with limited capacity for control.

Even today, when we no longer have to fear being eaten alive by a lion, this level of awareness remains focused on immediate needs and self-preservation (ME). However, it lacks complex thought or moral consideration. Eventually, around 3-4 million years ago, consciousness gained slower, more sophisticated processing abilities, enabling logical thinking and language. This advancement allowed for a broader sense of time—connecting past, present, and future—and a shift from a purely self-focused perspective to an emerging

121

awareness of ME and I. This new level of consciousness supports the integration of social rules and moral considerations, giving it moderate to high control over actions and the capacity to reflect on collective needs (WE), thus advancing beyond mere survival.

In the heart of the Amazon rainforest near Iquitos, I felt the primal energy of life itself—a relentless rhythm pulsing through the dense greenery. The air was heavy with the hum of insects and the cries of unseen creatures, a symphony of survival that echoed the instincts buried deep within my mind. My PRIMAL brain thrived here, urging vigilance and stirring subconscious fears shaped by years of conditioning. Yet, amid this wilderness, there was a surprising stillness. Sitting by ancient trees during my morning walks, I focused on my breath. With each inhale, the chatter of my PRIMAL mind softened, allowing my RATIONAL brain to emerge like sunlight filtering through the thick canopy above. This wasn't about silencing one to elevate the other but fostering dialogue between both.

Healing from SADAT—stress, anxiety, depression, addiction, and trauma—requires this reset of consciousness. It demands that we acknowledge the PRIMAL within us without judgment while cultivating the RATIONAL as our guide. The jungle reminded me of this truth: harmony is not in dominance but partnership. Just as the rainforest thrives on interdependence, so too can we find our balance within.

The Central Nervous System

Examining the intricate workings of the central nervous system (CNS) is essential to better understanding how the PRIMAL and RATIONAL systems influence our behaviors and conditioning. This core network governs our body's responses and processes to external and internal stimuli.

The Central Nervous System (CNS)

The Central Nervous system comprises the brain and the spinal cord. The brain is the control center for virtually all body functions and processes. Its functions are diverse and complex, encompassing voluntary actions like movement and speech and involuntary actions like breathing, temperature regulation, circadian rhythms, and heart rate. It processes and interprets sensory information from the environment and generates thoughts, emotions, and decision-making (higher cognitive functions).

The brain's high metabolic rate underscores its reliance on a continuous energy supply of oxygen and glucose for proper functioning. A typical brain has about 86 billion neurons, the fundamental units of the nervous system responsible for transmitting

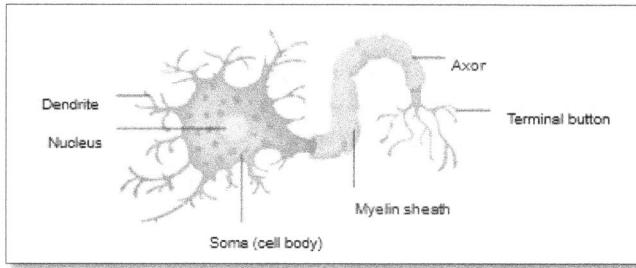

Figure 42: A Neuron Cell Body (Istock)

information throughout the body. Figure 42 shows a neuron, a brain cell that sends and receives messages to control body functions and thoughts. Key parts include dendrites that receive signals, the axon that transmits them, and a protective myelin sheath that helps signals travel quickly. Neurons become more efficient at firing together over time. This is best known as the Hebbian theory, summarized by the phrase "cells that fire together, wire together" (34).

The continuous strengthening of neuron connections underlies many cognitive and motor skills. Practice and repetition reinforce the connections that help us recall information or perform tasks more easily. This principle is central to our ability to reprogram many of our thoughts and neuropathways via the brain's plasticity. This principle alone can also explain why the brain may reprocess SADAT emotional memories in incessant loops.

Meanwhile, oxygen and glucose maintain cellular activities and regulate the chemical balance necessary for the propagation of nerve signals. So is proper hydration to support the brain's structure and function. Additionally, getting enough sleep is crucial for cognitive functions, memory consolidation, and removing brain waste products. Finally, mental activities, social interactions, and learning stimulate the brain and contribute to its health and plasticity. The brain's health is closely linked to the body's overall health, and any significant disruption in these requirements can adversely affect brain function.

By now, you should begin to understand the fantastic biological symphony the brain orchestrates every millisecond to keep us alive. Let us discuss the brain system that plays a critical role in the onset and persistence of SADAT: The peripheral nervous system. While the CNS serves as the command center for processing and interpreting information, the Peripheral Nervous System (PNS) acts as its communication network, relaying signals between the body and the brain to execute actions and responses

The Peripheral Nervous System (PNS)

The peripheral nervous system represents a network of nerves extending from the spinal cord to the extremities. It is divided into the *somatic nervous system* (SNS) and the *autonomic nervous system* (ANS). The SNS controls voluntary movements and transmits sensory information from our sensory organs to the CNS. Our sensory organs are specialized structures that allow us to perceive and respond to the environment by inhibiting or exciting sensory neurons. They include our eyes, ears, nose, tongue, and skin.

During my Amazon retreat, I experienced a profound awakening to the role of my peripheral nervous system through the heightened sensitivity of my senses. Walking barefoot in the jungle, feeling the earth's texture beneath my feet, and breathing in the humid air taught me to tune into the present moment in ways I had never experienced before. This visceral connection underscored the PNS's role in translating the world into physical sensations that profoundly shape our emotional responses. Meanwhile, our sensory systems gather data at astonishingly high rates (1 billion bits/s) compared to the throughput of conscious human behavior, which lags significantly at just 10 bits per second (35). This stark difference raises profound questions about brain function and efficiency. Why does the brain, equipped with billions of neurons, process such a minuscule amount of data for decision-making? The brain seems to operate in two modes: a high-speed "outer" mode for processing rich sensory and motor signals and a slow, deliberate "inner" mode for guiding behavior. While the neural complexity of the outer brain is well understood, the purpose of the inner brain's limitations remains elusive, pointing to a need for innovative research to uncover the mechanisms behind this bottleneck.

Eyes (*sight*)

By far, the visual sense is the most dominant channel through which we receive critical information. The visual system processes around 10 million bits per second, but only 40–50 bits reach conscious awareness, focusing on details like motion and contrast to aid immediate perception (36). Our visual processing function is the most energetically demanding system in the brain (37). It occupies almost 40% of the RATIONAL brain. Since our eyes and visual thoughts constitute a central portal through which we process our fears, it explains why so much of the SADAT suffering and healing implicates visual pathways. It also explains why so many healing modalities I will discuss later often require or suggest closing your eyes. In the jungle, the dense green canopy and intricate patterns of vines awakened my sense of sight. I learned that visual stimulation isn't just passive; it actively reshapes how the brain perceives its environment. This became a metaphor for seeing life with greater openness and clarity, a concept I embraced during the retreat.

Ears (*hearing*)

The auditory system handles about 1 million bits per second, with a small fraction consciously perceived, emphasizing meaningful sounds like speech while ignoring background noise. Sound localization is essential for survival, so it is processed in the oldest section of our PRIMAL brain, the brainstem. The *amygdala* (a tiny structure also nestled in the PRIMAL brain) is a sensory megaphone involved in the emotional processing of music and sounds. At the same time, the hippocampus (also PRIMAL) plays a central role in recording both sounds and their emotional significance. Music can undoubtedly play a significant role in our ability to excite or relax our nervous system. Likewise, disturbing sounds and excessive noise can contribute to stress and anxiety problems, and prolonged exposure to loud noise can damage hair cells in the inner ear, leading to hearing loss. The symphony of jungle sounds—chirping birds, rustling leaves, and distant rain—was overwhelming and grounding during my retreat in Peru. It reminded me that listening, whether to nature or our inner thoughts, is an act of presence. This auditory immersion demonstrated how tuning into our senses can foster deeper connection and flow.

Nose (*smell*)

We can recognize and act swiftly when our sense of smell detects dangerous chemicals like smoke or gas. Olfactory information flows roughly 100,000 bits per second, with only strong or novel odors reaching consciousness, particularly when associated with memory or environmental signals. Evolutionarily, the olfactory sense is regarded as one of the oldest (PRIMAL) senses (38).

The development of the olfactory system predates other sensory systems like vision and hearing in evolutionary history, indicating its primary importance in the early stages of life's development on Earth. Specific scents can also elicit calming or stress-reducing responses since they can positively influence mood and stress levels. The jungle's rich, humid aroma of damp earth, flora, and ancient trees—stimulated my sense of smell in ways I hadn't anticipated. It taught me how scents can evoke robust emotional responses, tying deeply into memory and transformation.

Tongue (*taste*)

The taste system processes around 1,000 bits per second, mainly bringing awareness to flavors critical for survival, such as bitterness, which may signal toxicity. Humans have approximately ten thousand taste buds, each containing 50–100 receptor cells. The human taste system can identify five basic tastes: sweet, salty, sour, bitter, and umami (savory). The sense of taste is essential for enjoying food, maintaining a balanced diet, and detecting potentially harmful substances.

For some individuals, stress can heighten taste sensitivity, while for others, it might dampen it. Stress activates the body's fight-or-flight response, releasing hormones like *cortisol* and *adrenaline* that affect taste perception. Stress can also either increase or decrease appetite. Some people may experience a heightened desire to eat (especially foods with strong tastes), while others may find their appetite suppressed because of stress. The bitter taste of Ayahuasca was a visceral reminder of the healing process—uncomfortable but transformative, as I experienced it in Peru. This sensory experience showed me how taste connects us to the present moment, encouraging us to embrace discomfort as part of growth.

Skin (*touch*)

Somatosensory processing, responsible for touch and body awareness, transfers about 1 million bits per second, though conscious awareness is limited to significant sensations like pain or sudden changes in pressure. Thanks to touch, we can feel when something is pressing against us and detect surface qualities (smooth, rough, etc.) and vibrations. Thanks to millions of touch receptors distributed throughout the skin and other tissues, we can sense heat and cold or be alerted to potentially damaging stimuli, signaling discomfort or pain. The gentle caress of Amazonian rain on my skin during a sudden jungle downpour was an unexpected revelation. Each droplet felt like a connection to something greater, reminding me of how touch grounds us in the present and opens the door to deeper emotional awareness.

While the five senses anchor us in the external world, other internal perception and cognitive functions like *intuition, balance, and proprioception* offer subtle yet essential insights into our body's state and movements. These *extra* senses deepen our self-awareness and fine-tune our interactions with the environment, linking the tangible world with a more intuitive, felt experience. While there is some debate among neuroscientists on defining intuition, balance, and proprioception as senses, many agree that they are crucial for understanding how we navigate and respond to physical spaces and nuanced internal signals. Essentially, they bridge our perception of the external and internal landscapes, allowing us to move through the world with conscious awareness and subconscious guidance.

Intuition

Intuition is often called our "sixth" sense. It is a subconscious synthesis of accumulated experiences, knowledge, and sensory inputs that can lead to an instinctual understanding or insight. Intuition is mainly informed by past experiences, knowledge, and learning, even though we may not recall these experiences actively. High levels of stress and anxiety can interfere with intuitive thinking. The prefrontal cortex, crucial for complex cognitive behaviors and decision-making, is particularly sensitive to cortisol (39). Also, anxiety often leads to overthinking and hypervigilance, which can overshadow intuitive thoughts. Intuitive insights typically occur in a relaxed awareness, free from emotional turmoil (40). Interestingly, the relationship between intuition and our gut –the *enteric nervous system* (ENS), is a subject of growing interest in neuroscience and psychology (41).

With 200 million neurons, the gut plays a significant role in our emotional and physical well-being, making it crucial to understand how it interacts with our moods and instincts. It also produces many neurotransmitters, including a large portion of the body's serotonin—nearly 80%—although only a tiny amount of this is used by the brain. This serotonin production highlights why our gut health is often linked to mental health and why probiotics can positively affect mood. Though research is ongoing, it is clear that the gut and brain communicate closely, suggesting that our gut instincts have an actual physiological basis, as the gut's influence on emotion and sensation plays a key role in what we experience as intuition (42). In the jungle, intuition became my unseen guide. Surrounded by unfamiliar terrain and experiences, I had to rely on a deep, instinctive knowing—beyond my senses. This inner voice reminded me that intuition is a subtle but powerful sense that connects us to our deeper selves and the flow of life.

Balance (vestibular system)

Our sense of balance, or *equilibrioception*, is essential for everyday activities like standing, walking, and navigating our environment. It is so fundamental that we often do not notice it until it is disrupted, such as when we feel dizzy or unsteady. Disorders of the vestibular system can lead to difficulties with balance but also generate stress and anxiety, emphasizing the importance of this sensory system in our daily lives.

Notably, the physiological effects of stress, such as increased muscle tension and altered breathing patterns, can also contribute to feelings of disequilibrium and dizziness. Additionally, there is a condition known as *functional dizziness*, where psychological factors, including stress and anxiety, play a primary role in the development and maintenance of dizziness and balance problems (43). Therefore, managing stress and anxiety can be an essential part of treating vestibular disorders, and conversely, addressing vestibular issues can help reduce stress and anxiety related to these symptoms. During Ayahuasca ceremonies, I often felt my physical balance shift—waves of dizziness and instability mirrored the emotional and mental turbulence I was navigating. These moments taught me that losing balance, literally and figuratively, is part of the process of realignment and growth, helping me find a deeper sense of grounding within myself.

Proprioception

Meanwhile, stress can impact body awareness or interoception (the sense of internal bodily states). Individuals under stress might be less attuned to their body's movements and positions. However, activities that enhance proprioception, such as yoga, Tai Chi, or Pilates, are known to be effective in reducing stress and will be discussed later in the book. Under Ayahuasca's influence, my awareness of my body's position and movements—my proprioception—became heightened yet strangely unfamiliar. I could feel the tension in my muscles as emotions surfaced and released, reminding me how deeply our physical awareness is intertwined with our emotional and spiritual state

As our senses interpret the world around us, the autonomic nervous system (ANS) operates quietly in the background, managing essential vital internal functions, helping the body maintain balance, and responding effectively to physical and emotional demands. Since the ANS directly orchestrates many physiological changes produced by emotions and stressors involved in SADAT, it is essential to understand its workings. For instance, the ANS regulates vital life support functions such as heart beating, sleep, and other critical biological functions. It also regulates involuntary functions like breathing and digestion, but not surprisingly, it operates mostly below your level of awareness.

During Ayahuasca ceremonies, my PRIMAL brain—the true boss of my Autonomic Nervous System—took center stage. I could feel its influence as my heart pounded, my breath quickened, and a wave of heat surged through my body. These involuntary reactions revealed their role in triggering survival responses. Over time, by focusing on conscious breathing, I learned to gently signal my ANS, inviting calm and reminding my PRIMAL brain that I was safe. This dance between my instincts and reasoning became a profound lesson in self-regulation and healing.

Let us now first move our attention to the *sympathetic* branch of the ANS (SNS), as it can generate the lion's share of the toxic effect of SADAT on your body and mind. The sympathetic nervous system prepares the body to respond to perceived threats or emergencies by raising the heart rate and redirecting blood flow to critical muscles (Fight or flight). Hormones like epinephrine (same as adrenaline), norepinephrine (same as noradrenaline), cortisol, and angiotensin II are integral to the functioning of the SNS. These hormones ensure the body can mobilize energy, maintain homeostasis, and protect itself in stressful situations. However, chronic activation, as seen in prolonged stress, can have adverse effects on health, contributing to all SADAT conditions and many more.

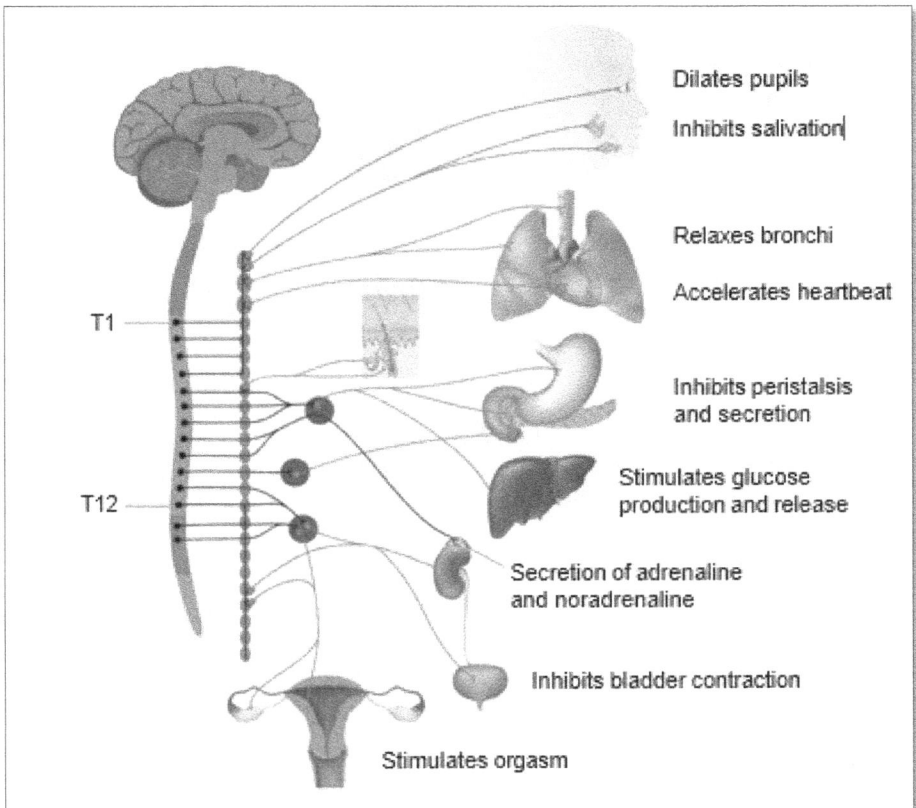

Dilates pupils

Inhibits salivation

Relaxes bronchi

Accelerates heartbeat

Inhibits peristalsis and secretion

Stimulates glucose production and release

Secretion of adrenaline and noradrenaline

Inhibits bladder contraction

Stimulates orgasm

T1

T12

Figure 43: The SNS (Istock))

Meanwhile, the *parasympathetic* branch of the ANS (PNS) promotes *"rest and digest"* activities, helping to conserve and restore energy. It performs opposite functions to the SNS, intended to quiet or relax our nervous system. The PNS uses the vagal system to produce inhibitory control—*the vagal tone*—over organs such as the heart. Vagal tone is often used to assess people's capacity to use "brakes" in stressful situations. High-performance athletes, especially, always strive to improve their vagal tone.

The parasympathetic nervous system helps the body relax, reduces heart rate, and lowers blood pressure by conserving and facilitating bodily functions during non-stressful times. The primary neurotransmitter associated with the PNS is acetylcholine, which binds to receptors that induce calming effects on various target organs. By doing so, the PNS influences various organ functions, such as reducing pupil size, stimulating bile production, increasing urination, and coordinating muscular contractions and movements in the rectum, the final

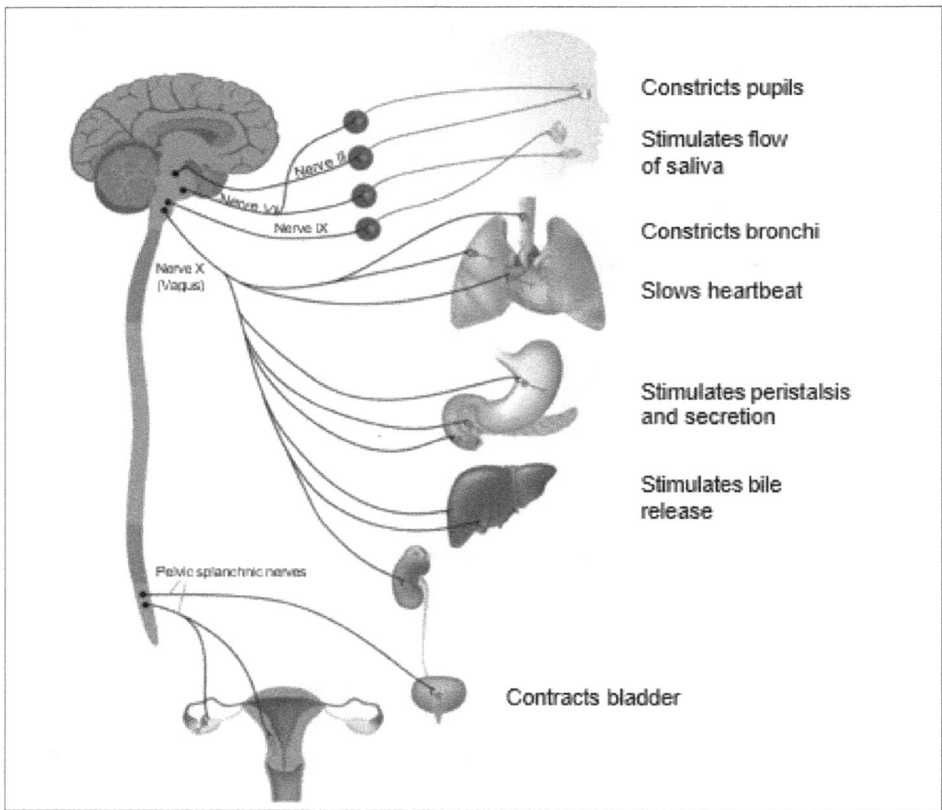

Figure 44: The PNS (Istock)

segment of the large intestine.

During my retreat, I vividly felt the profound effects of non-ordinary states of consciousness on my peripheral nervous system. In the intense moments of an Ayahuasca ceremony, my sympathetic nervous system surged, making my heart race and my breath quicken as unresolved fears surfaced. But as the plant medicine guided me deeper, my parasympathetic system took over, flooding me with a profound sense of peace. My body relaxed, my breathing slowed, and I felt an almost mystical connection to the rhythm of life as if my nervous system was recalibrating itself, opening me to calmness and presence I had never experienced.

Next, we will discuss how our nervous system manages essential internal functions with or without conscious control. Functional systems like attention, emotion, memory, and cognition allow us to process, respond to, and interpret experiences actively. They create an integrated support network that harmonizes unconscious regulation with conscious engagement in our daily lives. Attention, the focal point of our energy, is deeply intertwined with our wakefulness and the foundational aspects of consciousness that govern our most basic survival instincts. By understanding how reflexive and voluntary attention are regulated by different brain structures—from the PRIMAL brain's automatic responses to the RATIONAL brain's deliberate focus—we can gain insight into how we navigate the constant influx of stimuli. Additionally, the role of wakefulness in maintaining attentional processes highlights the critical balance between sleep and cognitive function. Finally, primal consciousness underscores the fundamental, often subconscious, ways our brain prioritizes survival and reward. Through this exploration, we can better appreciate the complexity of human consciousness and the factors that influence our mental health.

Attention, Wakefulness, and Primal Consciousness

The speed at which we react correlates highly with the degree to which we can control our attention willingly. Because of the voluntary or automatic qualities of attention, different neuronal pathways are at play during the production of higher or lower states of attention. For instance, the pre-frontal cortex (RATIONAL brain) is highly involved in selective or conscious attention. In contrast, reflexive or automatic attention is managed directly by the brain stem and the limbic system of the PRIMAL brain.

The amygdala plays a critical role in amplifying our attention. The ancient brain structure can hijack our entire body in milliseconds to help us respond to a threat. The bottom line is that attention is one of the most important portals through which we receive and store enormous stressors that can waste massive amounts of precious neurobiological energy. Correcting these patterns is possible once they are first recognized and rewired, which I will discuss when presenting the 3-step process in Chapter THREE.

Meanwhile, wakefulness provides the necessary level of arousal for the brain to engage in attentive processes. When fully awake, the neural circuits responsible for attention are more effectively activated and can process information more efficiently. The *Reticular Activating System* (RAS) in the brainstem (PRIMAL brain) plays a crucial role in regulating wakefulness and sleep-wake transitions. Insufficient sleep can lead to reduced attention span, slower cognitive processing, and impair the RATIONAL brain's *executive function*. While it is not entirely clear why we sleep, we do know that lack of sleep can seriously impair brain functioning or even cause death. If our brain needs to rest to maintain its healthy activity, of which consciousness is a necessary byproduct, then clearly, the quantity and quality of wakefulness matter.

Wakefulness produces an electrical current that can be plotted by an *Electroencephalogram* (EEG) as waves called *BETA brain waves*. BETA waves have the highest number of cycles per minute compared to waves produced during sleep. This pattern indicates the degree to which brain areas are firing multiple streams of neuropathways to attend to numerous cognitive activities. Thus, the deeper we fall asleep or the less conscious we become, the more synchronous the brain waves become, suggesting that the brain is firing less when cognitive functions are less recruited. Imagine the brain as an orchestra. When fully awake and engaged, every instrument plays its unique part in a complex, dynamic symphony, representing active cognitive functions and varied brain wave patterns. But as we drift into deep sleep, the orchestra plays a simple, unified melody, with all the instruments moving in sync. This synchronization reflects the brain firing less and working more cohesively, as it no longer needs the complexity required for higher cognitive functions.

As strange as it appears, we enter and leave many states of consciousness with or without awareness. However, the brain's attention system is biased towards using, first and foremost, what I call *PRIMAL consciousness* to attend salient events that may represent threats or opportunities. One triggers the urgency to survive, the other a craving to seek a *self-relevant* or desirable reward. Self-relevance is the degree to which an experience is attractive to us because it brings value or purpose. Therefore, the more relevant an event is to us, the more conscious our experience appears.

In a way, consciousness is overrated because the PRIMAL brain dominates many processes without requiring us to think about it. Never forget that our brain's efficiency is related to our ability to engage the fewest neurons to stay alive! Too much thinking is the worst enemy of our evolutionary future. It turns out that grey matter loss is correlated to cortical expansion, a measure of our cognitive capacity (44).

Gray matter loss refers to a reduction in the brain's gray matter, which comprises neurons (brain cells) and is crucial for processing information, decision-making, and controlling movements. Gray matter is found in areas like the cerebral cortex, which handles memory, emotions, and sensory perception. When gray matter decreases, it can lead to problems with thinking, memory, and emotional regulation. For example, gray matter loss is linked to conditions like Alzheimer's disease, depression, or schizophrenia, where people may struggle with focus, memory recall, or managing their emotions. It can also occur naturally with aging but may progress faster with chronic stress, substance abuse, or trauma.

Understanding how attention, wakefulness, and primal consciousness interplay with emotions, memory, and cognition is essential to decoding SADAT. The correlations highlight the layered processes that guide our decision-making and self-perception. Attention and primal consciousness anchor us in real-time, ensuring responsiveness to immediate needs. Still, emotions, memory, and cognition allow for a richer interpretation, connecting current experiences to past knowledge and future implications. This interaction shapes our ability to make thoughtful, context-sensitive decisions, which is essential for SADAT.

Emotions, Memory, and Cognition

The purpose of *emotions* is to make decisions or movements that help us restore balance or homeostasis. This view aligns with Antonio Damasio's work, particularly in *The Feeling of What Happens* (45), where he emphasizes the role of emotions in maintaining homeostasis through decision-making and adaptive behavior. Lisa Feldman Barrett's research, as discussed in *How Emotions Are Made* (45), also highlights how emotions act as predictive tools, drawing on past experiences to influence choices that align with physiological and psychological stability. Her perspective underscores the critical role emotions play in survival and well-being. They function not as mere byproducts of the brain but as integral mechanisms for maintaining equilibrium in the face of life's challenges. By understanding emotions through this lens, we can better appreciate their influence on our decision-making processes and overall health.

Remarkably, emotions originate from the PRIMAL brain, so we have little direct awareness of how emotions impact our moods when neurotransmitters and hormones are released throughout the rest of the brain and our entire body. Some emotions raise our heart rate while others lower it. The word *emotion* comes from the Latin word *movere*, which means to move. Indeed, emotions influence our ability to approach or withdraw from any situation in just a few milliseconds based on the emotional value we assign to an event (46). Inspired by Charles Darwin, evolutionary psychologists think emotions have evolved over millions of years to help us survive challenging moments (47). On the other hand, feelings, the conscious interpretation of emotion, may be considered more evolved than emotions, especially those elaborated in the RATIONAL brain, often responsible for excessive rumination.

Emotions became my greatest teachers during the retreat, unfolding in raw, unfiltered waves. In the structured chaos of the Amazon, I encountered a spectrum of feelings—from intense fear and sadness to unexpected moments of profound joy and peace. These emotions, often magnified during Ayahuasca ceremonies, were fleeting reactions and doorways to deeper understanding. Each one carried a story, a memory, or an unresolved tension, demanding my attention and inviting me to sit with it rather than resist.

The retreat showed me that emotions are not obstacles to overcome but signals from the PRIMAL brain, guiding us to unresolved areas within ourselves. Fear taught me about my need for control, sadness unearthed buried grief, and joy reminded me of my capacity for gratitude. In those moments, emotions were not just felt but processed, rewired, and released.

Through this emotional alchemy, I learned that vulnerability is a strength. Allowing myself to fully experience and express my emotions allowed neuroplasticity to flourish. By embracing these intense feelings, I began to reprogram my mind, letting go of old patterns and creating new pathways to self-awareness and healing. The jungle became a mirror, reflecting the profound power of emotions in shaping our lives.

Interestingly, one of the fascinating aspects of emotions is that they change our faces similarly, regardless of race, gender, or age. The study of facial expressions is growing among psychologists and media researchers because it is possible and reliable to reverse engineer distinct facial expressions and correlate them to the same emotions anywhere on this planet. This phenomenon is based primarily on the work of Paul Ekman (48).

Considered one of the most influential psychologists of all time, Ekman co-discovered that specific micro-expressions, tiny movements on our faces that create facial expressions, have a universal basis. He initially identified seven universal emotions producing the same facial response anywhere in the world. These emotions map to seven PRIMAL emotions: Anger, Disgust, Fear, Surprise, Happiness (Joy), Sadness, and Contempt.

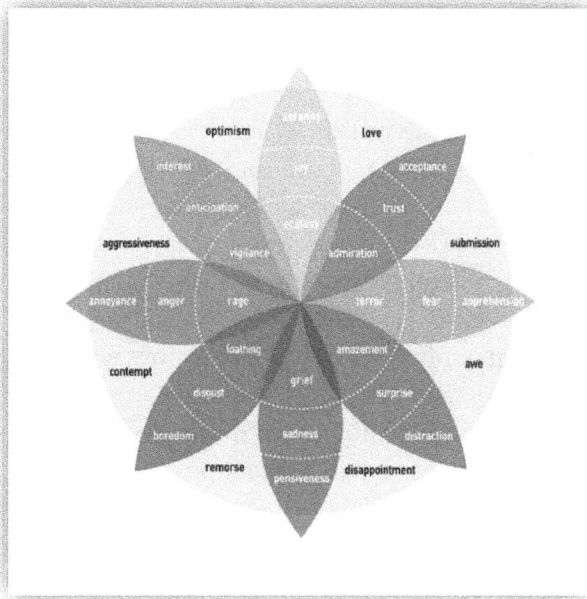
Figure 45: Plutchik Wheel of Emotions (Istock)

According to renowned psychologist Robert Plutchik, humans may have a range of 18,000 emotions (figure 45 features the top 28), which can be individually considered negative or positive (49). He suggested that more complex emotions have dynamic relationships, creating, in effect, a constellation of thousands of related feelings. While Plutchik's work has received praise from many psychologists, the logic of how emotions interact is challenging to prove. Since we only have about 7000 words in English describing emotional states, it explains why humans are so limited at verbalizing their feelings.

Additionally, emotions directly affect our memories, which is a critical byproduct of the effect of plasticity on the marking of traumatic events. In other words, emotions help us make decisions but also allow us to remember the context that motivated us. This is especially true of fear. That is why fear can be considered the most potent negative emotion we can ever experience. Ledoux, the foremost scientific expert on anxiety today, argues that fear has a unique evolutionary objective. He claims:

"To experience fear is to know that YOU are in a dangerous situation, and to experience anxiety is to worry about whether future threats may harm YOU" (50).

However, while Plutchik's model explains why emotions exist, it does not fully address how emotions are generated, experienced, or regulated in the brain. This is where Lisa Feldman Barrett's *Theory of Constructed Emotion* provides a groundbreaking shift in our understanding of the neurological basis of emotions (51).

Unlike Plutchik, Barrett argues that emotions are not innate, pre-programmed responses but instead constructed by the brain in real time based on bodily sensations (interoception), past experiences, and social context. Her research shows that the brain functions as a prediction engine, constantly generating and adjusting expectations about the world through sensory input. In this framework, emotions emerge not from a fixed set of neural circuits but from the brain's continuous efforts to interpret bodily signals and environmental cues. This theory also aligns well with Dr. Karl Friston's *Free Energy Principle*, which suggests that the brain's primary function is to minimize uncertainty by predicting and adjusting its internal models (52). Therefore, emotional experiences are not just responses to stimuli but dynamic constructions shaped by how the brain resolves prediction errors.

Understanding emotions in this way has profound implications for mental health, neuroplasticity, and self-transformation. If emotions are constructed rather than hardwired, they can be intentionally reshaped—through mindfulness, cognitive reframing, and spiritual practices. This perspective bridges neuroscience and self-awareness, offering a path to greater emotional intelligence, resilience, and, ultimately, a more profound understanding of consciousness.

Meanwhile, the role of the PRIMAL brain is key to help us navigate the world by always being vigilant, creating a default emotional charge we often need to mediate before it gets out of control. Incidentally, most people who suffer from SADAT tend to be hypervigilant. According to Stanislas Dehaene, an expert in the neuroscience of consciousness, when vigilance rises (as threats increase), the brain gradually recruits more brain areas in a bottom-up process, pushing more cerebral blood flow to our frontal lobes (53). In other words, we do not even begin to think about the seriousness of a threat until we first process the alert in our PRIMAL brain. Brain energy (glucose and oxygen) radiates from the subconscious ruler (PRIMAL brain) to the more conscious cognitive areas of our brain (RATIONAL brain). That explains why anxiety is part and parcel of who we are. Without it, we would take too many risks, and most likely, our species would not have survived.

Meanwhile, Sigmund Freud (1856-1939), a famous Austrian neurologist and the founder of the psychoanalysis movement, believed that anxiety is the source of all mental disorders (54). He claimed that humans could not tolerate living without a precise idea of what would cause us harm in the future. The word *anxiety* comes from the Latin word *anxieta*, itself related to the Greek word *angh*. Thousands of years ago, a*ngh* described uncomfortable physical sensations like tightness or dread. All this may explain why fear-based disorders like anxiety or traumas are the most prevalent of all psychiatric problems in the United States, if not the world.

Therefore, it is critical for anyone suffering from SADAT to recognize that emotions affect their decisions and behaviors in profound ways. Also, emotions directly affect what and why we remember anything, so traumatic events are difficult to erase from our memories. Research performed by Jim McGaugh confirmed long ago that emotional arousal enhances the storage of our memories (55). With emotional spikes, we tend to remember many details of what we do on a particular day or week. However, we may recall traumatic events decades later without effort or intention.

Retaining information is essential to survival and may explain why adverse events tend to be remembered more than positive ones (56). It is as if a "record" button in our brain automatically turns on during noteworthy events. Brain *memory* organization is a dynamic and complex process involving many structures and mechanisms. *Short-term memory (STM)* is a temporary holding area for information, while *long-term memory (LTM)* stores information more permanently, requiring complicated and enduring neural changes. Understanding this organization helps comprehend various aspects of human cognition and is crucial in addressing memory-related deficits associated with SADAT.

Our memories are stored in different sections of our brains. To simplify, older layers of the PRIMAL brain tend to be involved in storing and retrieving long-term memories (LTM), many of which are encoded without awareness. Interestingly, short-term memories (STM) are organized and maintained in the prefrontal cortex of the RATIONAL brain. Without rehearsal or active maintenance, the information in STM may fade in just a few seconds. Also, learning and memory formation are intimately related functions; one cannot happen with the other.

Memories surfaced like waves during my retreat, often unexpected and vivid. The Ayahuasca ceremonies were key, unlocking long-forgotten moments deep within my subconscious. Some were tender and joyful—a childhood moment with my parents, the laughter of my children. Others were heavy with unresolved pain—a harsh word, a feeling of abandonment, or a decision I wished I could undo. These memories were not random; they were threads woven into the fabric of who I had become, influencing my thoughts, emotions, and actions in ways I had not fully realized.

In the jungle, I learned to embrace these memories without judgment. Each one carried a lesson, an opportunity to understand myself more deeply. Ayahuasca revealed how the primal brain stores experience as emotional imprints, often bypassing the RATIONAL brain. Memories of fear and loss, for instance, shaped my patterns of anxiety and control, while joyful moments reminded me of my capacity for love and connection.

As I revisited these memories, I began to see them not as burdens but as opportunities for healing. By integrating them into my present consciousness, I could reframe their meaning and release their hold on me. This transformative process allowed me to rewrite my story with greater openness and grace.

To close our discussion of key brain systems, *cognitive functions* are complex and multifaceted, involving various processes enabling us to perceive, speak, think, and interact. Evolutionarily, cognition refers to developing and refining mental processes in living organisms over millions of years. This perspective suggests that cognitive abilities, such as perception, memory, language, and problem-solving, have been shaped by natural selection because they offer survival and reproductive advantages. The term *executive function* (EF), which I briefly introduced earlier, is often used to describe cognitive processes necessary for rational control of behavior, problem-solving, planning, decision-making, reasoning, flexibility in thinking (mental flexibility), and inhibitory control. Cognitive skills also include our ability to process and use spatial information about the environment and objects within to help us navigate spatial relationships and manipulate objects.

More specifically, *perception* is the process of acquiring, interpreting, selecting, and organizing sensory information, which involves primary and secondary sensory areas (visual, auditory, somatosensory cortices) and association areas of the RATIONAL brain. Perception enables us to recognize and interpret environmental stimuli through our senses. The cognitive process of selectively concentrating on one aspect of the environment while ignoring other things is considered an evolved skill. As you may remember, most cognitive processes follow a top-down route. For instance, the prefrontal cortex, parietal lobes, and thalamic nuclei enable our focus on specific stimuli or thoughts so that we can shift this focus as needed.

My retreat in the Amazon challenged my executive function in unexpected ways. As the part of the brain responsible for planning, decision-making, and self-regulation, executive function typically thrives in structured environments. But the jungle's unpredictability, coupled with the deep introspection induced by Ayahuasca, disrupted my usual reliance on logic and control. During ceremonies, I felt my executive function temporarily fade, leaving me in a state where emotions and primal instincts dominated. At first, this felt disorienting—my ability to rationalize, analyze, and control outcomes seemed out of reach. However, this surrender revealed something profound: the overreliance on executive function can suppress the deeper, intuitive wisdom of the primal brain and emotional centers.

As the retreat progressed, I found a new balance. Moments of calm reflection outside the ceremonies allowed my executive function to reengage, helping me process and integrate powerful emotional and sensory experiences. This interplay between my brain's RATIONAL and PRIMAL parts taught me that executive function is not just about control; it's also about flexibility— learning when to lead and when to let go. This realization became a cornerstone of my journey toward greater openness and flow.

Meanwhile, the ability to understand and produce *language* is specialized in the left temporal lobe of the cortex. Damage to these areas compromises our ability to understand or use language. It can cause people who suffer from such damage considerable stress and anxiety by making it challenging to use grammatical constructions or name objects. Language and social cognition helped us interact and understand others using empathy as we evolved. *The theory of mind* (*TOM*) describes our innate capacity to grasp other's beliefs, desires, and intentions that are different from our own (57). TOM also encompasses the knowledge that others have: beliefs, desires, and intentions different from one's own. This ability is crucial for human social interactions as it allows individuals to predict and interpret the behavior of others. TOM develops in early childhood and is considered a key aspect of social cognition. Deficits in TOM are often associated with certain neurological and developmental conditions, such as autism spectrum disorders. While SADAT does not explicitly reduce empathy, it can restrict the brain's capacity for balanced emotional processing, potentially leading to diminished empathetic responses due to stress and survival-mode activation.

To conclude, it is essential to note that the brain's *default mode network* (*DMN*) plays a significant role in regulating cognitive processes. It is like your brain's idle mode, like a car engine running in the background when you are parked or waiting at a stoplight. Just as the car's engine hums quietly, ready to move when needed, the default mode network is active when your brain is at rest—not focused on a specific task. The DMN contributes to mind-wandering and daydreaming, self-referential thinking, planning, and emotional regulation, all highly related to our psyche's ego drive. Altered activity and connectivity in the DMN have been observed in various psychiatric and neurological disorders, including depression, Alzheimer's disease, and autism spectrum disorder (58-60). On the other hand, transient states during which the DMN's activity is reduced (contributing to ego suspension) can create opportunities for increased levels of plasticity, which will be discussed in the next chapter.

In summary, emotions catalyze memory, influence cognitive decision-making, and disrupt cognitive functions when negative emotions become chronic. This interplay is key to understanding how emotional well-being is vital for healthy cognitive processes. Building on the foundational role of emotions, memory, and cognition in shaping our experiences, we can explore the next layer of awareness: meta-consciousness. This state allows us to step back from our thoughts and emotions, observe them objectively, and foster a deeper understanding of how they influence our behavior and sense of self. By engaging with *meta-consciousness*, we gain the tools to consciously navigate, reframe, and even reshape the mental patterns formed through our emotional and cognitive experiences.

Meta-consciousness

For centuries, philosophers, religious scholars, and scientists have researched the nature of *consciousness*. However, the working definition of consciousness has been the subject of ongoing discord. Consciousness is used or considered a substitute for many other words like life, awareness, attention, mindfulness, alertness, wakefulness, and even morality. The debate on what consciousness is will not end anytime soon if there is no universal adoption of a clear and operational definition. In the middle of the 17th century, French philosopher René Descartes argued that consciousness was generated by the soul, which he claimed was managed independently from the body. Descartes proposed that the soul could reside in the pineal gland, a small structure within the brain uniquely situated, as it is not divided into two hemispheres like most other brain components. His views gave birth to a critical philosophical movement called *dualism*, the separation of the body and the mind. Dualism continues to influence many thinkers today but is rejected by most neuroscientists, for whom the brain produces the mind. For example, Antonio Damasio believes we must recognize biological and psychological processes as interconnected and interdependent to construct an integrated view of consciousness (61).

Since it is commonly accepted that what makes humans unique, if not superior to other species, is our ability to be more conscious, it is not surprising that the scientific debate about consciousness has been raging for thousands of years. Most recently, however, brain science and quantum physics have helped elevate this challenging dialogue beyond rhetoric and ideology. Given the complexity of the topic, I aim to make this discussion approachable and balanced.

To start, I will discuss meta-consciousness from a *material perspective*, which is the reductionist view of scientists in general. The materialistic view of consciousness is like seeing the brain as a computer. Just as a computer's processes depend on physical circuits, hardware, and electricity, the materialistic view holds that all aspects of consciousness—thoughts, feelings, and awareness—are products of the brain's physical structures and chemical reactions. On the other hand, the non-materialistic view of consciousness is like seeing the brain as a radio rather than a computer. In this analogy, the brain is not the source of consciousness but more of a receiver or a tuning device for something beyond itself, like radio waves. This perspective suggests consciousness could persist even if the brain is altered or ceases to function.

Materialistic Perspective on Meta-consciousness

Figure 46: Materialistic Perspective (AI)

Since brain scientists are just beginning to map neuronal activity that correlates with states of awareness, the scientific study of consciousness is evolving rapidly. Often referred to as ***physicalism*** in the context of a philosophy of mind, the materialistic view posits that consciousness and mental states are entirely the result of biological processes within the brain. This view holds that all aspects of consciousness, including thoughts, emotions, and perceptions, are the product of neural activity and can be explained in terms of cellular, molecular, and biochemical processes. As a result, the brain is considered the sole creator and regulator of consciousness since changes in mental states are correlated with changes in brain states.

Materialism rejects any dualistic notion that the mind and the body are fundamentally different substances. The mind is not seen as separate from the physical world but as a product of it. Therefore, materialistic scientists only approach the study of consciousness through empirical research, primarily in neuroscience, where brain activity is measured and correlated with subjective experiences. In theory, if consciousness is solely a product of physical processes, it could be possible to recreate these processes in artificial systems, leading to artificial consciousness. For decades, I have aligned with this materialistic perspective, finding its reliance on empirical evidence and measurable phenomena compelling and comforting. In theory, if consciousness is solely a product of physical processes, it could be possible to recreate these processes in artificial systems, leading to artificial consciousness. However, as my journey unfolded, I began to see the limitations of this view and its inability to fully account for the depth and breadth of subjective human experience.

Meanwhile, a prominent neuroscientist, Christof Koch, proposed that there may even be a central brain structure responsible for consciousness called the *claustrum* (62). Koch's research focuses on the biophysical mechanisms underlying consciousness and how they are embodied in the complex neural networks of the brain. His work often explores the question of which specific physical systems give rise to feelings of conscious awareness. The claustrum is a thin layer of PRIMAL gray matter adjacent to the amygdala and is believed to have evolved from the ancient fear-centric almond-size structure. The claustrum receives input from almost all cortical regions and returns information to nearly all RATIONAL brain areas. This makes it a good candidate for being the center of consciousness.

For cognitive neuroscientists, consciousness refers to the levels of awareness we have of our immediate experience of sensations and feelings (63). Therefore, they believe states of consciousness can be measured and identified as representative of specific neuron firing patterns. As we discussed earlier, attention is a key factor in assessing the quality of varying states of consciousness. Our focus changes based on the self-relevance of an event, but more importantly, the energy committed to it. In other words, we remain conscious thanks to an infinite combination of those factors. As a result, a dominant view is that consciousness is a spectrum of energy constantly oscillating from PRIMAL to RATIONAL (53). The more relevant or salient an event is to us, the more conscious energy our experience seems to recruit. *f*MRI studies have confirmed that, even during the resting stage of the brain, the part with the highest metabolic activity is where we generate self-thoughts, an activity correlated to our ability to develop self-relevance (64).

To perform all these computations, the brain requires a continuous supply of energy, primarily glucose and oxygen, to function optimally. Therefore, variations in the availability of these resources can affect our level of consciousness. For example, low blood sugar (hypoglycemia) or oxygen deprivation (hypoxia) can lead to decreased consciousness or even loss of consciousness. When energy levels are high, the brain can sustain complex and energy-demanding processes like attention, problem-solving, and memory. Conversely, when energy is low, the brain may downscale to more basic, less energy-intensive functions, leading to a reduced state of consciousness. During sleep, especially in non-REM stages, the brain's metabolic rate decreases, allowing for restoration and the removal of metabolic byproducts. Conditions that affect brain metabolism, such as neurodegenerative diseases, brain injury, or strokes, often disrupt the brain's energy utilization and can lead to states of altered or reduced consciousness.

Finally, certain substances can change the brain's metabolic activity, thus impacting consciousness. Stimulants, for instance, can enhance consciousness by increasing neural activity and energy consumption, while depressants can reduce consciousness by decreasing brain activity.

The Role of Subjective Awareness

While you may believe your awareness is based on observable, measurable facts and data rather than *subjective feelings*, the evidence suggests otherwise. You are mainly influenced by your unique representation of a given moment as produced by your sensations or feelings. It is only accessible to you and not to an external observer. It is the essence of your private and personal experience of life. The term *qualia* is often used to describe this fascinating property of your consciousness as it only represents how things appear (65). Many philosophers argue that qualia make the study of consciousness more an art than a science. However, materialistic scientists remain unmoved by the philosophical argument and continue researching ways to map or measure manifestations of any awareness, from objective to subjective.

Qualia took on an entirely new dimension during my retreat. Under the influence of Ayahuasca, the ordinary sensations of sight, sound, and touch transformed into vivid, almost otherworldly phenomena. Colors became richer, sounds carried layers of meaning, and textures conveyed emotions in ways that defied explanation. In one ceremony, I recall the sensation of holding a simple leaf. Its veins seemed to pulse with life, and its earthy scent evoked memories I could not quickly identify. This heightened awareness of qualia reminded me of the unique lens through which each of us experiences the world. Ayahuasca stripped away the filters of routine perception, allowing me to see, hear, and feel with unparalleled clarity.

These moments underscored the importance of presence in my life. Often, we rush through experiences, missing the richness of their qualia. The retreat taught me that these deeply firsthand experiences of perception are not trivial; they are the essence of how we connect with the world and ourselves. By becoming more attuned to qualia, I cultivated a profound appreciation for the present moment, enhancing my journey toward greater openness and flow.

Because of the availability of neuroimaging technologies such as MEG and *f*MRI, it has become routinely possible to record brain activity spatially and temporally to look at gaps between objective measures of awareness (i.e., brain activity) and the subjective manifestation of consciousness as reported by subjects. This investigation, however, depends on people's ability to observe their qualia moments, a skill that does not appear to be innate. Surprisingly, studies have revealed significant latency between the time we decide (cognitive awareness) and the time we begin to form the decision in our brain, presumably because emotions influence our choices below our threshold of awareness and ahead of our subjective awareness (66). I know from conducting many neuroscientific studies that synchronizing data from brain activity to brain tasks is challenging but depends on the counter-intuitive steps through which the brain goes to adapt and regulate our behavior.

As we enter and leave multiple states of consciousness with or without awareness, we can also infer that consciousness is not produced by external stimulation alone. Internal thoughts, ideas, beliefs, and even dreams can generate multiple states of consciousness. Since brain scientists are just beginning to map neuronal activity that correlates with many of those states, the neurobiological study of consciousness is evolving slowly. While I have moved away from being a materialistic scientist for many years, I remain partial to the taxonomy proposed by Dehaene et al. (67). Dehaene, a renowned neuroscientist and cognitive psychologist, has significantly contributed to consciousness studies and understanding of human cognition. As a professor at the Collège de France and director of *NeuroSpin*, one of the world's most advanced brain imaging centers, Dehaene's research spans topics like numerical cognition, language, and the neural basis of consciousness. Dehaene's classification system came at a critical time when consciousness researchers were producing conflicting results discussing the neural basis of varying states of awareness.

In their *Global Neuronal Workspace Theory* (*GNWT*) model, Dehaene and his colleagues propose that the level of vigilance associated with each state appears to unify different states of consciousness, a point I already discussed to explain the nature of the bottom-up effect characterizing many of our decision under stress. Vigilance is a psychological and cognitive state characterized by sustained attention and alertness over a prolonged period, particularly when monitoring for infrequent or unpredictable events. This concept is crucial in various fields, including psychology, neuroscience, and occupational health, as it pertains to maintaining focus and reacting promptly to potential threats or important stimuli. Vigilance is implicated in wakefulness, sleep, coma, anesthesia, and, as discussed earlier, triggers PRIMAL consciousness below the awareness radar. Studies confirm that early consciousness is controlled mainly by the thalamus and the brain stem (68), while some point to the primary visual cortex (69, 70). However, Dehaene's GNWT theory also suggests that consciousness arises when information is integrated and made globally available to various cognitive systems. This means that conscious awareness is not tied to a specific location in the brain but to the dynamic interaction of widely distributed networks. An example is when you first see a word in a foreign language, it will most likely remain subconscious. But once you learn its meaning, it enters your global workspace and becomes part of your conscious thinking.

On the other hand, Giulio Tononi, another renowned neuroscientist for his significant contributions to consciousness studies, proposed a different model of consciousness called the *Integrated Information Theory (IIT)* (71). This comprehensive theoretical framework explains consciousness, how it can be measured, how it correlates with brain states, and why it fluctuates between states such as awake, dreaming, and dreamless sleep. Like the Dehaene model, IIT proposes that consciousness is a product of integrating information into the brain. It suggests that consciousness can be quantified as a value of interconnectivity and information integration within a neural system. A brain with many highly interconnected networks has more integrated information than a simple circuit, meaning it is more conscious.

This explains why a human has more consciousness than a thermostat, even though both process information. IIT is supported by various neuroscientific methods, including neuroimaging, transcranial magnetic stimulation (TMS), and computer modeling. Tononi's research offers a unique perspective that blends empirical investigation with theoretical modeling. His work has been influential in deepening our understanding of how consciousness arises from the physical structure of the brain and its dynamic activities.

Recently, The Templeton World Charity Foundation organized a contest on theories of consciousness involving the Global Neuronal Workspace Theory (GNWT) and the Integrated Information Theory (IIT) (72). The challenge, however, yielded inconclusive results. An international team led by Dr. Lucia Melloni conducted experiments through six independent labs. While IIT fared slightly better, neither theory was definitively proven or disproven. The study demonstrated activity in the posterior cortex aligning with IIT and some aspects of consciousness predicted by GNWT in the prefrontal cortex. Still, the inconclusive results left most materialistic scientists stuck on the *hard problem* of scientifically measuring consciousness.

The Emergence of Spiritual Consciousness

At this point in our discussion, it is important to stress that material scientists mostly disregard the notion that spiritual states exist or play a significant role in our understanding of consciousness. Spiritual states often involve subjective experiences that are challenging to quantify or study with standard scientific methods. This lack of measurable evidence leads many materialistic scientists to focus on physical explanations. A core principle of scientific research is reproducibility – the ability for an experiment or study to be repeated with the same results. However, spiritual experiences are often unique and personal, making them difficult to reproduce in a controlled setting. This makes it challenging to study these experiences in a manner that aligns with scientific methodology. They are often described in non-physical terms, so they do not fit neatly into a rational framework. Some consider that a scientific approach is not necessarily a rejection of the existence of spiritual or supernatural aspects but rather a focus on natural explanations that can be empirically investigated.

For instance, introduced earlier in Chapter ONE, the American physicist, novelist, and essayist Dr. Alan Lightman believes that materialistic perspectives of the universe can coexist with transcendent, spiritual experiences (16). Yet, Lightman asserts that everything is made of atoms and molecules governed by physical and quantum laws. Therefore, spiritual experiences, such as feelings of awe, connection, and beauty, must arise from the complex interactions of neurons in the brain. Additionally, the notion that they are *neuronal correlates of consciousness* or *NCC* is central to Lightman's argument and many other materialistic scientists (73). According to them, NCC enables neural complexity, supports an integrated nervous system, and manages our mental maps, mechanisms of selected attention, and even memory storage. Lightman argues that one way to demonstrate the existence of organized NCC is by considering that older adults who have compromised grid cells essential to the functioning of NCC cannot sustain many emergent properties of consciousness. Furthermore, psychiatrist and neurologist Todd Feinberg and Jon Mallatt argue that NCC emerged as a primitive form of consciousness nearly 500 million years ago (74).

Therefore, Lightman claims that the recognition of levels of consciousness in other animals further confirms that consciousness is rooted in the material brain and that the human brain and its capacities are not qualitatively different from the brains of other animals. However, his views ultimately combine respect for the laws of science with an appreciation for the spiritual aspects of human experience, emphasizing that some human experiences may not be reducible to mere physical explanations.

Interestingly, Lightman recognizes the gap between the materialistic view of the universe and the rich, subjective experiences that define our spirituality and humanity. For instance, Lightman acknowledges that some aspects of these experiences, particularly their deeply personal and subjective nature, may not be fully explainable in purely material terms. This perspective, which Lightman refers to as *spiritual materialism*, suggests that even with an understanding of nature's workings, we can experience transcendence through the conscious awareness that emerges from our physical brains.

Many other leading theories of spiritual consciousness attempt to explain consciousness's nature, origin, and experiences. For instance, *Transpersonal Psychology* (TP) studies transcendent or spiritual aspects of human experience. It integrates aspects of modern psychology with insights from spiritual traditions. Known for his hierarchy of needs, Abraham Maslow, a founder of the TP movement, introduced the concept of self-actualization and later self-transcendence as the pinnacle of human experience (75). Another prominent Transpersonal Psychology scholar, Dr. Stanislav Grof, explored non-ordinary states of consciousness through psychedelic therapy initially and later developed the proprietary method he named *Holotropic Breathwork*. Grof believes non-ordinary states of consciousness can lead to spiritual insight and healing (76). Over the last couple of decades, I have attended many workshops and retreats with Dr. Grof and found his views compelling, especially given his medical background and impressive research and books.

Meanwhile, the American philosopher Ken Wilbert boldly attempted to synthesize all human knowledge into a comprehensive framework that includes the spiritual dimension. He proposed a model called *AQAL* (All Quadrants, All levels), which provides for various dimensions of reality and stages of development and is based on the idea that consciousness spans a range from pre-personal to personal, lastly from personal to transpersonal stages (77). In this context, *pre-personal* refers to stages of consciousness before developing a distinct sense of self or personal identity. These stages are often associated with infancy, early development, or primal states of awareness where the individual has not yet formed a coherent or reflective sense of *I*. Pre-personal consciousness is typically characterized by instinctual, sensory, or subconscious processes that operate without self-awareness or higher-order thought.

Wilbert's model suggests a continuum of consciousness development, where pre-personal stages transition into *personal stages* (marked by self-awareness and identity) and, ultimately, *transpersonal stages*, which transcend the ego and connect to universal or collective aspects of experience. This perspective is often used in psychology, spirituality, and transpersonal studies to explore the evolution of human consciousness across biological, psychological, and spiritual dimensions.

Also, many Eastern philosophies and spiritual practices emphasize the cultivation of spiritual consciousness through meditation, mindfulness, and other practices. I am particularly interested in the Advaita Vedanta, a non-dualistic school of Hindu philosophy and spiritual practice. It is one of the most well-known and influential interpretations of the Vedanta, which refers to the end part of the Vedas, the ancient sacred texts of Hinduism. Advaita means non-dual signifying the belief in the fundamental oneness of all existence. According to Advaita Vedanta, there is no absolute separation between the individual self (Atman) and the ultimate reality (Brahman) (78).

To conclude, the perspective of most neuroscientists is that the brain does not switch on and off but instead operates on a continuum of consciousness. This material view contrasts with *dualist or spiritual perspectives*, which posit that non-physical entities or processes (such as a soul or mind) are fundamental to understanding consciousness. Critics of materialism (I hope it is obvious by now that I have become one!) argue that the view struggles to fully account for the subjective, qualitative aspect of conscious experience.

Non-materialistic Perspective on Meta-consciousness

Figure 47: Non-materialistic Perspective (AI)

Non-materialistic views assert that physical processes or brain activity alone cannot fully explain consciousness. These perspectives argue that consciousness has qualities or aspects that are fundamentally non-physical or transcend material explanations. Such views often encompass philosophical positions like dualism, which posits a distinct realm of mental phenomena separate from the physical, or idealism, which considers consciousness the primary or sole reality. These viewpoints challenge materialistic or physicalist theories that equate consciousness solely with brain function or material processes.

Ancient cultures like the Egyptians and Greeks conceptualized consciousness with the soul or spirit, often linking it to life forces or divine aspects. Philosophers like Aristotle and Plato pondered the nature of consciousness, considering it a fundamental element of the human experience intertwined with reasoning and the essence of life. These early perspectives laid the groundwork for later philosophical and scientific explorations into the nature of consciousness. René Descartes, the French philosopher, scientist, and mathematician, established the tenets of dualism in the seventeenth century. According to Descartes and his dualistic model, mental phenomena are non-physical, meaning the mind and body are distinct and separable. Meanwhile, most organized religions support dualistic models to the extent that they integrate consciousness with broader spiritual, ethical, and metaphysical concepts.

Top Religious Views on Consciousness

While investigating the religious views on consciousness is way beyond the scope of this book, it is nonetheless helpful to be reminded of the core principles that guide the narrative on the origin and role of consciousness in leading theological belief systems.

Christianity (1.2 billion Catholics, 625 million Protestants, 293 million Orthodox, 75 million Anglicans)

In Christianity, consciousness is often intertwined with the concept of the soul, which is believed to be a unique, immaterial essence granted by God. This soul is considered the seat of consciousness and personal identity, surviving beyond physical death. Christians generally view the soul as an aspect of a human being's ability to have moral judgment and a relationship with God. This perspective emphasizes consciousness's spiritual and ethical dimensions, aligning it with eternal destiny and moral responsibility.

Catholics

For the 1.2 billion Catholics worldwide, consciousness is closely tied to the soul as the immaterial essence created by God. Catholic doctrine emphasizes that the soul is immortal, holding personal identity and moral responsibility even after physical death. This belief is central to Catholic theology, which views consciousness as a gift enabling humans to discern right from wrong and to maintain a relationship with God, ultimately determining their eternal destiny.

Protestants

Protestants similarly understand the soul as the source of consciousness and moral agency. However, Protestant traditions often strongly emphasize personal accountability and a direct relationship with God through faith. Consciousness in this context reflects one's ability to respond to God's grace, make ethical decisions, and seek salvation through individual faith and scriptural understanding.

Orthodox Christians

Orthodox Christians approach consciousness through a deeply spiritual lens, often emphasizing the transformative journey of the soul toward *theosis* or union with God. In Orthodox thought, consciousness involves the spiritual awareness necessary to align one's life with divine will. It is cultivated through prayer, sacraments, and ascetic practices that refine the soul and draw it closer to God's presence.

Anglicans

For Anglicans, consciousness is considered integral to the soul's relationship with God, reflecting Protestant and Catholic influences. Anglican theology highlights consciousness's ethical and spiritual dimensions, balancing individual moral responsibility with communal worship and tradition. Consciousness, as a facet of the soul, is seen as a means of engaging with God's grace while navigating moral and spiritual challenges.

Islam (1.5 billion Sunnis, 215 million Shias)

In Islam, consciousness is closely linked with the soul (*nafs*) and is considered a gift from Allah. It encompasses self-awareness, moral and ethical discernment, and the capacity for spiritual understanding. Islam emphasizes the soul's accountability and responsibility in choices and actions, viewing consciousness as central to the human experience and the individual's relationship with God. This perspective underscores the importance of using one's conscious faculties for moral growth and spiritual development.

Sunni Muslims

For Sunni Muslims, who represent much of the Islamic faith, consciousness is seen as an integral aspect of the soul (nafs) and a divine gift from Allah. It includes self-awareness, moral discernment, and spiritual comprehension. Sunni teachings emphasize the soul's accountability for its actions, with consciousness guiding individuals toward fulfilling their obligations to Allah and striving for moral growth through the Quran and Sunnah.

Shia Muslims

Among Shia Muslims, consciousness is also profoundly connected to the soul (nafs) but often carries a heightened focus on justice and the moral implications of actions. Shia Islam emphasizes the soul's role in discerning ethical truths and maintaining a conscious alignment with the teachings of the Prophet Muhammad and the Imams. This view sees consciousness as central to achieving spiritual enlightenment and cultivating a deeper relationship with Allah.

Sufi Muslims

For Sufi Muslims, consciousness takes on a mystical dimension, serving as a pathway to spiritual enlightenment and union with Allah. In Sufism, the soul (nafs) undergoes a process of purification, moving beyond the ego-driven self to achieve a higher state of awareness. Consciousness is cultivated through meditation, dhikr (remembrance of God), and devotion, aiming to transcend the material world and experience the divine presence.

Hinduism (1 billion Hindus)

In Hinduism, consciousness is considered a core aspect of existence. At the individual level, it is embodied in the *Atman* (soul), which is seen as eternal and beyond physical existence. At the universal level, consciousness is represented by *Brahman (*as briefly mentioned when I discussed Advaita Vedanta), the ultimate, all-encompassing reality. This perspective posits that individual consciousness (*Atman*) and universal consciousness (*Brahma*n) are fundamentally the same, and realizing this unity is a key aspect of spiritual enlightenment. The understanding of consciousness in Hinduism is profoundly spiritual and philosophical, intertwining with concepts of reality, self, and the universe.

Buddhism (334 million Mahayana Buddhists, 185 million Theravada Buddhists)

Buddhists believe in immortal and nonmaterial consciousness. In Buddhism, consciousness (*Vijñāna* in Sanskrit) is considered a fundamental aspect of existence, deeply interconnected with the physical and mental phenomena. It is viewed as a continuous process rather than a static entity, constantly changing and evolving. Buddhism emphasizes the concept of "no-self" (*Anattā*), suggesting that consciousness is not a permanent, unchanging self or soul but a stream of interconnected and impermanent mental events. This perspective encourages mindfulness and meditation to understand consciousness's transient nature and alleviate suffering by detaching from the illusion of a permanent self.

Theravāda Buddhism

In Theravāda Buddhism, consciousness (Vijñāna) is viewed as a fleeting process arising and ceasing based on conditions such as sensory input and mental formations. It is impermanent and lacks any unchanging essence. Central to this tradition is the doctrine of Anattā (no-self), which denies the existence of a permanent soul or self. Instead, what is perceived as "self" is a dynamic interplay of the five aggregates: form, sensation, perception, mental formations, and consciousness. This understanding helps practitioners avoid the illusion of a stable identity, leading to liberation from suffering and attaining Nirvāṇa, the cessation of rebirth.

Mahayāna Buddhism

Mahayāna Buddhism expands the understanding of consciousness with models like the Yogācāra school's eight layers (79). These include sensory consciousness, ego-awareness (Mano-Vijñāna), and the deeper Ālaya-Vijñāna (storehouse consciousness), which holds karmic imprints and past experiences. Despite this complexity, Mahayana adheres to the concept of Anattā, rejecting a permanent soul while exploring consciousness as a dynamic process. This layered approach provides a path toward transcending ego and ordinary perception, guiding practitioners toward profound awareness and compassion. The ultimate goal remains to achieve Bodhisattva-hood or liberation, with a focus on alleviating suffering for all beings, reflecting Mahayana's universalist ethos

Shinto (121 million practitioners)

Shinto, Japan's traditional religion, does not explicitly articulate a defined concept of consciousness like Western philosophies and religions might. Instead, it focuses more on rituals and practices honoring *Kami* (spirits or gods) that inhabit all aspects of nature. Shintoists view humans as part of the natural world, with a strong emphasis on harmony and balance. While it does not explicitly address consciousness philosophically, Shinto implies a kind of interconnected awareness through its reverence for nature and the spirits within it.

The church of Jesus Christ of Latter-day Saints (17 million members)

The concept of eternal progression in LDS theology suggests that individuals continue to grow, learn, and develop forever, including in their consciousness and understanding. This eternal perspective on growth and learning highlights the expansive view of consciousness as an evolving attribute of the soul. LDS beliefs about the afterlife include continued consciousness, relationships, growth, and progression. This suggests a view of consciousness as not only continuing after death but as an integral part of one's eternal identity and purpose.

Judaism (15 million Jews)

In Judaism, consciousness is also often linked with the soul (*Neshama*), viewed as a divine spark given by God. This perspective emphasizes moral and ethical responsibility, aligning consciousness with self-awareness, free will, and the ability to choose between good and evil. In Judaism, the soul is seen as the essence of a person, central to their identity and relationship with God and others. This view underscores the importance of conscious choices and actions in life's moral and spiritual journey.

Even from this brief review, it becomes evident that the common denominator in the views of significant religions on consciousness is the belief in an immaterial, enduring aspect of the self, often linked with the soul or spirit. This aspect is fundamental to human identity and moral responsibility, transcending physical existence. Religions typically view consciousness as integral to one's relationship with the divine, ethical actions, and pursuing spiritual understanding or enlightenment. None of the religions I just reviewed support the notion that consciousness is a material, transient property of the brain.

For many years, I had cultivated this feeling—a deep sense that there was *something more*, something just beyond the grasp of materialistic science. Yet, I hadn't fully understood its limitations, nor did I recognize what I was truly missing. I was immersed in the frameworks of empirical research, neuroscience, and cognitive theory, yet an undeniable gap remained—a missing dimension that science alone couldn't explain. Immersed in the jungle's vibrant tapestry of life, guided by ancient rituals and sacred songs, I experienced a significant paradigm shift. One night, under a canopy of stars, I sat with the visceral energy of the rainforest enveloping me, my rational mind struggling to grasp the intangible forces at work. Yet, in that liminal space, something clicked. The intricate songs of the shamans felt like neural pathways rewiring, urging me to let go of my need for empirical control and to fully embrace the unseen, the spiritual, and the interconnected. I understood then that, though powerful, materialistic science could only take me so far. I had to open myself to perspectives that blended the measurable with the mystical to understand transformation and healing fully. This epiphany inspired me to explore a more holistic science that bridges neuroscience, neuroplasticity, and ancient spiritual wisdom. It was not about rejecting science, but rather about expanding it—embracing a more holistic understanding that integrates neuroscience, neuroplasticity, and ancient spiritual wisdom. This moment of clarity guided me toward an approach that honors both the measurable and the mystical, the rational and the transcendent.

After all, science is not just about what we can measure but about how we make meaning of our experiences. And in that space—where knowledge meets the unknown—I found the inspiration to bridge the worlds of modern brain science and timeless spiritual insight by writing *OPEN*.

Meanwhile, I must address the views of the millions of people who do not follow a particular religion or support a god-centric model of consciousness in this discussion. Since Indigenous approaches to consciousness largely inspired this book and my prior book on the neuroscience of serenity, this section is especially important to explain non-materialistic views of consciousness from shamanic, pagan, and well-established spiritual frameworks.

Shamanic, Pagan, and Spiritual Views of Consciousness

Shamanistic and *Pagan* traditions are our planet's oldest self-organized spiritual movements. Consciousness is generally viewed as deeply interconnected with the natural and spiritual realms. These beliefs often emphasize that every element of nature possesses a spirit or consciousness and that humans can interact with these spirits through altered states of consciousness.

Shamanism

Shamanism is considered one of the earliest forms of spiritual practice and religious belief, dating back to prehistoric times. While it is challenging to pinpoint an exact date for its emergence due to the lack of written records from that era, it is generally believed to have developed during the Paleolithic era over 10,000 years ago.

Some researchers even suggest that Neanderthals (>100,000 years ago), who buried their dead and used red ochre, might have engaged in proto-shamanistic practices, although this remains speculative. Additionally, evidence such as symbolic artifacts, burials, and cave art indicates that spiritual and shamanistic practices were present among early Homo sapiens in the Upper Paleolithic (40,000+ years ago). Objects like the Lion-Man of Hohlenstein-Stadel (Germany, c. 40,000 BCE), a carved ivory figure blending human and animal traits, are interpreted by some archaeologists as representations of spiritual or shamanic concepts.

These artifacts indicate that early humans engaged in rituals involving altered states of consciousness, communication with spirits, and the use of symbolic objects. Shamanism is not confined to a specific region or culture. It has been practiced by Indigenous peoples worldwide, from Siberia and Mongolia to the Americas, Africa, and Australia. This widespread presence suggests that shamanistic practices emerged independently in various cultures as a response to the human need to understand and interact with the natural and spiritual world.

Shamanism encompasses a specific set of practices within a cultural or religious context. It is often tied to distinct ethnic groups and their traditional spiritual practices (80). Shamanism involves practices aimed at entering trance states to communicate with and gain knowledge from the spirit world. This perspective sees consciousness as a human attribute and a universal quality shared by all elements of nature.

Indigenous communities often view consciousness as deeply connected to the natural world and the community. This perspective sees consciousness as more than individual awareness, extending to the land, animals, ancestors, and spiritual entities. Shamanistic views often consider consciousness a shared attribute, linking humans with nature and the spiritual realm and central to maintaining harmony and balance in life. These positions emphasize the interdependence of all life and the importance of respecting and understanding these connections for the well-being of individuals and the community.

Native Americans also think consciousness is the source of a deep connection between the self, nature, and the spiritual world. It is not limited to individual awareness but is interwoven with the land, community, ancestors, and various spiritual entities. This perspective recognizes a universal life force or spirit that flows through all beings and elements of nature, underscoring the interconnectedness and interdependence of life. Respect, balance, and harmony within this interconnected web are central themes in their understanding of consciousness (81).

Shipibo Views of Consciousness

Meanwhile, Amazonian Indigenous communities also view consciousness as deeply interconnected with the natural and spiritual worlds. They typically believe that all elements of nature, including plants, animals, and geographical features, possess a spirit or consciousness. These communities often use rituals and practices to connect with and understand these various forms of consciousness, viewing them as integral to maintaining harmony and balance in their individual and social lives. This perspective emphasizes a holistic and interconnected understanding of consciousness, transcending individual human experience.

Living a month in a Shipibo village in the middle of the Peruvian Amazon gave me a deep sense of the rich and complex understanding of consciousness intertwined with their cultural and spiritual beliefs. Ancient traditions, rich shamanic practices, and regular use of plant medicines like Ayahuasca guide their life. The Shipibo believe everything in the universe, including plants, animals, and even inanimate objects, possesses a spirit or consciousness. This perspective fosters a deep sense of interconnectedness and respect for all life forms. They also recognize multiple levels or dimensions of reality and consciousness. The physical world is just one aspect, with other spiritual dimensions existing alongside it. These dimensions are accessible through certain rituals, especially those involving plant medicines like Ayahuasca.

Indeed, Ayahuasca plays a central role in Shipibo spirituality and their understanding of consciousness. Ayahuasca facilitates access to higher states of consciousness and enables communication with the spiritual world. This is crucial for healing, as it allows shamans to identify and treat illnesses they perceive as having spiritual origins. Shamans can navigate different levels of consciousness and interact with the spiritual world. Through their knowledge and skills, they guide healing ceremonies, offer wisdom, and balance the physical and spiritual realms (11).

As I discussed in Chapter ONE, icaros, the sacred songs believed to carry spiritual energy and guide consciousness during ceremonies, particularly ayahuasca rituals, are central to their healing practices. Icaros are used to harmonize the participants' energy, connect them to higher states of awareness, and facilitate healing by interacting with the spiritual dimensions of plants and the universe. These songs are considered a bridge between the material and spiritual realms, shaping and elevating consciousness to foster transformation and alignment.

Meanwhile, Shipibo's cosmology and art deeply reflect their views of consciousness. Their intricate designs and patterns, often seen in their textiles, are believed to represent the fundamental structures of the universe and the interconnected nature of all things. These patterns are more than art; they are expressions of their understanding of the cosmos and consciousness. They believe that wisdom about the universe and human existence has been passed down through generations and is embedded in their cultural practices (28). For the Shipibo, exploring consciousness is about individual understanding or enlightenment and the community's well-being. Their practices and beliefs emphasize the importance of harmony, balance, and collective growth.

Paganism

Paganism is considered a broad category of religions and spiritual paths that often emphasize connection to nature, recognize and worship multiple deities (polytheistic beliefs), and ancient rituals. In contrast to shamanism, paganism can include reconstructed or newly created traditions that may not be tied to a specific culture.

It generally refers to religious traditions and practices outside the world's main religions, especially those that predate monotheistic religions like Christianity, Islam, and Judaism. It often includes polytheistic and nature-based faiths. Paganism includes many traditions and beliefs, including ancient Greek and Roman religions, Norse and Celtic spirituality, and modern Neo-Pagan movements like *Wicca*.

Wicca is a contemporary Pagan religious movement that draws upon a diverse set of ancient pagan and 20th-century hermetic motifs for its theological structure and ritual practices. It was developed in England during the first half of the 20th century and was introduced to the public in 1954 by Gerald Gardner, a retired British civil servant. Wicca is now practiced in various forms worldwide and is one of the most well-known forms of modern Paganism (82).

Each pagan tradition has its deities, beliefs, and practices. A common thread among them is reverence for nature and the belief in multiple gods and goddesses, each with different aspects and domains. Celebrations and rituals often align with natural cycles like the seasons and lunar phases. Many pagan traditions draw on ancient or Indigenous cultures' mythology, folklore, and practices. They often seek to revive or reinterpret these traditions in a modern context. Modern paganism is usually characterized by individualized practice and an eclectic approach, where practitioners may blend elements from various pagan and sometimes non-pagan traditions to create a personal spiritual path.

The views on consciousness within Paganism are as diverse as the traditions and paths that fall under the Pagan umbrella. Paganism includes many beliefs, from reconstructed ancient religions like Hellenism and Asatru to modern movements like Wicca and Druidry. Pagans often see consciousness as a quality shared by all living beings, sometimes even inanimate objects and natural phenomena. This view fosters a sense of interconnectedness between nature and the environment.

Spiritual Movement

People who identify as spiritual hold views on consciousness that blend elements of philosophy, religion, mysticism, and sometimes science. These beliefs can vary widely due to the diverse nature of spirituality, encompassing a range of traditions, cultures, and personal interpretations. The term *spiritual* is subject to many definitions and debate. Generally speaking, it means "being connected to something bigger than yourself," according to 74% of people surveyed by the Pew Research Center. In comparison, 70% of those who responded specified that "spiritual" meant "being connected to God (83)."

Non-materialistic views of consciousness dominate the general field of spirituality and do not depend on the belief in God. In the US alone, 22% of adults consider themselves spiritual but not religious, according to the Pew Research study. However, you may be surprised to learn that 83% of all US adults believe they have a soul in addition to their physical body, and 81% say there is something spiritual beyond the natural world. Many spiritual perspectives view consciousness not as a byproduct of physical processes but as a fundamental, pervasive quality of the universe. This view often aligns with the concept of a universal consciousness or a unifying spirit that connects all beings. A common belief in spiritual circles is that all consciousness is interconnected and part of a larger whole. This idea can manifest as unity with other living beings, the environment, and the universe.

Spiritual views often include the belief that consciousness can transcend the physical limits of the brain and body. This might consist of beliefs in out-of-body experiences, astral projection, or the existence of a soul or spirit that exists independently of the body. Many spiritual beliefs emphasize a holistic view of consciousness, where the mind, body, and spirit are profoundly interconnected and affect one another. This perspective often incorporates practices such as meditation, yoga, or prayer to harmonize these aspects. Beliefs in an afterlife or reincarnation are also common from a spiritual perspective. These beliefs suggest that consciousness continues in some form beyond physical death, whether in a spiritual realm or through rebirth in a new body. Spirituality often places a strong emphasis on the evolution or development of consciousness. This can involve a journey toward enlightenment, self-realization, individuation, or higher awareness, often through personal introspection and spiritual practices.

Some spiritual traditions, particularly those influenced by Eastern philosophies like Hinduism and Buddhism, advocate a non-dualistic view of consciousness for which the individual self and the universe are seen as an illusion, with proper understanding coming from recognizing the oneness of all existence. These spiritual views focus on the experiential and subjective aspects of consciousness, often emphasizing personal experience, introspection, and the pursuit of a deeper understanding or connection with the cosmos. They differ from more materialist or scientifically oriented views in that they often incorporate mystical, transcendent, and non-empirical elements.

Panpsychism

Panpsychism is a philosophical theory about the nature of consciousness and its relationship to the physical world. It suggests that consciousness is a universal feature like its physicalist counterpart, but it often imbues this consciousness with spiritual or divine qualities (84). It posits that consciousness, a mind-like aspect, is the universe's fundamental and ubiquitous feature. In other words, panpsychism suggests that all things, not just humans or animals, but also plants, inanimate objects, and even elementary particles, possess some form of consciousness or subjective experience. The movement has recently gained popularity; many consider it a hybrid model between materialist and non-materialist extremes. This offers us a beautifully simple, elegant way of integrating consciousness into our scientific worldview, of marrying what we know about ourselves from the inside and what science tells us about matter from the outside (85).

The work of Donald Hoffman, a cognitive scientist, provides a unique perspective that resonates with certain aspects of panpsychism: the view that consciousness is a fundamental and ubiquitous feature of the universe (86). Hoffman challenges the traditional materialist assumption that consciousness arises solely from complex arrangements of matter. Instead, he argues that consciousness is fundamental and that what we perceive as the physical world is a construct that conscious agents generate. In this view, the material world does not exist independently of observers; instead, it emerges from interactions between conscious entities. This aligns with the panpsychist idea that consciousness is not limited to humans or complex organisms but is instead an intrinsic feature of reality, present at all levels of existence.

Hoffman's theory suggests that our perceptions are akin to a "user interface" designed by evolution to help us navigate the world effectively rather than to reveal objective reality. This interface hides the true nature of reality, which, according to Hoffman, is structured by consciousness itself. By positing that conscious agents are the fundamental building blocks of reality, his theories imply that what we consider "physical objects" are merely representations of the conscious experiences of these agents rather than independent entities. Thus, Hoffman's ideas contribute to a broader philosophical discourse that sees consciousness not as an emergent property of complex systems but as a foundational aspect of the cosmos, aligning with the core tenets of panpsychism.

While Hoffman's theories overlap with spiritual ideas, he tends to present his views within cognitive science, evolutionary theory, and philosophy rather than directly engaging with spirituality as traditionally defined. His work is more focused on challenging the scientific understanding of perception and reality, though it naturally leads to questions often explored within spiritual contexts.

To sum up, panpsychism claims that consciousness is a fundamental quality in the universe embedded in all matters. It challenges the traditional materialist view by suggesting that the mental and physical are deeply intertwined at every level of existence. This holistic perspective sets the stage for our discussion on quantum consciousness, which builds on the interconnectedness and non-locality observed in quantum mechanics.

In quantum consciousness, the enigma of observation and its role in collapsing quantum states offers a compelling bridge to panpsychism ideas. The act of measurement—a phenomenon central to quantum mechanics—raises questions about the relationship between an observer and the observed, hinting at a participatory universe where consciousness may play an intrinsic role. This scientific lens provides a potential mechanism through which the foundational tenets of panpsychism could manifest at the quantum level.

This next section invites a deeper inquiry into the origins and nature of consciousness, leveraging both philosophical insights and the groundbreaking implications of quantum theory to uncover a more integrated understanding of reality. This interplay highlights the expanding frontier where metaphysics and physics converge.

Quantum consciousness is a term used to describe a group of hypotheses that suggest quantum mechanics plays a role in explaining consciousness. It is a speculative and controversial field, combining concepts associated with atoms and subatomic particles with neuroscience. While the field is very challenging to wrap your head around, there are simple principles that are important to consider. Roger

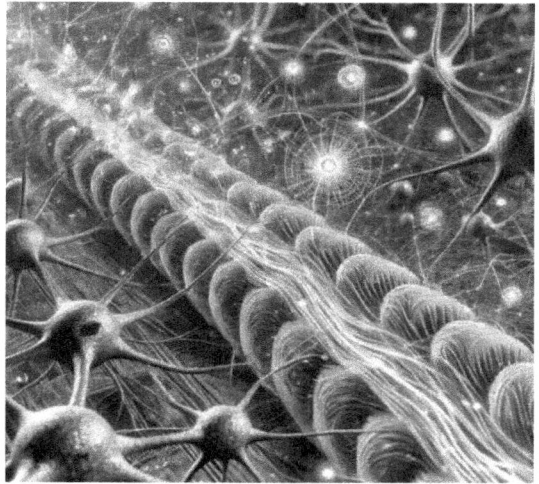

Figure 48: Quantum Consciousness (AI)

Penrose, a mathematician and theoretical physicist, and Stuart Hameroff, an anesthesiologist and consciousness researcher, proposed the first theory in this domain over twenty years ago (87). They suggested that consciousness arises from quantum computations occurring in *microtubules* within brain cells. Microtubules are cylindrical structures within the cytoplasm of eukaryotic cells. These dynamic structures are pivotal in cell division, ensuring accurate chromosome segregation. Additionally, microtubules are crucial for cell locomotion and fluid movement in tissues.

More recent studies have confirmed the role of microtubules in consciousness through potential quantum biology mechanisms that are thought to be fundamentally different from classical neural processing (88). The idea is that quantum states, such as *superposition* and *coherence*, could be involved in brain function. Superposition means that particles exist in multiple states simultaneously, which have been measured and observed in experiments. Coherence describes a situation where quantum entities are correlated and synchronized (89). Proponents of quantum consciousness suggest that these properties could explain some of the complex, non-linear properties of brain activity and consciousness.

For instance, Roger Penrose and Stuart Hameroff proposed a model called *Orchestrated Objective Reduction* (Orch-OR), which combines quantum mechanics with neuroscience and posits that quantum computations in neuronal microtubules are the basis of consciousness. This model suggests different levels of consciousness based on the coherence of these quantum processes, with greater coherence leading to higher levels of consciousness (87).

Meanwhile, some theories draw on the *quantum observer effect*, meaning that a quantum system is not describable without including the impact of observation. This has led to philosophical and speculative discussions about the role of the conscious observer in shaping reality. Additionally, there is intense speculation about whether *quantum entanglement*—a phenomenon where particles remain connected so that actions performed on one of them affect the other, regardless of distance—could play a role in the brain's functioning, potentially contributing to aspects of consciousness connecting with other beings millions of miles, if not light-years away from each other.

Finally, emerging theories suggest that consciousness may arise from neural activity and the brain's electromagnetic fields (90). These fields are generated by synchronized neural firing, particularly during gamma oscillations, closely associated with conscious perception and attention. Research indicates that these oscillations might integrate information across the brain, addressing the "binding problem" by creating a unified framework for subjective experience.

Experimental evidence supports this idea; transcranial magnetic stimulation (TMS), which applies external electromagnetic fields, can modulate or disrupt conscious awareness. This suggests that electromagnetic fields are more than mere by-products of neural activity—they may actively participate in cognitive processes. The Global Workspace Theory aligns with this view, proposing that electromagnetic fields could serve as the medium for this integration.

While the hypothesis is compelling, challenges remain in establishing causality between electromagnetic fields and consciousness. Current studies often highlight correlations, but further research is needed to verify whether these fields directly generate conscious awareness. Advances in computational modeling and brain-imaging technologies hold promise for exploring the mechanisms by which these fields might shape the human experience. Such work bridges traditional neuroscience with emerging paradigms, inviting interdisciplinary efforts to unravel the nature of consciousness.

Thus, despite its intriguing and promising premises, quantum consciousness is not widely accepted in the scientific community. Critics argue that the brain's warm, wet, and noisy environment is not conducive to quantum phenomena typically only observed at near absolute zero temperatures. Additionally, there needs to be more empirical evidence directly linking quantum mechanics to brain processes. Therefore, while it presents an innovative approach to understanding consciousness, quantum consciousness remains a theoretical and highly debated concept in neuroscience and physics.

Towards a Neurospiritual Model of Consciousness (NMC)

Modern science and philosophy have studied and debated the nature of consciousness for millennia, providing a rich and complex understanding of this fundamental phenomenon. As I have presented in *OPEN* so far, consciousness exists at the intersection of various disciplines—chemistry, biology, neuroscience, psychology, quantum physics, and spirituality, each offering unique insights into nature, origin, and evolution. In reflecting on this vast body of knowledge, I offer the following model with great humility and reverence to the many scientists, philosophers, and scholars who have dedicated their work to advancing our understanding of consciousness.

While I recognize the boldness, if not the presumption, of proposing yet another model of consciousness, I felt compelled to contribute to the conversation. I hope this model will inspire meaningful dialogue and reflection on the topic. Therefore, I present a framework based on the metaphor of an inverted pyramid. In this model, consciousness is conceptualized as evolving from primal biological awareness at birth, progressing through various emotional and cognitive development levels, and ultimately culminating in higher-order spiritual states like hyper-consciousness. The aim is to integrate the neurological foundations of consciousness and its spiritual dimensions, creating a comprehensive framework for understanding human awareness.

The Inverted Pyramid Structure

The *Neurospiritual Model of Consciousness* (NMC) conceptualizes consciousness as a multi-tiered structure, starting with foundational, primal biological states and moving up through increasingly complex cognitive and spiritual levels. This structure is depicted as an inverted pyramid to illustrate how consciousness broadens as it ascends from its base. At the foundation, consciousness is rooted in survival-focused awareness, where PRIMAL drives and instincts primarily operate. These primal states, which include basic responses to immediate needs and threats, form the narrowest, most focused layer of the pyramid, necessary for fundamental survival but limited in scope.

As one moves up the pyramid, the model describes transitions to higher levels of consciousness characterized by rational thinking, reflection, and eventually, spiritual awareness. In the RATIONAL layer, cognitive processes such as logical reasoning, self-reflection, and social understanding become dominant, expanding awareness beyond survival to abstract thinking and problem-solving. Finally, at the top of the pyramid lies the SPIRITUAL layer, where consciousness can encompass unity, interconnectedness, and spiritual transcendence. This model suggests that as consciousness expands, it grows in complexity and inclusiveness, ultimately allowing for states of unity that transcend the Self, merging individual awareness with a sense of collective and universal consciousness.

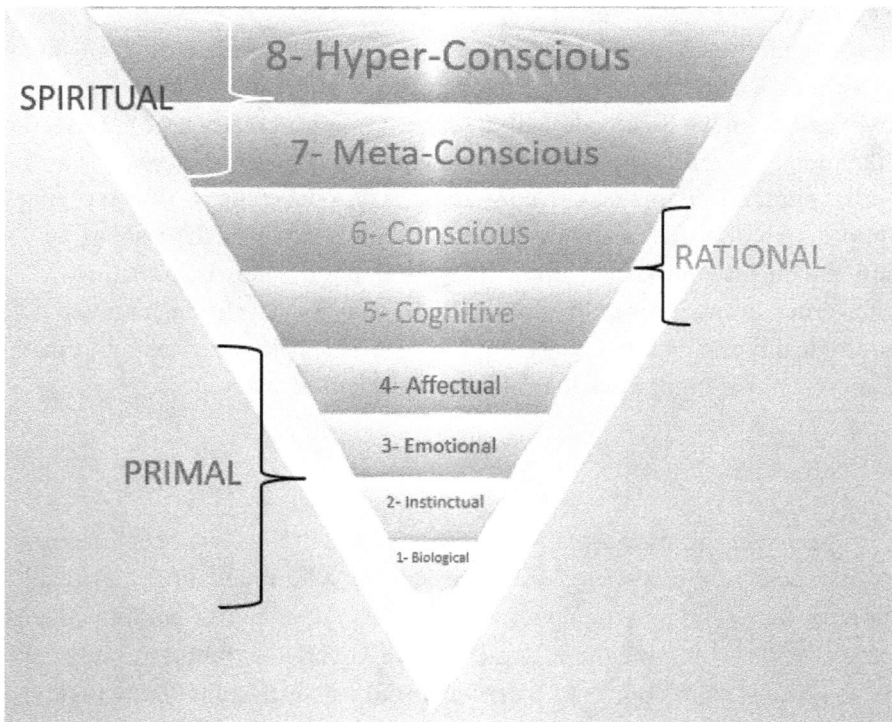

Figure 49: The Neurospiritual Model of Consciousness (NMC)

The 8 Levels of the NMC

1. **Biological: Primitive Awareness**
   At the foundation of NMC, the brainstem and medulla govern essential survival functions, including breathing, heart rate, and reflexes. This level reflects fundamental awareness focused purely on maintaining life.

2. **Instinctual: Sensory Processing**
   Expanding upon primitive awareness, sensory processing involves the thalamus, which relays sensory input to the primary sensory cortex. Here, consciousness interprets immediate environmental stimuli and establishes essential exchanges with the external world.

3. **Emotional: Reactivity**
   Moving beyond instinctual perception, the amygdala and the limbic system introduce an emotional layer to consciousness. This level is dominated by autonomous responses to emotions like fear, pleasure, and aggression, which shape reactions to environmental and social cues.

4. **Affectual: Basic Social Awareness**
   The anterior cingulate cortex (ACC) is essential for developing social awareness at this level. This involves empathy and social bonding, marking a shift from individual survival to understanding social dynamics.

5. **Cognitive: Self-awareness and Ego**
   The default mode network (DMN), which includes the medial prefrontal cortex and posterior cingulate cortex, becomes central at this level, generating self-referential thoughts and a stable sense of self. This level of consciousness is characterized by self-reflection and ego-centric processing, where the mind is occupied with self-identity, personal narrative, and introspection. While useful for self-understanding, prolonged DMN activity can keep individuals cognitively focused inward, limiting transcendence to higher states of consciousness.

6. **Conscious: Abstract Thinking**
   The dorsolateral prefrontal cortex (DLPFC) enables higher-order cognitive functions, such as planning, reasoning, and abstraction. At this level, consciousness expands to encompass complex problem-solving, hypothetical scenarios, and the ability to envision long-term goals beyond immediate needs.

7. **Meta-conscious: Moral and Ethical Reasoning**
   This level engages areas within the prefrontal cortex and the temporoparietal junction, fostering moral and ethical reasoning. Meta-consciousness transcends self-interest, incorporating empathy, altruism,

and consideration for the well-being of others. It allows for judgments aligned with principles of fairness and justice. At this state, consciousness is observable, creating states that provide more agency and flow.

8. **Hyper-conscious: Spiritual Unity**

   At the pinnacle of the model, spiritual unity is attained through the ability to suspend the default mode network (DMN) at will, enabling a transcendence of self-focused awareness. Enhanced activity in the parietal cortex supports a sense of interconnectedness and unity beyond the individual self. This state represents the ultimate expansion of consciousness, where boundaries dissolve, and an experience of universal or collective consciousness is realized.

How does NMS align with other prominent models of consciousness?

In my model, consciousness is seen as a complex, evolving phenomenon that bridges scientific and spiritual realms, recognizing it as influenced by neurobiological processes and transcendent experiences. This perspective integrates adaptable neuroplasticity and higher awareness achieved through spiritual practices. Comparing it to models proposed by Freud, Jung, Maslow, Grof, Wilber, Dehaene, Tononi, Newberg, Penrose, Bohm, Goswami, and Hoffman reveals how neurospirituality merges insights from neuroscience, psychology, physics, and spirituality into a cohesive and dynamic framework.

Sigmund Freud (1856–1939)

Freud's early 20th-century model of consciousness is foundational in psychology. He divided the mind into conscious, preconscious, and unconscious domains, with consciousness serving as a control over the hidden impulses of the id. Freud's model does not incorporate spirituality and focuses instead on ego-driven self-regulation. NMC, however, considers consciousness a mental regulator and a pathway to higher states of awareness. Unlike Freud's mechanistic approach, neurospirituality introduces practices that expand consciousness beyond rational control, integrating spirituality as a means of achieving mental health and harmony.

Carl Jung (1875–1961)

Jung expanded on Freud's ideas by introducing the collective unconscious, a reservoir of shared archetypes and symbols. Jung believed consciousness could connect to this universal source, which shapes individual experience and collective understanding. Neurospirituality builds on Jung's spiritual lens, suggesting that through neuroplastic rituals like meditation and prayers, people can connect with universal themes, deepening self-awareness. Neurospirituality also extends Jung's view by incorporating neuroscientific principles, proposing that neuroplasticity allows individuals to reshape and elevate personal archetypes and beliefs.

David Bohm (1917–1992)

Physicist David Bohm proposed the holographic model, suggesting that all elements of reality, including consciousness, are interconnected within an "implicate order" that underlies physical phenomena. In the neurospiritual model, consciousness similarly connects to a deeper universal structure. Practices like meditation are believed to help individuals access this implicate order, promoting a sense of oneness. Neurospirituality aligns with Bohm's view by exploring consciousness as both individual and cosmic, although it adds neuroplastic insights to explain how spiritual practices can foster lasting changes in consciousness.

Abraham Maslow (1908–1970)

Maslow's humanistic model (1940s–1960s) introduced self-actualization and self-transcendence as the ultimate stages of psychological development. He proposed that individuals reach higher consciousness by satisfying basic needs and seeking self-fulfillment. Neurospirituality parallels Maslow's hierarchy but expands it by emphasizing that achieving higher states of awareness requires intentional neuroplastic engagement. This model integrates Maslow's goal of self-transcendence through brain-altering habits and positioning practices like meditation and mindfulness as pathways toward holistic growth.

Stanislav Grof (1931–)

In the 1960s, Grof developed transpersonal psychology, studying consciousness through non-ordinary states such as those induced by LSD, especially. Grof saw these states as gateways to expanded consciousness and deep healing, often transcending personal ego and facilitating spiritual awareness. Neurospirituality incorporates these principles, recognizing altered states to access higher consciousness and foster neuroplasticity. Both models advocate for the spiritual use of psychedelics, rituals, and meditation to tap into the self's healing potential. However, neurospirituality also emphasizes the role of brain adaptability in sustaining these changes.

Ken Wilber (1949–)

In the late 20th century, Wilber proposed the integral theory, positing consciousness as an evolving spectrum from basic sensory awareness to transpersonal stages. Like neurospirituality, Wilber's model recognizes that consciousness can expand through spiritual practice. Neurospirituality aligns with this by suggesting that meditation and self-reflection can raise consciousness, moving it through various stages of awareness. However, neurospirituality further emphasizes neuroplasticity as a key mechanism for maintaining growth and supporting sustained states of compassion, empathy, and mindfulness.

Roger Penrose (1931–)

As discussed earlier, Penrose's quantum model challenges the idea that consciousness is purely biochemical. The neurospiritual model shares Penrose's openness to unconventional explanations, suggesting that consciousness might be fundamentally non-material and influenced by quantum forces. However, neurospirituality emphasizes accessible neuroplastic changes achieved through daily rituals, highlighting practical pathways to expand awareness and interconnectedness.

Amit Goswami and Donald Hoffman (1940s–)

With his quantum consciousness theory, the physicist Amit Goswami suggests that consciousness is fundamental, giving rise to the physical world rather than emerging from it. Neurospirituality resonates with this non-materialistic view, seeing consciousness as foundational and transformative. Donald Hoffman's theory of "conscious agents" proposes that our perception of reality is shaped by consciousness, implying that our physical world is a "user interface" for a deeper reality. Neurospirituality aligns with Hoffman's view, suggesting that we can reshape our perceptions to "tune in" to expanded layers of reality, highlighting neuroplasticity as a mechanism for altering perceptual habits and enhancing self-awareness.

Jean-Pierre Changeux (1936---) and Stanislas Dehaene (1965–)

In the 1990s, Dehaene and Changeux developed the global workspace theory, proposing that consciousness emerges when information is shared across neural networks. As discussed earlier, this view presents a mechanistic basis for consciousness but doesn't fully account for subjective experience. Neurospirituality appreciates this structural model but goes beyond it, suggesting that while brain connectivity is crucial, practices that promote neuroplasticity can facilitate experiences of higher consciousness and interconnectedness, extending beyond mechanistic frameworks.

Giulio Tononi (1960–)

In the early 2000s, Tononi introduced the integrated information theory (IIT), positing that consciousness corresponds with the brain's ability to integrate information. While IIT provides insights into the neural complexity underlying consciousness, neurospirituality further emphasizes subjective experience and the transformative effects of practices like meditation. Neurospirituality, unlike IIT, considers spiritual practices as catalysts for neuroplastic changes, shaping consciousness by expanding emotional, cognitive, and spiritual understanding.

To conclude, the neurospiritual model of consciousness synthesizes insights that align with many prominent consciousness thinkers. It adopts Jung's archetypal richness, Maslow's self-transcendence, Grof's non-ordinary states, and Wilber's spectrum of consciousness while integrating quantum and informational perspectives from Penrose, Bohm, Hoffman, and Goswami. Unlike purely mechanistic models (e.g., Dehaene, Tononi), neurospirituality views consciousness as both plastic and transcendent, suggesting that neuroplastic practices and spiritual rituals can evolve consciousness toward empathy, interconnectedness, and profound self-awareness. This holistic approach bridges science and spirituality, emphasizing that neuroplasticity offers a practical and transformative path toward higher consciousness.

AI, Consciousness, and Neuroplasticity

My journey to creating a productive partnership with AI has been nothing short of transformative. As I returned from Peru, I realized how generative AI could become a collaborative force to enrich my writing process and connect ideas from my early draft of *OPEN* in remarkable ways. From the beginning, I approached AI as more than just a tool; I saw it as a creative partner. I quickly learned to fine-tune a custom GPT with my research, prior writings, and key concepts explored in *OPEN*. From the start, I ensured that the insights it generated would always engage with the depth of my work. This move to partner with AI wasn't about relinquishing control but augmenting my reflective capabilities—allowing AI to catalyze deeper connections and explore broader perspectives. I often started with a single idea, and through carefully crafted prompts, I guided the AI to examine critical concepts explored in the book, like consciousness, neuroplasticity, and flow, in greater depth. It allowed me to cross-reference fields I am passionate about, like neuroscience, neurobiology, quantum mechanics, spiritual practices, and psychology. Armed with quick and contextually relevant information, I was able to weave together threads I might not have otherwise explored. The process felt like a rich, iterative dialogue where each response from the AI sharpened my thinking and elevated my work.

One of the most rewarding aspects of working with AI was its capacity to accelerate the research process. I used it to generate relevant work I could read and cite, checking the references each time, knowing too well it can "hallucinate." Also, I recruited it to review and improve the writing of early drafts, explore alternative phrasings, and refine complex scientific ideas into more accessible language, all aligned with my tone, experience, and credentials. When I felt stuck, AI provided fresh angles, often surprising me with perspectives that deepened my understanding of the topic. I realize that some people may not necessarily be pleased or impressed by my bold move here to engage with AI, but I believe it has improved the book enormously!

For *OPEN*, this collaboration with AI meant the ideas didn't just remain isolated in my head; they evolved dynamically as I worked with them. It helped me refine the book's style, ensuring it resonated with a broad audience while staying true to the science and spirituality that underpin the framework. One key skill I acquired for this partnership to flourish was mastering prompt engineering—a topic that became the foundation for unlocking AI's potential.

Prompt engineering is the art and science of crafting precise and intentional instructions to guide AI's responses. During my early work with AI, I quickly realized that the quality of its output was directly tied to the clarity and focus of my input. This meant thinking critically about what I wanted to achieve with each query, whether generating a metaphor, summarizing research, producing visuals, or exploring a new narrative angle. Refining my prompts, I created a productive, creative, and responsible workflow. I learned how to balance specificity with openness—giving the AI enough structure to stay on track while allowing space for its generative capabilities to shine. For instance, when exploring a complex topic like neuroplasticity, I would structure prompts to ensure the response incorporated scientific rigor and accessible language I could use to augment or improve my original writing.

AI also became instrumental in bringing *OPEN* to life visually. I leveraged generative AI tools to create illustrations illustrating the book's stories and themes. The ability to describe concepts—like the interplay of neuroplasticity and consciousness—and see them rendered visually was both efficient and inspiring. I have always believed in the power of visualizing ideas, especially when communicating complex concepts like those in *OPEN*. This belief is grounded in decades of research I have conducted on the brain's response to visuals. Visuals activate multiple areas of the brain, deepening understanding and emotional resonance. Complex ideas, such as rewiring neural pathways or achieving flow states, can feel abstract when conveyed only through text. Visuals serve as cognitive shortcuts, simplifying complexity and fostering stronger memory connections. For example, using images to depict the interplay of brain functions or the dynamics of stress and transformation makes these concepts more intuitive and engaging. Finally, the brain processes images faster and retains them longer than text, making visuals indispensable for explaining intricate topics like neuroplasticity, consciousness, and the neurospiritual framework central to *OPEN*.

The visuals in *OPEN* are more than embellishments; they are integral to its mission. Unlike most books in this genre, which rely heavily on text, *OPEN* uses visuals as a central narrative element. I brought abstract concepts to life with AI-generated illustrations, making them more relatable and inspiring. This approach enriches the reader's experience, turning complex theories into accessible and memorable insights. They make the abstract tangible, helping readers visualize their transformation. By merging visuals with text, *OPEN* creates a dynamic, engaging journey rooted in my long-standing research on the brain's preference for visual learning.

Meanwhile, throughout this process, I remained mindful of the ethical considerations inherent in working with AI. I ensured the collaboration was a dialogue rather than delegating my creative and writing responsibilities. AI didn't replace my insights; it enhanced them. This commitment to creative integrity was essential in preserving the authenticity and originality of *OPEN*. By embracing AI as a partner, I created a book reflecting my vision and synthesizing disciplines and ideas that might have taken more years to discover. The iterative process allowed me to explore connections between neuroplasticity, spirituality, and personal growth with a depth and clarity that would have been harder to achieve otherwise.

Ultimately, this partnership has reshaped how I approach creativity. Working with AI was not about shortcuts but amplifying my thinking, writing, and creating capacity. It allowed me to transcend the boundaries of my perspective and produce a book as innovative as the process that created it. This experience has taught me that when approached thoughtfully and ethically, AI can be a powerful ally in shaping ideas, deepening understanding, and bringing visions to life.

Meanwhile, at the peak of using generative AI while writing *OPEN,* I experienced a synchronicity that felt almost uncanny. This coincidence aligned perfectly with the trajectory of my work. Amid my deep dive into neurospiritual concepts and the interplay of human and machine intelligence, I was offered an opportunity to design and teach courses on Prompt Engineering and Generative AI at Johns Hopkins University. The timing was remarkable, as if the universe recognized my immersion in this transformative technology and responded by expanding my platform to share its potential. This serendipitous opportunity connected with my passions: exploring consciousness, pushing the boundaries of creativity, and integrating technology into human growth. It affirmed the role of AI not just as a tool for productivity but as a collaborator in reshaping education, innovation, and self-discovery. Creating these courses has become more than an academic pursuit; it has allowed me to guide others in navigating the promise and challenges of generative AI and inspire a responsible and transformative partnership with technology. This synchronicity reminded me of the interconnectedness of purpose and opportunity, showing that when we align our intentions with action, the paths we need often unfold in unexpected and meaningful ways.

Building upon the *Neurospiritual Model of Consciousness*, which promotes growing our awareness to reach greater flow and happiness, AI has, then, emerged as an ally in this pursuit. Since the neurospiritual approach emphasizes inner transformation through neuroplastic practices and spiritual rituals, AI offers tools to enrich and expand this journey through each level of consciousness. By augmenting cognitive processes, enhancing information access, and enabling more profound insights, AI can support and accelerate our understanding and experience of consciousness at each stage—from the foundational levels of perception and awareness to the higher realms of abstract thought and spiritual unity.

Let me be clear: I believe AI is a remarkable extension of human intellect, not a substitute for it. It operates on algorithms and vast datasets, processing information at speeds and scales far beyond human capability. However, despite these impressive capabilities, AI still lacks subjective experience—the defining characteristic of consciousness. Consciousness, as we understand it, involves self-awareness, intentionality, and the ability to experience emotions and sensations, elements that AI does not possess yet.

Nonetheless, AI's most significant potential lies in expanding and enriching human consciousness. More importantly, AI can enhance our awareness, understanding, and interaction with the world by serving as a tool that augments our cognitive processes (91-93). This expansion can occur in several ways:

- AI can analyze vast amounts of information and present it in an easily digestible way for the human brain. This ability enables us to access new knowledge, insights, and perspectives that would otherwise remain beyond our cognitive reach. As a result, our capacity to learn, adapt, and think critically can be significantly enhanced.

- AI-driven tools can stimulate neuroplasticity by offering tailored cognitive exercises, feedback, and stimuli that promote optimal brain function. These interventions can help maintain and enhance neuroplasticity throughout life, supporting cognitive health and resilience.

- AI can help expand our awareness of internal and external environments. For instance, AI-powered meditation apps and biofeedback devices can help individuals become more attuned to their physiological and mental states, fostering a deeper understanding of their consciousness. This heightened awareness can improve emotional regulation, mindfulness, and overall mental well-being.

- In therapeutic contexts, AI can significantly facilitate healing and recovery, particularly for individuals recovering from neurological injuries or disorders. AI-based rehabilitation programs can be designed to promote neuroplasticity, helping patients regain lost functions and optimize brain health. These AI-driven interventions can accelerate recovery and improve outcomes by continuously adapting to the patient's progress.

- As I demonstrated with the 100+ images I generated for *OPEN*, Generative AI has the potential to visualize and analyze narratives from altered states of consciousness, such as those experienced during psychedelic journeys, dreams, and meditation. This capability enriches our understanding of altered states of consciousness and supports emerging therapeutic practices by providing insights into the subconscious mind, facilitating deeper self-awareness and healing.

- AI can translate complex and abstract experiences into visual representations through advanced machine learning algorithms, offering us a tangible way to explore and interpret their inner worlds. Moreover, AI can provide sophisticated analysis of these experiences, identifying patterns, emotional undercurrents, and symbolic content that might otherwise go unnoticed.

Thus, the intersection of AI and consciousness offers immense potential benefits for human development. As AI tools become more integrated into our daily lives, they can help us stay "awake" metaphorically—alert, engaged, and conscious of our actions and thoughts. This heightened awareness and cognitive engagement is crucial for maintaining neuroplasticity, as it encourages continuous learning, adaptation, and growth. Furthermore, AI can support mental and physical health by providing personalized interventions that enhance our capacity for self-care, stress management, and cognitive maintenance. These tools can guide individuals in maintaining their maximum neuroplasticity level, promoting lifelong cognitive health and resilience. AI may never achieve consciousness in the human sense, but its role in expanding our consciousness and supporting neuroplasticity is profound and transformative. By harnessing the power of AI, we can unlock new dimensions of human potential, enabling us to live more aware, adaptive, and fulfilling lives. This symbiotic relationship between AI and human cognition promises to enhance individual well-being and advance our collective understanding of consciousness.

While the potential of artificial intelligence to enhance human intelligence is profound, it is equally critical to acknowledge the responsibility that comes with this expansion. Integrating AI into society offers obvious opportunities for progress but carries significant risks if not guided by ethical principles and caution. If misused or left unchecked, AI can exacerbate existing inequalities, threaten privacy, and even perpetuate harm through bias or misuse of power. AI's rapid advancement is already threatening many jobs by automating tasks, reshaping roles, and potentially displacing workers, making it crucial to address workforce transitions with education, reskilling, and thoughtful policy. Additionally, the massive computational power required to train and operate advanced AI systems consumes significant energy, contributing to carbon emissions and raising concerns about its impact on climate change, underscoring the need for sustainable AI development practices.

Thus, as we push the boundaries of human-machine collaboration, we must remain vigilant to ensure these technologies serve humanity's best interests. This requires robust regulatory frameworks, ethical oversight, and ongoing dialogue about AI's societal implications. Embracing AI responsibly means harnessing its benefits and confronting and addressing its potential threats, ensuring its integration fosters equality, transparency, and the greater good.

Above all, the rise of AI calls for humanity to cultivate a deeper, more spiritually attuned consciousness that transcends ego-driven pursuits and embraces our interconnectedness with each other and the planet. This awakening is essential not only to navigate the uncertainties of an unknown future with caution and hope but also to engage with AI in ways that honor the soul of humanity. By aligning technological innovation with ethical principles, compassion, and a higher purpose, we can transform AI into a tool for elevating consciousness, fostering global harmony, and enriching the sacred experience of being human.

It should be clear now that consciousness is intimately related to our happiness and mental health. With deep respect for the many scholars and thinkers who have contributed to our understanding of consciousness, the neurospiritual model of consciousness seeks to offer a theoretical framework that integrates complementary biological and spiritual dimensions. The inverted pyramid metaphor captures the expansion of consciousness as it ascends through various levels, from instinctual awareness to spiritual unity. By building on the insights of earlier models and offering a new perspective, I hope this model will continue to inspire thoughtful dialogue and exploration of consciousness in its neurobiological and spiritual forms.

While the potential for consciousness to expand and foster deeper awareness is inherent within us all, conditions that distort or suppress this growth often hinder it. Stress, anxiety, depression, addictions, and trauma (SADAT) create maladaptive patterns in the brain, restricting our ability to remain present and open. These conditions also interfere with neuroplasticity by anchoring us in fear, doubt, and emotional turmoil. As we delve into the neurophysiological basis of SADAT, it will become clear that these challenging states limit our consciousness and pose significant barriers to healing and transformation. Therefore, we can dismantle their hold by understanding their roots and pave the way for greater openness and flow.

The Impact of SADAT On Consciousness and Plasticity

This section presents the current scientific views on the direct effect of stress, anxiety, depression, addictions, and traumas (SADAT) on our nervous system and especially our ability to achieve meta-consciousness and recruit neuroplasticity under such conditions. I intend to help you understand and appreciate the neurobiological mechanisms that are disrupted by the toxicity caused by exacerbated or prolonged SADAT symptoms. In Chapter ONE, I shared a deeply personal account of how I have confronted and, for the most part, overcome the debilitating effects of stress, anxiety, depression, addiction, and trauma in my own life. I chose to reveal deeply personal details of my struggles so that, through my experience of each condition, I would encourage more people to consider the strategies and mindsets that helped me mitigate these challenges, providing readers with a firsthand perspective on the path to recovery and resilience.

Please note that the information presented in this book is intended for educational and informational purposes only. I must remind the readers here that *I am not a medically licensed professional and do not diagnose, treat, or prescribe medications for mental health disorders*. The content herein is based on current scientific research and theories related to the effects of stress, anxiety, depression, addiction, and trauma on the brain, as well as their impact on consciousness and neuroplasticity.

Thus, this book should not be used as a substitute for professional medical advice, diagnosis, or treatment. If you are experiencing symptoms of a severe mental health condition or have concerns about your mental well-being, I strongly encourage you to seek the guidance of a licensed healthcare provider, such as a psychiatrist, clinical psychologist, or other qualified mental health professional.

Stress is a sensation of pressure from various external factors, including psychological, physiological, and economic conditions.

Figure 50: Stress (Istock)

Events such as wars or a global pandemic can induce significant stress through direct exposure or watching them on the news media. Stress arises when you perceive that the demands placed upon you— whether for mental or physical effort—exceed your capacity to cope. Life events such as poverty, relocation, divorce, or job searching cause considerable stress due to the perceived imbalance between demands and coping abilities. The American Psychological Association annually publishes a report (94) titled "*Stress in America: A Nation Recovering from Collective Trauma Identifies Several Major Stressors.*" Conducted on a sample of 3185 people, the survey confirmed that stress is primarily caused by health-related issues, with 46% explicitly citing mental health as their top stressor. As for the specifics behind stress, education was the first cause for 75%, while finances affected 40% of people, the economy 32%, and family responsibilities 22%.

Meanwhile, 45% reported a mental illness in 2023 compared with 31% in 2019. Alarmingly, younger adults ages 18 to 34 still reported the highest rate of mental diseases at 50% in 2023. Physiologically, stress activates a response system that increases adrenaline and cortisol production, affecting blood pressure, immunity, digestion, and muscle tension (see Figure 54 depicting the stress system and its effect on multiple organs). Prolonged or intense stress can lead to more serious issues like depression and substance abuse. Chronic stress can also lower ***testosterone*** production, affecting energy levels, mood, and physical attributes like muscle mass. In men specifically, testosterone deficits can produce numerous symptoms, including weight gain, feelings of depression, moodiness, and low self-esteem.

The Effect of Stress on Consciousness and Plasticity

Stress may impact consciousness to varying degrees depending on its intensity, duration, and individual factors. High stress levels can reduce attention and focus, leading to difficulty concentrating on tasks or staying present. This results in a narrowed awareness of surroundings and decreased responsiveness to stimuli (95). Stress may also interfere with the brain's ability to encode, store, and retrieve information, leading to forgetfulness, confusion, and cognitive decline. Additionally, stress can disrupt emotional regulation, leading to

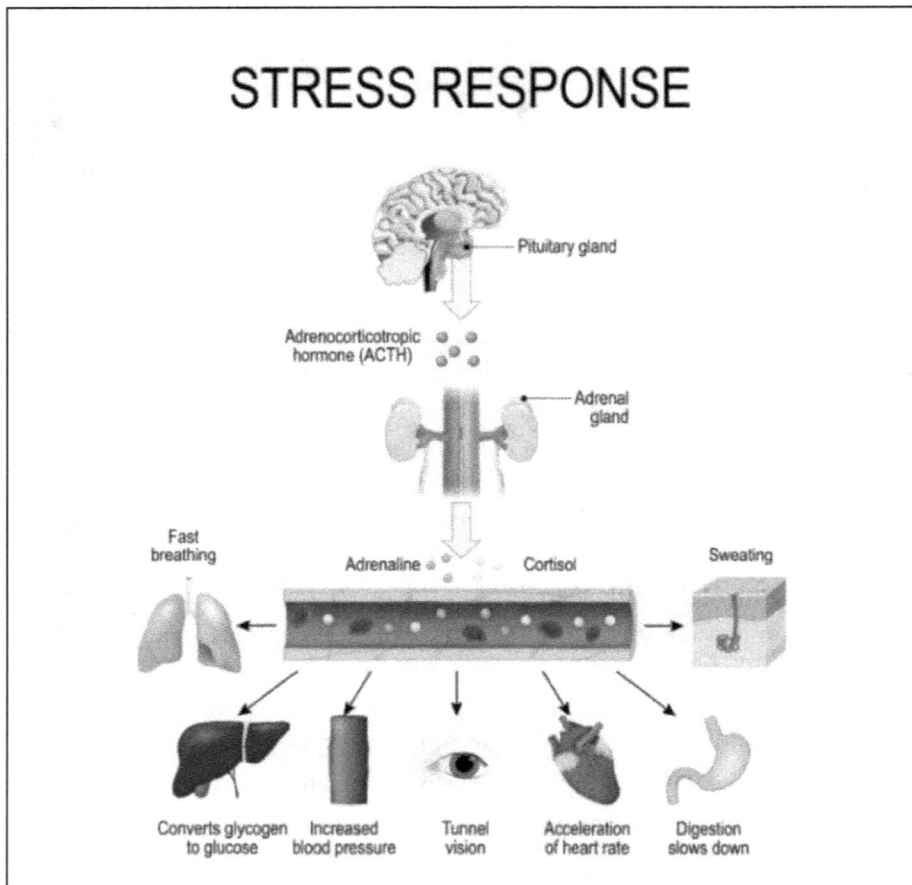

Figure 51: Stress Response (Istock)

heightened emotional reactivity, mood swings, and difficulty managing emotions. This alters the quality of consciousness by influencing the subjective experience of emotions and affecting one's perception of reality.

During my time in the Amazon jungle, I experienced firsthand the transformative power of stepping away from the stress-inducing patterns of modern life. Immersed in nature's raw beauty and simplicity, I was confronted with the stark contrast between the unrelenting pace of my usual environment and the unhurried rhythm of the jungle. The constant connectivity and demands of daily life, which had often amplified my stress and anxiety, were suddenly replaced by the soothing presence of the natural world.

Living among the trees, rivers, and wildlife, I noticed a profound reduction in my stress levels. The absence of external distractions allowed my nervous system to recalibrate, facilitating deep relaxation and clarity. Practices such as breathwork, mindfulness, and Ayahuasca ceremonies further supported this process, helping me identify and release deeply held tensions I had carried unknowingly for years.

This experience highlighted the importance of physically and mentally creating space for healing and restoration. The jungle became more than just a setting; it was an active participant in my transformation, reminding me of the healing potential inherent in nature and the profound impact of stepping into a space free from the pressures of modern life.

Thus, prolonged or intense stress can result in a diminished sense of self-awareness and coherence in consciousness. In extreme cases, severe stress or trauma may even induce altered states of consciousness, such as *dissociative states* or temporary lapses in awareness. These experiences can involve detachment from reality, distorted perceptions of time and space, or fragmented self-awareness.

Meanwhile, connections between reduced neuroplasticity and heightened stress levels are complex and multifaceted, involving various neurological and psychological factors (96, 97). Remember that this adaptability is crucial for learning, memory, and recovery from brain injury. Unfortunately, chronic stress has been shown to negatively affect the brain, particularly areas like the hippocampus (PRIMAL brain), which is vital for memory and learning.

Recent research indicates that high stress levels can also reduce connectivity between the hippocampus and the dorsolateral prefrontal cortex (RATIONAL brain), a key stress mediation area. Excessive release of stress hormones like cortisol can be neurotoxic, inhibit *neurogenesis* (forming new neurons), and reduce *synaptic plasticity*. This can impair cognitive functions such as learning, memory, and decision-making.

Finally, comorbid conditions (co-existing conditions) such as anxiety and depression are often related to high stress; both conditions are also linked to reduced neuroplasticity, as I will discuss next.

Anxiety is considered an internal response to stress, typically characterized by a persistent sense of dread or apprehension, even in the absence of immediate threats. This constant state of worry profoundly impacts our quality of life, especially our consciousness and neuroplasticity levels.

Figure 52: Anxiety (Istock)

In the United States, anxiety disorders are the highest among High school students, as 70% consider anxiety and depression significant problems, with academic pressure being a primary cause. A staggering 31% of college students have received a lifetime diagnosis of an anxiety disorder, with 34% showing signs of generalized anxiety disorder (GAD).

Anxiety disorders affect over 40 million individuals and have remained the top mental health disorder for many years (98). Their impact is widespread, influencing individual well-being, economic productivity, and public health. The increasing awareness and diagnosis rates highlight the importance of addressing this significant public health concern through effective treatment and support systems.

Diagnosing anxiety alongside other conditions (dual diagnosis) can be challenging, as many anxiety disorders are known to elevate stress levels. Moreover, about 20% of individuals with excessive anxiety engage in substance use, including prescription drugs, alcohol, or illegal substances. The data underscores the immense challenge of coping with high anxiety levels worldwide.

Interestingly, moderate levels of anxiety can be considered normal and even functional, as they are part of our basic neurological wiring, serving a protective purpose. Much of the anxiety experienced might stem from our tendency to view it negatively, overlooking its evolutionary role. As will be evident in this section, my interest in anxiety has spanned decades, partly fueled by my experiences with excessive rumination. Although I am not a clinical psychologist, my research on the brain over the last twenty years motivated me to write *The Serenity Code,* highlighting the role of anxiety in sustaining a state of flow and harmony between PRIMAL, RATIONAL, and SPIRITUAL brain systems.

Earlier, you learned that our PRIMAL brain, which connects directly to our spinal cord, constantly works to eliminate fears, threats, and risks for survival. It is the guardian of our safety, and its vigilance constantly influences our decision-making. A certain level of tension is necessary for survival, as its absence could lead to excessive risk-taking and endanger our species. Sigmund Freud, the renowned Austrian neurologist and father of psychoanalysis, even considered anxiety to be the root of all mental disorders. Thus, it is essential to recognize that, unlike stress, anxiety appears to be deeply embedded in our DNA, forming a vital part of our survival mechanism.

Meanwhile, high anxiety and stress are strongly associated with memory deficits. Numerous studies across psychology and neuroscience support this relationship. For instance, as we discussed, stress activates the hypothalamic-pituitary-adrenal (*HPA*) axis, leading to elevated levels of cortisol that can impair the function of the hippocampus, a brain region critical for memory formation and retrieval (99). High anxiety and stress can also disrupt the balance of neurotransmitters such as serotonin, norepinephrine, and dopamine, all of which play roles in memory (100). Both stress and anxiety can also negatively impact the brain's ability to adapt and reorganize. This can lead to structural changes in the hippocampus and prefrontal cortex, further impacting memory processes (101). For example, high-stress situations can impair working memory, which is crucial for temporarily holding and processing information (102).

Finally, long-term studies have indicated that people suffering from chronic stress or high anxiety are at greater risk for developing sleep issues, memory, cognitive impairments, and even dementia later in life. This confirms a cumulative detrimental effect of both conditions on mental health (100). The link between sleep and memory is important to understand. Most of what we remember happens during certain stages of sleep, particularly *REM sleep* (103). As strange as this may sound, recording our memories and even what we learned is done when we are unconscious. It also explains why a poor sleeping routine can damage your ability to store and retrieve information (104, 105). Meanwhile,

Therefore, the problem with anxiety is not just its genetic basis. Excessive levels of mind chatter lead to difficulties in coping with life. Additionally, recent research has found that chronic anxiety sufferers exhibit permanent anatomical changes in their brains, particularly in areas related to emotional regulation, like the ***anterior cingulate cortex*** (ACC). The ACC connects both the PRIMAL and the RATIONAL brain but is a relatively new brain structure (106). Rumination is also linked to the hyperactivity of the **default mode network** (DMN), which can interfere with the brain's ability to regulate attention and lead to cognitive deficits over time.

Anxiety typically begins when emotional stimuli (Figure 53) originate from external or internal sources that are transmitted to the brain through our sensory system. For instance, an external trigger could be hearing about a friend's cancer diagnosis, while an internal trigger might be realizing you cannot

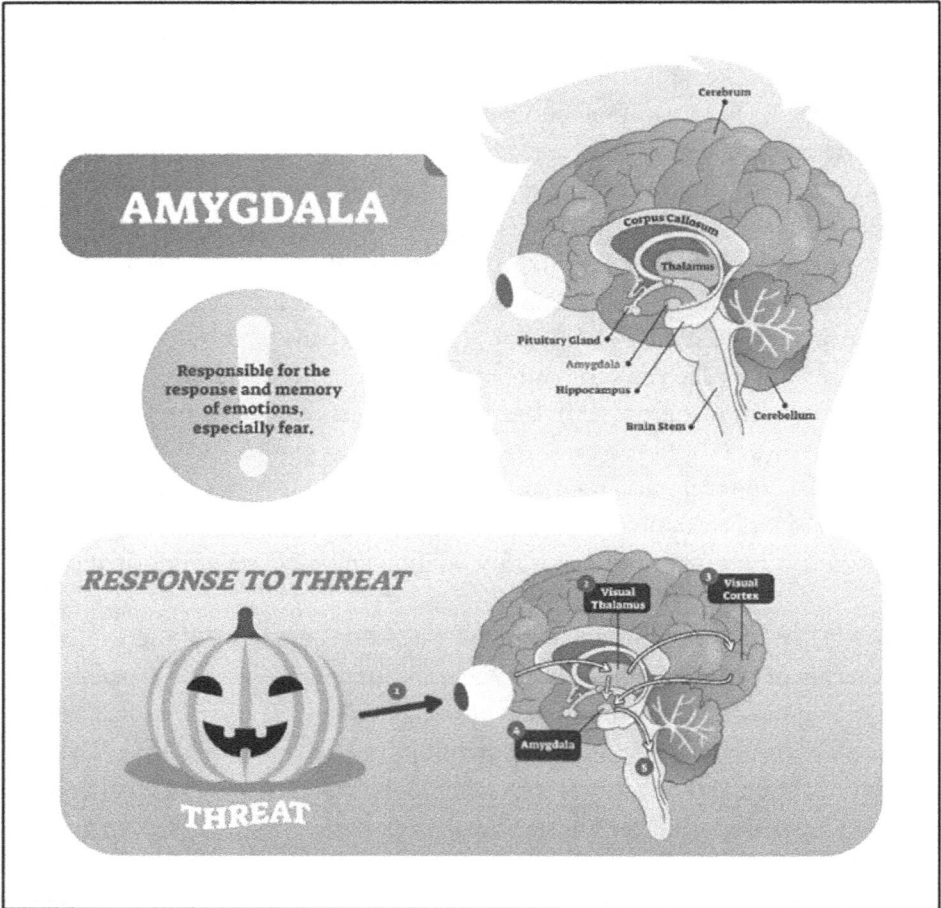

Figure 53: Anxiety Response (Istock)

pay your rent. In both cases, the information will likely be first processed by the amygdala, a small structure in the PRIMAL brain and part of the limbic system, acting as the amplifier of salient emotional events.

While the amygdala is often associated with fear processing, it is also essential to recognize its role in handling positive stimuli. The amygdala's remarkable speed in transmitting information to the RATIONAL brain (approximately 13 milliseconds) contrasts sharply with the slower response time of the neocortex (RATIONAL brain), which takes about 500 milliseconds to recognize a threat. For individuals with chronic anxiety, such as those with ***Generalized Anxiety Disorders*** (***GAD***), the amygdala activation triggers the release of neurotransmitters like norepinephrine and dopamine as well as stress hormones like cortisol and epinephrine for balance restoration.

Joseph Ledoux, a foremost expert on anxiety, distinguishes between anxiety and fear, proposing that fear is a cognitive interpretation arising from the PRIMAL survival circuits (50). He points out that animals may not experience fearful events in the same way as humans do since they cannot elaborate their cognitive logic. Although understanding animal emotions is challenging due to communication barriers, brain scans can provide insights into how animals might engage both PRIMAL and RATIONAL circuits in response to fear (107).

The RATIONAL brain, which eventually processes the importance of emotional events, can sometimes exacerbate stress and anxiety through negative thoughts. Although capable of mitigating stress and anxiety, as mentioned earlier, the RATIONAL brain may be overwhelmed by the PRIMAL brain's rapid and intense responses, potentially amplifying adverse effects.

During my time in the Amazon jungle, I observed a dramatic shift in my own neurobiological state, directly impacting my experience of anxiety. Far removed from the hyper-stimulating environments of urban life, the jungle provided a profound opportunity to witness how natural settings can recalibrate the brain's stress-response system.

Immersed in the serene yet unpredictable rhythms of the jungle, I felt my amygdala—the brain's fear center—quiet down while my parasympathetic nervous system engaged more fully. The absence of constant stimuli and long periods of deep silence punctuated by the sounds of nature allowed my body to reconnect with its innate rhythm. My anxiety, often exacerbated by overactivation of neural circuits tied to worry and hypervigilance, began to diminish sharply.

This personal experience reinforced the concept that anxiety has a deeply embodied neurobiological basis, often heightened by the pressures of modern living. Nature's capacity to downregulate stress hormones and activate calming neural pathways demonstrated a profound truth: our environment can significantly influence the neurobiological underpinnings of anxiety, offering pathways to healing and resilience.

The Effect of Anxiety on Consciousness and Plasticity

Meet Sarah, a 35-year-old marketing professional who has battled high anxiety for most of her adult life, but recent stressors like work demands and family responsibilities caused it to escalate. She began noticing changes in her ability to function. Focusing during meetings or retaining critical details became difficult at work, and decision-making felt overwhelming as her mind fixated on worst-case scenarios. Colleagues observed her forgetfulness and missed deadlines, which added to her growing sense of inadequacy.

Emotionally, Sarah felt consumed by fear and worry, often triggered by minor stressors. Her heightened vigilance amplified her perception of threats, like interpreting a manager's tone as dissatisfaction. These emotional spikes clouded her thinking, leaving her feeling trapped in a cycle of fear and helplessness. Meanwhile, her body mirrored her inner turmoil with a rapid heartbeat, constant muscle tension, and shortness of breath, making it nearly impossible to stay present. Occasionally, the intensity of her symptoms caused dissociation, where she felt detached from her surroundings and even herself.

Chronic anxiety began to take a toll on Sarah's brain. Her hippocampus and amygdala became dysregulated, and elevated cortisol levels impaired her brain's ability to adapt, learn, and form new connections. She struggled to shift her mindset, and her diminished cognitive flexibility made even small changes feel insurmountable. Realizing the impact on her life, Sarah sought help. Through cognitive-behavioral therapy, mindfulness, and lifestyle changes, she began to address her anxiety. Gradually, these interventions helped her restore balance and improve her neuroplasticity. Exploring psychedelic-assisted therapy also allowed Sarah to reframe her thought patterns and regulate her emotions quickly. Over a remarkably short time, she regained her ability to focus, make decisions, and approach life's challenges more resiliently. Her experience illustrates how anxiety narrows consciousness and limits plasticity—and how intentional interventions can lead to recovery and growth.

There is strong evidence showing that anxiety can significantly narrow consciousness, affecting cognitive functioning, emotional regulation, and overall awareness. Anxiety can also impair memory, attention, and decision-making (108). Individuals experiencing high anxiety find it difficult to concentrate on tasks, remember information, or make rational judgments. This can lead to difficulties in understanding one's surroundings. Anxiety often leads to a heightened state of arousal and vigilance, which can lead to being overly focused on perceived threats or worries. This heightened attentional focus can result in a limited awareness of other stimuli and a reduced ability to stay present in the moment.

Meanwhile, high anxiety is often accompanied by intense emotional experiences such as fear, worry, or panic. These emotions can overwhelm individuals, leading to difficulties in emotional regulation and a sense of being consumed by one's feelings. Emotional turmoil clouds consciousness, making maintaining a clear and balanced perspective on thoughts and actions challenging.

Anxiety can manifest in various physical symptoms, such as Sarah's account of rapid heartbeat, shortness of breath, and muscle tension. These bodily sensations can further contribute to lowered consciousness by causing discomfort and distraction. In extreme cases, high anxiety may trigger dissociative experiences where individuals feel detached from reality or themselves.

When chronic, anxiety can impact plasticity in several ways. Chronic anxiety is known to affect brain regions like the hippocampus and amygdala. As noted earlier, the hippocampus is vital for memory and learning and may exhibit reduced neurogenesis and lower synaptic plasticity (109). The amygdala can also become hyperactive, further contributing to anxiety, which, like stress, often leads to the prolonged release of hormones like cortisol. High cortisol levels harm neuroplasticity, inhibiting the growth of new neurons and affecting synaptic regulation. Additionally, high anxiety can lead to a decrease in cognitive flexibility, making it harder for individuals to adapt to new situations or modify their thoughts and behaviors as the brain becomes less adept at forming new connections and pathways.

Meanwhile, research has shown that chronic anxiety can lead to structural and functional changes in the brain. These changes can manifest as alterations in brain volume in specific regions that impact the brain's plasticity (110). Fortunately, the effects of anxiety on plasticity are not necessarily permanent. Cognitive-behavioral therapy (CBT), medication, stress management techniques, and lifestyle changes can mitigate these effects and improve plasticity. So can psychedelic modalities, as I discuss in Chapter THREE.

Thus, the interaction between the PRIMAL and RATIONAL brains highlights the challenge of managing anxiety since control over the PRIMAL brain's unconscious threat response is limited.

Globally, depression affects approximately 380 million individuals, with about 21 million cases reported in the United States as of 2021. This condition often leads to significant impairments in daily life activities.

The DSM-5, a key reference manual used by mental health professionals to diagnose and classify mental health conditions, defines depression as a period of at least two weeks marked by a depressed mood or a loss of interest or pleasure in daily activities, accompanied by a range of symptoms like sleep disturbances, changes in eating habits, energy levels, concentration, or feelings of self-worth.

Figure 54: Depression (Istock)

Major depressive disorder is notably prevalent in the United States, with higher incidence rates in adult females (10.3%) compared to males (6.2%) (111). Young adults aged 18-25 exhibit the highest prevalence rate (18.6%). However, over one-third of those suffering from depression do not seek treatment.

Meanwhile, both medium and high levels of anxiety are found to be more likely to escalate into depression than comparable levels of stress. This may be explained by the fact that depression involves PRIMAL brain regions like the hippocampus and amygdala but also RATIONAL brain areas like the frontal lobes. The hippocampus of depressed people can decrease in size due to excessive cortisol, impacting long-term memory organization and retrieval. Additionally, *serotonin* plays a pivotal role in regulating anxiety, depression, and mood (112). An imbalance in this crucial neurotransmitter, either due to reduced production or insufficient receptor sites in the brain, can lead to depression.

Interestingly, as noted earlier, the primary site of serotonin production is our gut, while only a small portion is metabolized in the brain. While the topic remains controversial, it is essential to recognize that depression can be considered a biological brain disorder with the potential to cause brain tissue damage akin to neurodegenerative diseases like Alzheimer's or Parkinson's. A growing body of research reinforces the biological underpinnings of depression and challenges outdated views that reduce the condition to merely a psychological or circumstantial disorder. Instead, depression appears to be a complex interplay of neurobiology, genetics, and environmental factors, requiring multifaceted treatment approaches. Also, studies show that some individuals with depression have fewer opioid receptors in the brain, which is crucial for regulating sensations of pain and pleasure. This finding opens possibilities for early detection of depressive risks using methods like positron emission tomography (PET) scans.

The Effect of Depression on Consciousness and Plasticity

Depression can significantly lower consciousness by affecting cognitive function, emotional regulation, motivation, and overall awareness (113). It often leads to cognitive deficits, such as concentration, memory, and decision-making difficulties. Depressed people may experience "brain fog" or a sense of mental sluggishness, which can impair their ability to think clearly and process information effectively, resulting in a lowered level of consciousness.

A persistent low mood and a lack of interest or pleasure in activities characterize depression. This emotional numbness can lead to a diminished awareness of others' feelings and experiences, making them feel disconnected from their surroundings. Additionally, depression can sap individuals' motivation and energy levels, making it challenging to engage with the world around them. This lack of motivation can lead to decreased self-efficacy, further lowering consciousness as individuals may withdraw from social interactions and activities.

During the COVID pandemic, I experienced a profound encounter with depression, one that revealed how deeply it can narrow awareness and drain the capacity for engagement. The inability to travel and advocate for my ideas—a central part of my identity—left me feeling unmoored, as if the foundation of my purpose had been pulled away. Even the simplest tasks became daunting, and my mind felt trapped in a fog of hopelessness. My ability to focus, make decisions, or connect meaningfully with others diminished, amplifying a sense of isolation and disconnection.

Physically, depression manifested as chronic fatigue, disrupted my sleep, and colored each day with a pervasive sense of heaviness that seemed to settle in my body. I began to recognize how depression impacts the brain's plasticity, hindering the ability to form new connections and locking me into rigid patterns of negative thinking. Without my usual avenues for growth and exploration, it felt like my mind had lost its flexibility and capacity for renewal.

Healing came slowly, through deliberate steps. Without the ability to travel, I immersed myself in practices that could be done from home—mindfulness, journaling, and rituals to reconnect with my body and spirit. Later,

I explored psychedelic modalities, which enabled me to access and reframe the emotions I had been suppressing for decades and had been stirred by the COVID crisis. Combined with neuroplasticity-focused strategies, these tools helped me rebuild resilience and rediscover purpose. The pandemic taught me that even in isolation and stagnation, there is a path to renewal when we commit to change and openness.

Depressive thoughts often revolve around themes of self-criticism, worthlessness, hopelessness, and dwelling on past failures or perceived shortcomings. Depression can also manifest in various physical symptoms, such

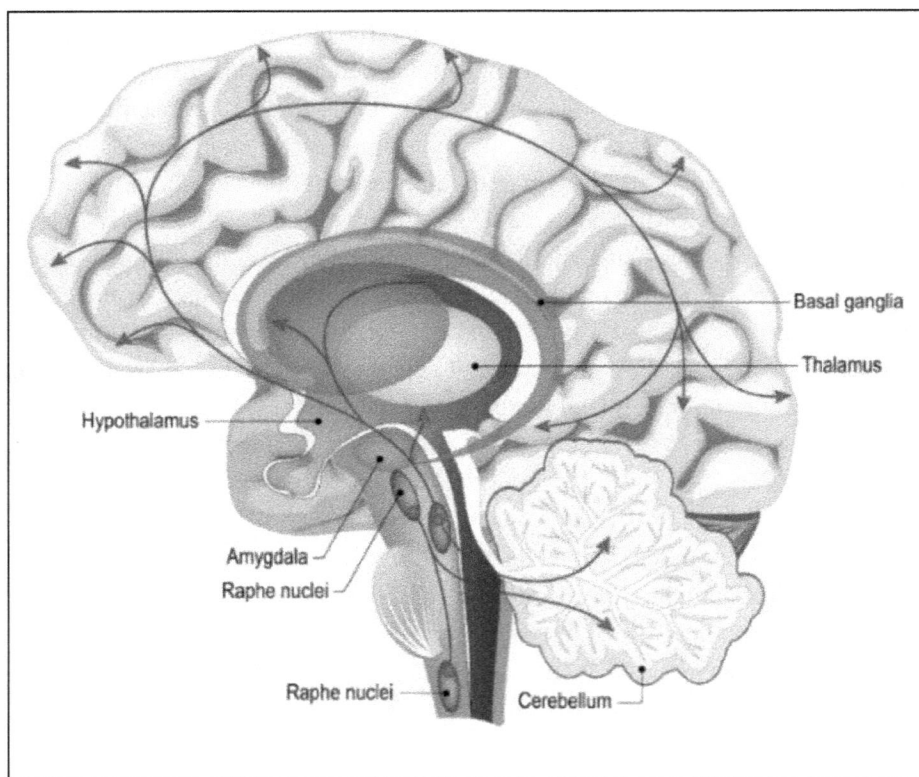

Figure 55: Depression Serotonergic Network (Istock)

as fatigue, sleep disturbances, and appetite changes. These physical symptoms can further decrease consciousness by causing discomfort, lethargy, and decreased alertness.

Meanwhile, consciousness deficits can also predict depression, making the relationship between the two states even more important to study (114). Brain structure and function changes have been observed, such as atrophy in RATIONAL areas like the prefrontal cortex and the PRIMAL regions like the hippocampus, which are crucial for emotional regulation and memory. These changes, along with reduced connectivity, impair neuroplasticity (115). Furthermore, lower levels of brain-derived neurotrophic factor (*BDNF*), an essential protein for neuroplasticity, are often measured in individuals with depression. These neuroplasticity impairments contribute significantly to the development and persistence of depressive symptoms.

Finally, depression frequently coexists with anxiety disorders. Nearly half of those diagnosed with depression are also diagnosed with an anxiety disorder. This comorbidity indicates a substantial overlap between the two conditions, suggesting that individuals who suffer from one are at a heightened risk of developing the other (116). Alarmingly, since the COVID-19 pandemic, there has been a significant increase in the prevalence of mental health disorders, particularly among young adults. Overall, more than two in five adults experienced symptoms of anxiety or a depressive disorder. This rise underscores the impact of external stressors on mental health. All these findings highlight the intertwined nature of depression, anxiety, and stress, illustrating how these conditions often co-occur and exacerbate each other.

Addiction

Research on addictions has been a health priority for a long time. The magnitude of the problem is extremely difficult to estimate and even define. Substance abuse disorder alone affects more than 10 million people in the US population (117). Seventy thousand drug overdose deaths are occurring annually. The average life

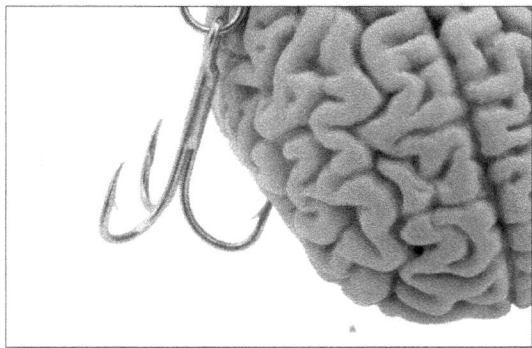

Figure 56: Addiction on the Brain (Istock)

expectancy in the US has declined due to opioid overdose deaths. The Center for Disease Control (CDC) has reported a sharp increase in drug overdose deaths, particularly from synthetic opioids like fentanyl. Overdose deaths in the U.S. increased from 91,799 in 2020 to 106,699 in 2021, a nearly 16% rise. These deaths have played a substantial role in reducing life expectancy, which dropped from 77 years in 2020 to 76.4 years in 2021 (118)

Over the last decade, though, our understanding of addictions has evolved significantly. In the US, for instance, the rise of the synthetic opioid fentanyl has significantly impacted the landscape of substance abuse, overdose deaths, and public health policies. The fentanyl crisis has been declared a public health emergency in many regions, claiming over 100,000 deaths in 2023 (119). Fentanyl is cheap to produce, up to 50 times stronger than heroin and 100 times stronger than morphine, making it one of the most dangerous opioids available on the streets. The crisis has overwhelmed emergency services, influenced healthcare policies, and necessitated a significant shift in resources towards combating opioid addiction and overdose prevention.

Meanwhile, most experts now recognize that all addictions have a behavioral component that is rooted in similar biological and psychological mechanisms. Substances like alcohol or behaviors like work or gambling—share common underlying patterns in how the brain and mind work. At their core, addictions involve behaviors that become hardwired through repetition, creating a strong connection between specific actions and feelings of relief, pleasure, or escape. These patterns are driven by biological mechanisms, such as the release of dopamine, a chemical in the brain that reinforces the behavior by making it feel rewarding. Psychologically, addiction often stems from using these behaviors to cope with stress, pain, or unresolved emotions. In other words, regardless of what someone is addicted to, the cycle of craving, reward, and reliance is built on similar foundations in the brain and mind.

Also, genetics and environmental factors play a significant risk in addiction risk, while co-morbid conditions like ADHD, OCD, stress, and anxiety often affect people with addictive behaviors. More importantly, the neuroscience of behavioral addictions has shed light on how certain habits or cravings related to substance abuse, gambling, internet use, or excessive use of social media can become toxic (120, 121). They activate the PRIMAL mesolimbic pathway, including structures like the *nucleus accumbens* (NAcc) and *ventral tegmental area* (VTA), stimulating the release of neurotransmitters like *dopamine*. Think of dopamine as a pleasure-seeking molecule that quickly motivates people to repeat rewarding behaviors. Over time, this can lead to changes in the brain's reward system, increasing the desire for the dopamine 'high.' Whether it is a substance or a behavior, the dopamine release during these activities reinforces them, making the person more likely to repeat them. Over time, this makes the brain more sensitive to addiction cues and less sensitive to natural rewards.

The ability to resist impulses or delay gratification is severely compromised in addiction. The prefrontal cortex, responsible for decision-making and impulse control, is often weakened by behavioral addictions. This impairment can lead to difficulty in controlling addictive behavior despite being aware of its negative consequences. It is also partly due to altered neurotransmitter activity in brain areas responsible for impulse control, leading to increased impulsivity and difficulty resisting the urge to engage in addictive behaviors.

As noted earlier, I was forced to confront long-standing patterns of overworking, reliance on alcohol, and constant worry that had previously gone unchecked during the COVID crisis. Without the ability to travel or engage in my usual routines, these toxic habits became glaringly apparent. Work had long been my refuge, where I could channel my energy and distract myself from underlying fears and uncertainties. When that outlet became restricted, I found myself turning more frequently to alcohol to cope with the mounting anxiety and isolation. The relentless cycle of worry only compounded these struggles, creating a sense of being trapped in my own mind.

This period highlighted how addiction—to work, alcohol, and worry—had become my way of managing stress and avoiding deeper emotions. The pandemic removed the distractions that had kept these patterns hidden, leaving me face-to-face with their impact on my mental and physical health. I realized that these behaviors were not solutions but symptoms of more profound struggles, eroding my ability to feel present and at peace.

Breaking this cycle required a profound shift. I began engaging in mindfulness practices, rituals to ground myself, and techniques to regulate my emotions more healthily. Exploring psychedelic modalities was another key step, as they allowed me to confront and reframe the fears and feelings that had been driving these patterns for years. Through this process, I understood that true resilience and clarity come not from avoidance but from embracing and transforming the underlying forces that shape our lives.

Most experts now agree that all types of addictions, whether to substances like alcohol or behaviors like work or gambling—share common underlying patterns in how the brain and mind work. At their core, addictions involve behaviors that become hardwired through repetition, creating a strong connection between specific actions and feelings of relief, pleasure, or escape. These patterns are driven by biological mechanisms, such as the release of dopamine, a chemical in the brain that reinforces the behavior by making it feel rewarding. Psychologically, addiction often stems from using these behaviors to cope with stress, pain, or unresolved emotions. In other words, regardless of what someone is addicted to, the cycle of craving, reward, and reliance is built on similar foundations in the brain and mind.

Thanks to many recent studies, we also understand that the brain's stress systems and inability to regulate emotions are also involved in addictions (122). This means that chronic engagement in addictive behaviors can alter these systems, leading to increased stress reactivity and emotional dysregulation, which can further exacerbate the cravings. Also, behavioral addictions can lead to tolerance: a need to engage in behavior more intensely or frequently to achieve the desired effect. As a result, many addicts experience withdrawal symptoms when not engaging in the behavior.

Finally, both genetic predispositions and environmental factors, such as exposure to specific stressors or cues, can also contribute to the risk of developing a behavioral addiction.

The Effect of Addiction on Consciousness and Neuroplasticity

Addiction can significantly lower consciousness by impacting various cognitive, emotional, and behavioral functions. Addiction alters the brain's reward system and impairs decision-making processes. Individuals may prioritize seeking and using the addictive substance or engaging in addictive behaviors over other essential aspects of life, leading to poor decision-making and a narrowed focus of attention. Addiction can distort individuals' perception of reality, leading to cognitive biases and irrational beliefs that reinforce addictive behaviors (123). This distorted perception may also lead to a diminished awareness of the negative consequences of addiction and a heightened focus on obtaining the substance or engaging in addictive behavior.

Addiction often co-occurs with emotional dysregulation, characterized by mood swings, irritability, and difficulty managing emotions. Individuals may use addictive substances or behaviors as coping mechanisms to regulate their emotions, which can lead to decreased consciousness as emotions may overshadow rational thought processes.

Furthermore, addiction is characterized by a loss of cognitive control over the addictive substance or behavior, leading to compulsive use despite negative consequences. This loss of power can result in a diminished sense of agency and self-awareness, as individuals may feel powerless to resist cravings and impulses associated with addiction (121). As a result, addiction can lead to social isolation and withdrawal from meaningful relationships and activities. Addicts often prioritize their addiction over interpersonal connections, leading to a decreased awareness of social cues and a narrowed focus on engaging in addictive behaviors.

Surprisingly, addictions are also caused by neuroplasticity. Changes in the brain's structure and functions happen because the brain adapts in response to repeated experiences. This often involves the reinforcement of pathways associated with the rewarding aspects of the addictive behavior, leading to an increased desire to engage in the behavior (124). At the same time, the neural circuits involved in self-control and decision-making might weaken. Therefore, chronic addictions can lead to a decreased ability to plan, organize, and solve problems effectively. This can be attributed to the impact of addictive substances or the effect of addictive behaviors on cognitive functions. It can even be related to the preoccupation with the addiction itself.

To conclude, repeated engagement in addictive behaviors leads to dysfunctional neuroplastic changes in the brain that can alter neural circuits related to reward, motivation, memory, and control over behavior. As a result, individuals may find it increasingly difficult to resist addictive behavior. Therefore, the impact of addiction on cognitive control is a crucial factor in the perpetuation of addictive behaviors and presents significant challenges in the treatment and recovery process. Understanding this link is crucial for developing effective treatments for any behavioral addiction.

Trauma

Defining the term trauma can be confusing. Additionally, it is often a controversial topic to discuss with so-called trauma experts. Some argue that only specific events meet the definition of trauma, like being involved in a car crash or a personal attack, arguing that events that do not cause physical pain but rather psychological turmoil, like witnessing an act of violence, should not be diagnosed as traumas.

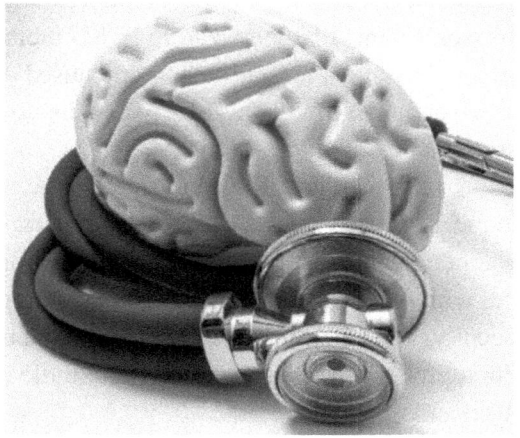

Figure 57: Trauma (Istock)

For instance, the Coalition for National Trauma Research defines trauma as "a bodily wound or shock produced by sudden physical injury, such as that from violence or an accident, including vehicle crashes, severe falls, gunshots or knives, blunt force, blasts, and burns" (125). Fortunately, significant progress is being made in defining trauma and developing improved treatment options for those affected by it. There is hope at the end of the tunnel. Hope comes from thought leaders from the medical, scientific, and psychological communities but also elders from Indigenous tribes who have long considered traumas as spiritual disorders.

The Perspective of Dr. Gabor Maté on Trauma

Dr. Gabor Maté's views on trauma are often considered unique since he believes that "trauma is not what happens to you but what happens inside of you" (126). Gabor Maté is a renowned physician and author known for his work on addiction, trauma, and mind-body health. His holistic approach emphasizes the role of early childhood experiences and emotional stress in the development of physical and mental illness. Maté argues that unresolved trauma, often stemming from adverse childhood experiences, is a significant driver behind addiction and many chronic conditions. His work integrates insights from neuroscience, psychology, and ancient healing practices, advocating for compassion and self-awareness in healing processes.

For Maté, understanding trauma starts when we consider the impact of a traumatic event, not just at the physical but also at the emotional and spiritual levels. Part of that impact is considered in the term *post-traumatic stress disorder (PTSD)*. According to the National Institute of Mental Health (NIH), PTSD can develop after exposure to a traumatic event that is beyond a typical stressor. Events that may lead to PTSD include, but are not limited to, violent personal assaults, natural or human-caused disasters (127), accidents, of course, but also many other forms of psychic disturbances related to emotional abuse, harassment, and more.

For example, the vicious dog attack I survived at 27 had a lasting impact on me, as I was gripped with fear and panic during an encounter with any dog, whether small or large, for decades. Thankfully, the deep psychological work I did, with Aya's guidance, gave me the courage to confront my trauma. More recently, it even empowered me to take the bold step of getting a puppy after my retreat in Peru. It has been a joy to rekindle my love for all dogs, including huskies!

Using a broader definition of trauma like the one suggested by Dr. Maté helps us recognize that traumas affect millions of people daily, if not all of us. The official data tracking trauma victims indicates that 50% of all U.S. adults will experience at least one traumatic event in their lives, but most do not develop PTSD. People who experience PTSD may have persistent, frightening thoughts and memories of the event(s), sleep problems, numb feelings, or may be easily startled. In severe forms, PTSD can significantly impair a person's ability to function at work, at home, and socially. An estimated 3.6% of US adults have PTSD, the prevalence of which is much higher among females (5.2%) than males (1.8%) (111). Most likely, these numbers underestimate the actual psychological impact traumatic events can have.

This is why Dr. Maté emphasizes the importance of understanding the deep-rooted causes of trauma and going beyond official definitions. According to him, this leads to an essential reframing of how trauma affects each of us. It is not just a function of what causes trauma but also a function of how the brain processes it. He believes, like most psychologists do, that trauma happens when "big events" shock the foundation of our psyche in such a way that they may leave unconscious markings in the emotional body as well as the brain.

As a result, trauma can cause alterations in the brain structure and function. For instance, it can cause amygdala hyperactivity, a heightened response to fear or stress. It can lead to hippocampal changes impairing memory and spatial navigation functions. It will likely alter the Prefrontal cortex, disrupting decision-making and impulse control. Changes in neurotransmitter levels are also likely to happen and affect mood and stress response (128).

Furthermore, trauma has been linked to increased inflammation in the body. Indeed, traumatic experiences, especially if chronic or severe, can activate the body's stress response system, leading to elevated levels of stress hormones like cortisol. This hormonal imbalance can subsequently affect the immune system, increasing inflammatory responses. These changes can all contribute to the development of co-morbid mental health disorders like PTSD, anxiety, addiction, and depression (129).

The Effect of Trauma on Consciousness and Neuroplasticity

Trauma can significantly impact your consciousness, affecting it in various ways that span immediate and long-term effects (130). The extent to which trauma affects consciousness can depend on the nature of the trauma, individual vulnerability and resilience factors, and the availability of support and therapeutic interventions. At worst, trauma can lead to dissociation, a disconnection from immediate surroundings, one's sense of self, thoughts, and feelings. This can range from mild detachment to more severe forms of dissociation, such as depersonalization or derealization, affecting an individual's consciousness and perception of reality.

Trauma can affect how memories are processed, stored, and recalled. This can lead to fragmented or intrusive memories, amnesia for specific events, or intense flashbacks that can disrupt current consciousness by making past trauma feel vividly present. Trauma can also alter perception, leading to hyperarousal (increased sensory sensitivity) or hypoarousal (numbing sensory experiences). Additionally, cognitive functions, such as attention, concentration, and executive function, can be impaired, affecting decision-making processes and the ability to focus and engage with the environment (131).

Trauma can impact the ability to regulate emotions, leading to heightened anxiety, fear, or emotional numbness. These emotional states can significantly affect an individual's conscious experience, influencing mood, behavior, and interpersonal interactions. They affect sleep patterns, leading to difficulties such as insomnia or nightmares. Sleep disruptions can influence overall consciousness, including alertness, attention, and cognitive function during waking hours. Finally, on a deeper level, trauma can impact one's sense of self and identity, influencing how individuals perceive themselves and their place in the world. This can lead to existential questions and concerns that affect one's overall consciousness and way of relating to the world (132).

Meanwhile, trauma can influence neuroplasticity in both negative and positive ways, depending on various factors such as the nature of the trauma, the timing of an individual's development, and the presence of supportive interventions.

The negative impacts of trauma on neuroplasticity

Traumatic experiences can lead to brain damage in areas responsible for stress response, emotion regulation, and memory (97). Such changes may contribute to the development of trauma-related disorders like PTSD, affecting an individual's ability to cope with future stressors and regulate emotions. Understanding this link is essential for developing effective treatments for trauma-related conditions.

The hippocampus, which is critical for memory and learning, can become smaller in volume, a change associated with PTSD. Similarly, the amygdala, involved in emotional processing, can become hyper-reactive, leading to heightened responses to stress and fear. The brain's altered structure and function due to trauma can impair cognitive and emotional functioning. As we have already seen with stress, anxiety, and depression, this includes difficulties in learning, memory, attention, and emotional regulation. Such impairments can reinforce negative behaviors and thought patterns, making it more challenging for individuals to cope with stress and manage emotions effectively.

Meanwhile, changes in the brain's stress response system can make you more susceptible to future stress and trauma. This heightened sensitivity can create a feedback loop, where exposure to stress further alters brain function, increasing the risk for mental health disorders.

The positive impacts of trauma on neuroplasticity

This aspect of neuroplasticity can help you recover from trauma, demonstrating the brain's capacity for healing and adaptation. In some cases, the experience of overcoming trauma can lead to enhanced coping mechanisms and personal growth. Individuals may develop stronger emotional resilience, greater self-efficacy, and improved problem-solving skills. These changes can be seen as positive adaptations facilitated by neuroplasticity in response to the challenges posed by trauma.

It is well known that neuroplasticity can regenerate, repair, and modify its structures in response to traumatic experiences, ultimately decreasing the negative effects of trauma and even facilitating recovery (133). However, recent research on neuroplasticity has highlighted how critical periods of plasticity can be influenced by trauma.

Indeed, certain types of traumas can reopen critical periods of neuroplasticity, allowing for significant changes in brain structure and function that might not occur otherwise. Critical periods are specific times during development when the brain is particularly receptive to environmental stimuli, leading to significant neuroplastic changes. These periods are typically associated with early childhood and certain conditions, such as trauma or significant changes in sensory input.

For instance, animal studies have shown that traumatic experiences can reopen these critical periods, enabling the brain to undergo changes usually restricted to early development. Studies on rats have demonstrated that damage or significant changes to sensory inputs can activate critical periods in the auditory cortex, allowing for new plasticity changes that facilitate adaptation and recovery (134).

This research suggests that carefully controlled trauma or stress could potentially be used to induce neuroplasticity in a way that promotes recovery from previous damage (135). However, it is crucial to note that while this concept shows promise in controlled animal studies, it is not yet a safe or ethical approach for human treatment. The idea of intentionally inducing trauma to enhance recovery is still far from being validated in clinical settings. Much more research is required to understand the potential risks and benefits fully. However, these studies underscore the potential for positive neuroplastic changes in response to trauma, highlighting the importance of supportive interventions and environments in facilitating recovery and enhancing mental health resilience.

The Role of Epigenetics in SADAT's Origin and Evolution

One of the most important byproducts of any SADAT condition is how it may cause or be caused by changes in our DNA. Scientists have proven that ***epigenetics*** affects a wide range of mental disorders, including depression, anxiety, schizophrenia, and bipolar disorder (136). Epigenetics is a branch of biology that studies changes in gene expression that do not involve alterations to the underlying DNA sequence—a change in phenotype without a change in genotype. Age, environment, lifestyle, and disease state can influence these changes. Epigenetics is fundamentally about how genes can be turned on or off and how the expression of genes is regulated.

Meet Emma, a 35-year-old teacher, who grew up in a household where her father, a police officer, struggled with chronic stress and anxiety due to his job risk and financial instability. Although Emma had never experienced severe hardships, she often felt an unexplained sense of tension and hypervigilance. She found it difficult to relax and noticed that her responses to stress seemed disproportionate to her situations.

After exploring her family history, Emma learned about epigenetics—how life experiences, like her father's prolonged stress, could influence the activation of specific genes in future generations. Her father's constant exposure to stress hormones as a cop likely altered the expression of genes related to anxiety and stress regulation, changes that were passed on to Emma. While these genetic changes didn't mean she was destined to feel anxious, they made her more sensitive to environmental stressors.

Emma worked with a therapist to develop mindfulness techniques and incorporate lifestyle changes like regular exercise and improved sleep, which helped regulate her stress response. Over time, these positive habits not only improved her mental well-being but also had the potential to influence her gene expression, breaking the cycle and setting a healthier foundation for future generations. This case highlights how epigenetics bridges biology and the environment in shaping our experiences.

Epigenetics provides insight into how environmental factors interact with genetic predispositions in these conditions. For example, significantly prolonged or severe stress can lead to epigenetic modifications affecting how stress response and mood regulation genes are expressed. These changes can influence an individual's susceptibility to mental health disorders and can even be passed down to future generations.

Trauma and lifestyle choices can also lead to epigenetic changes. These modifications can affect brain development, stress response systems, and neurotransmitter functioning, contributing to the emergence or progression of mental health conditions. Some epigenetic changes can be inherited, potentially influencing mental health across generations. This area of research is critical to understanding the complex interplay between genes and the environment in mental health.

In this section, I have considered only materialistic views on the cause and effect of SADAT conditions. Unfortunately, western societies tend to ignore or dismiss their "spiritual" basis. On the other hand, ancient healing traditions often regard stress, anxiety, depression, addiction, and trauma through a holistic lens, emphasizing the importance of balance and connection between mind, body, and spirit. Indigenous cultures essentially believe that parts of the soul can leave the body due to trauma or distress, leading to a diminishment of vitality and well-being. Practitioners of these traditions might say that soul loss affects consciousness by causing disconnection or a sense of being "incomplete."

In the Jungian perspective, *soul loss* refers to fragmentation or disconnection from one's essential self, often resulting from traumatic experiences or unresolved psychological conflicts (137). Carl Jung, the founder of analytical psychology, did not explicitly use the term "soul loss." However, the concept is closely related to his ideas of the "shadow" and the "anima/animus."

According to Jung, soul loss can be understood as a state in which a person loses touch with essential aspects of their psyche, leading to a diminished sense of self and purpose. This disconnection often manifests as a lack of vitality, meaning, or direction in life and can contribute to various forms of psychological distress. As such, soul loss may explain the maladaptive responses of stress, anxiety, depression, addiction, and trauma through the lens of Jungian psychology by emphasizing the disintegration of the psyche. When an individual experiences trauma or chronic stress, parts of the self may become repressed or split off, leading to an inner fragmentation that manifests as these psychological symptoms. For example, addiction can be seen as an unconscious attempt to fill the void created by this disconnection, while depression may reflect the soul's cry for reintegration and wholeness. Anxiety and stress, in this context, could be understood as the psyche's response to unresolved conflicts and unacknowledged parts of the self that have been exiled from conscious awareness.

Jung believed that healing required the process of individuation, where the person reintegrates these lost parts of the soul, leading to a more balanced and unified self. In therapeutic contexts that draw on these traditional concepts, soul loss might be used metaphorically to describe experiences of dissociation, depression, or identity fragmentation. The idea is that recovering or reintegrating these "lost" parts of the self can lead to a more cohesive consciousness and improved mental health.

Meanwhile, almost all organized religions and spiritual belief systems accept that the quality of our connection to a divine source, is central to humans' physical and mental well-being. As a result, many religious and spiritual traditions consider spiritual harmony central to mental health. They often see these conditions as imbalances or disruptions in one's life force or energy, and healing rituals are geared towards restoring this balance. Practices that include prayer, meditation, energy healing, and plant medicine all promote the ability to connect to something bigger than our ego. Therefore, non-materialist views stress the importance of understanding the underlying spiritual factors contributing to these conditions.

- Two central brain systems regulate our survival function (PRIMAL) and decision-making and planning function (RATIONAL). The PRIMAL is fast and automated, while the RATIONAL is slow and programmed over a long period.
- The human nervous system protects and commands the body to navigate external and internal stimuli. The ultimate objective is to maintain or restore homeostasis. Our sensory organs are channels through which we acquire and interpret environmental information. The autonomic nervous system (ANS) regulates our fight-and-flight, rest, and digest functions with minimal cognitive contribution.
- Neurons are the nerve cells we use to transmit and receive electrochemical signals, enabling us to react, feel, think, and decide. Using more neurons strains our energy; therefore, many processes seek to save or shortcut the use of glucose and oxygen.
- Early attention protects and controls many of our behaviors. We do not even need to be awake to react to threats. PRIMAL consciousness ("ME" level) is the first layer of awareness we recruit to stay alive and navigate threats and opportunities.
- Emotions are chemical signals that help us approach or avoid a situation. Our brain runs a critical saliency network to detect and respond to noticeable events within milliseconds. Our memory system relies on an emotional cocktail to improve the marking of significant events.
- Our memory banks are distributed throughout the brain. A large majority of the encoding process is completed during our sleep. Long-term information is mainly organized and stored in the PRIMAL brain, while the frontal lobes of the RATIONAL brain manage short-term memory.
- Cognition is a set of complex processes the brain has evolved to perform over hundreds of thousands of years. Thanks to trillions of connections in our RATIONAL brain, we can perceive, communicate, plan, decide, or self-reflect.
- RATIONAL consciousness ("I level) helps us understand the power of our agency, the ability to act independently and exercise free will.
- Meta-consciousness ("WE" level) is our unique ability to observe our behavior and thoughts without biases or influence from others' opinions while recognizing that we are more than the small self we see at the ME and I levels.

- Hyper-consciousness represents the ultimate expansion of consciousness, where boundaries dissolve, and an experience of universal or collective consciousness is realized.
- Quantum consciousness is a growing movement of scientists, philosophers, and scholars seeking to include the teachings of our understanding of how sub-atomic particles organize and behave. While quantum theories are still hotly debated, they offer innovative frameworks to explain emergent properties of consciousness.
- The Neurospiritual Model of Consciousness (NMC) offers a theoretical framework that integrates biological and spiritual dimensions. The inverted pyramid metaphor captures the expansion of consciousness as it ascends through various levels, from instinctual awareness to spiritual unity
- SADAT conditions all originate in maladaptive responses from our nervous system. While the brain is designed to maintain and restore homeostasis, excessive stress, anxiety, persistent negative moods, addictions, and traumas negatively impact our bodies, emotions, and cognitive resources. The toxicity caused by SADAT lowers consciousness and impairs our innate ability to rewire neuropathways implicated in the normal functioning of our autonomic nervous system. Understanding the unique link between plasticity and SADAT conditions points to effective healing methods described in the following sections.

Now that you have a good understanding of SADAT and its correlation to both consciousness and plasticity, you are ready to learn the 3-step process through which you can mitigate or even erase its symptoms:

1. **OPEN your Mind:** Raise your consciousness to change your perspectives on the impact of SADAT on your life.
2. **OPEN your Self**: Reframe the narrative of your life journey to enjoy it without SADAT.
3. **OPEN your Life**: Reprogram your life to adopt rituals that nurture and strengthen your ability to maintain awareness and openness.

CHAPTER THREE

OPEN Your Mind

"The unexamined life is not worth living"

--**Socrates**, Philosopher

The topic of raising consciousness is complex, considering the multitude of views that define our ability to observe our thoughts and behaviors. As I have discussed, there are many models of consciousness; some follow strict scientific frameworks (materialistic), while others are inspired by religious, quantum, para-scientific, and spiritual belief systems (non-materialistic). Regardless of the model used, there is a broad consensus that greater consciousness brings us more inner balance and joy.

Some may wonder if being *'woke"* is an important focus of the book. "Woke" originally referred to being aware of social injustices, particularly race-related, and staying vigilant against systemic oppression. Over time, it has evolved to encompass broader social awareness of issues like gender inequality, LGBTQ+ rights, and climate justice. While the term has positive connotations for those advocating social progress and consciousness, it has also been criticized and politicized by opponents who view it as overzealous or divisive.

Therefore, the clear answer is that *OPEN* is not about being "woke". Raising consciousness is not limited to becoming aware of important issues like social justice or climate change, even though studies do point to stress and anxiety caused by injustice among adolescents (138). This is partly due to the constant bombardment of negative news and the pressure to remain constantly vigilant and responsive to calls for action.

So, to be clear, "raising consciousness" in this book refers to:

"Your ability to access the highest number of paradigms of mental awareness through which you can feel and understand your inner and outer world."

Unfortunately, as we just reviewed, SADAT conditions impair both consciousness and neuroplasticity because of the toxic and lasting impact they have on our bodies and emotions. When you are in a chronic state of feeling overwhelmed, anxious, depressed, or resisting cravings, you compromise your ability to self-diagnose or rewire toxic patterns. Yet, think of the power of identifying, pulverizing, or transcending the forces of resistance that block your ability to be more conscious and find the path to self-healing. That is the promise of raising consciousness.

First and foremost, I will discuss *the barriers to consciousness (BC)* that make raising awareness (STEP 1) challenging. Table 3 illustrates the situation you may find yourself in today, especially if you suffer from any of the SADAT conditions. So, first, ask yourself what level of consciousness you access the most frequently: PRIMAL, RATIONAL, or SPIRITUAL (META). Each level has pros and cons, but expanding consciousness is an accrual process, which means you cannot expect to easily access level 2 if you are stuck at level 1

(PRIMAL). More importantly, you cannot reach level 3 (SPIRITUAL/META) if trapped in level 2 (RATIONAL) or 1 (PRIMAL).

Consciousness Level	Priorities	Access	Perspective
PRIMAL	Survival	PRIMAL only	ME
RATIONAL	Survival and cognitive functions	PRIMAL and RATIONAL	ME and I
SPIRITUAL (META & HYPER)	Integrate inner and outworld streams of consciousness	PRIMAL, RATIONAL, and SPIRITUAL	ME, I, and WE

Table 3: Consciousness Matrix

A story might help further explain the meaning of the table.

Figure 58: Apollo 13 Mission (AI)

Despite a critical mid-mission failure, Apollo 13's triumphant return to Earth in 1970 is a remarkable story of multidisciplinary collaboration and problem-solving. After an oxygen tank exploded, jeopardizing the spacecraft and the astronauts' lives, NASA teams on the ground and the astronauts in space had to work together to resolve multiple complex challenges. Under immense pressure, this crisis required aerospace engineering and physics expertise and quick, innovative thinking. The mission's ground team, consisting of engineers, technicians, and flight controllers, worked tirelessly to devise solutions to conserve power, re-route oxygen supply, and ensure the crew's safe return. Simultaneously, the astronauts implemented these solutions, showing extraordinary composure and skill. As they all journey into the heart of the danger, they relied on each other's strengths to solve unexpected challenges. Their collaboration revealed solutions to the crisis and a deeper understanding and respect for their differing viewpoints. Their journey is a testament to the power of raising consciousness and plasticity. Their combined efforts exemplify how diverse expertise, perspectives, and teamwork can overcome seemingly insurmountable challenges, even in the most hostile environments.

Thus, I see raising consciousness as your opportunity to recruit as many "internal" scientists, engineers, technicians, and ground workers as possible to investigate and solve your mission-critical health issues. Therefore, I will discuss a broad range of techniques to help you build an army of allies, friends, confidants, consultants, and muses who will rally to open your heart, offer wisdom, guidance, and encouragement, and ultimately lead you to victory over SADAT.

First, I will discuss the benefits and value of clearing as much toxicity from your emotional body as possible through cathartic methods. While Western doctors rarely present this topic as a valid or even credible therapeutic option, it is a prominent tradition in most Indigenous cultures. Then, I will present *natural* and *entheogen-based* approaches that help you reach new levels of awareness.

By "natural approaches," I mean practices that do not involve using psychoactive substances. These include techniques like meditation, mindfulness, breathwork, yoga, spending time in nature, and engaging in creative or reflective activities. These rely on the body's innate abilities to cultivate focus, clarity, and emotional balance, making them accessible and sustainable tools for personal growth.

By contrast, most entheogens are derived from plants, even though they may be only available in synthesized forms like LSD (ergotamine) and MDMA (sassafras). Entheogens are primarily employed in various spiritual, religious, or shamanic practices to elicit changes in consciousness. These substances are utilized for spiritual exploration, therapeutic insights, or divination, facilitating profound experiential understanding and self-awareness.

As I have shared already, I am biased towards the responsible use of plant medicine because of the speed and effectiveness at which this modality can raise consciousness and provide a short-term path to long-term SADAT relief. However, I also recognize and honor that many natural "consciousness-raising" modalities have been practiced for thousands of years without being rooted in the use of mind-altering substances.

So, let us first examine how SADAT's effect on consciousness and other barriers to consciousness may trap us in mental states that ultimately compromise our ability to self-heal.

Barriers to Consciousness (BC)

In a state of low consciousness, you are not fully aware of your behaviors and emotions but also the forces of resistance that limit your capacity to self-reflect and self-heal. This lack of awareness can lead to automatic, habitual reactions instead of deliberate responses. Such patterns often perpetuate negative behaviors and emotional states, like stress and anxiety, since you cannot even recognize the need or opportunity for change. Being trapped in this cycle hinders personal growth and the ability to adapt to new situations, as there is little to no reflection on past actions or consideration of their consequences. A restricted state of awareness keeps individuals stuck in maladaptive behaviors, making it challenging to break free and adopt healthier coping strategies.

For decades, I was trapped in the invisible grip of my Savior Complex, a pattern of behavior that operated far below the surface of my awareness. I couldn't see how deeply it shaped my relationships, decisions, and self-worth. Even worse, I never believed it could happen to me. The idea that I might be driven by such a compulsive need to rescue and fix others was entirely at odds with the image I held of myself, carefully crafted through years of academic and professional success.

The thought of confronting this truth was terrifying. My pride, rooted in my achievements and the identity I had built, would have been utterly shattered. Growing up as a child of parents without formal education, I forged my identity through academic and professional accomplishments. My self-worth became inextricably tied to earning advanced degrees—an MBA, a Master's, and a Ph.D. in psychology—writing books and touring the globe to deliver teachings to thousands of CEOs and senior executives.

My achievements were not just accomplishments; they were a fortress protecting me from vulnerability. The idea that my seemingly noble efforts to help others were driven by an unconscious need for validation and control was a reality I was not ready to face. Yet, the more I ignored this truth, the more imprisoned I became by my patterns. It wasn't until I began the painful work of self-reflection that I saw how much this complex had dictated my life, holding me back from genuine connection and inner peace.

I will discuss two categories of barriers to consciousness: unconscious and conscious. You would think that being conscious of being unconscious or close-minded would be impossible or make you crazy, but the material I share will surprise you!

Unconscious Barriers to Consciousness

The list of unconscious headwinds I have identified below is by no means exhaustive. However, it includes *coping mechanisms* you may have used or continue to use to face daily challenges. A coping mechanism is a psychological strategy to manage stress and emotional discomfort in life-critical situations.

These mechanisms can be *adaptive*, helping effectively manage emotions and situations, or *maladaptive*,

Figure 59: Unconsciousness Barriers (AI)

leading to difficulties. While unconscious barriers are presented under the unconscious section, they can also be partially or fully conscious. For instance, over time or with the help of a therapist, we can become more aware of how we are "triggered" by what others say or how they respond to our behavior.

Ego Resistance

It is puzzling to think that our creator (or evolution) would have found it beneficial to let the ego restrict our consciousness. However, it is a fact, and I will present multiple interpretations that may shed light on its mysterious origin. Let us first define the role of your *"ego."*

Freud's theory posits that the ego is a crucial element of our mental structure (139). The ego acts as an intermediary, balancing the innate impulses of the *"id."* For Freud, the id is the most primitive part of the human psyche. It is the source of basic impulses and drives, such as hunger, sex, and aggression. The id seeks immediate gratification without consideration for morality or social norms and is entirely unconscious. Freud believed that the id is present from birth and drives an individual's fundamental behaviors and instincts.

I suggest that the PRIMAL brain is the force behind the id, while the *"superego"* originates in the RATIONAL brain. According to Freud, the superego is the seat of our ethical conduct and morality. Essentially, the ego bridges our unconscious and conscious realms. It is tasked with assessing reality and fostering a sense of self-identity. Simply put, the ego oversees transitioning from the "ME" level of our self-awareness to the 'I' level, which is crucial in orchestrating our thoughts, emotions, and behaviors in response to the world around us.

Given that, the ego often resists acknowledging limitations or weaknesses, including having lower levels of awareness. Fear of vulnerability or judgment from others might prevent your egos from admitting you are not fully conscious or self-aware. Additionally, societal norms often do not encourage deep introspection or the admission of not understanding oneself fully. This can create a barrier to seeking help or exploring personal growth opportunities. It is a big issue for people working in organizations based on building emotional resilience, like the military or police force (140). Admitting a lack of consciousness can blow your ego, the fear of being vulnerable or weak, and cultural stigmas like the impression that you cannot handle a problematic situation.

The ego's resistance drives low consciousness as a defense mechanism to protect our *self-image*. Self-image, a mental picture of oneself, is crucial for the ego. A positive self-image is essential for psychological well-being. It encompasses perceptions, beliefs, and attitudes about our abilities, appearance, and worth. The ego invests heavily in maintaining a positive self-image, central to one's identity and self-esteem. When this image is threatened, the ego may use defense mechanisms to protect it. Still, an overly rigid or unrealistic self-image can lead to difficulties adapting to change, acknowledging faults, and personal growth.

My ego resisted fiercely during the retreat, clinging to the familiar frameworks of logic and control that had defined my career. In the face of the Shipibo rituals, my inner skeptic questioned everything, seeking to dismantle the intangible with rationality. But as the nights unfolded, it became clear that this resistance was a defense mechanism, a barrier protecting me from the unknown and, paradoxically, from a more profound understanding. The struggle wasn't against the experience but within me, a dance between surrender and the need to hold on. Only by acknowledging and softening this resistance could I be open to the profound lessons offered.

As this book will continue to insist, balancing self-acceptance with a realistic understanding of oneself is vital to a healthy ego. However, to maintain a positive self-concept, the ego may resist acknowledging shortcomings or areas for growth, even if it involves your health. This resistance is a natural human tendency to avoid discomfort and preserve self-esteem.

Repression

In psychology, *repression* is a defensive coping mechanism, as also identified by Sigmund Freud (141). It involves unconsciously blocking unwanted or distressing thoughts, memories, or desires. Repression is a way for the mind to protect itself from psychological distress by keeping unpleasant thoughts or experiences hidden from the psyche. This mechanism can play a role in shaping your behavior and can lead to psychological issues if unresolved repressed material surfaces or continues to influence you unconsciously. The concept of repression is relatively abstract; therefore, examining cases for which this unconsciousness counterforce is acute and chronic is helpful. Here are general patterns observed in clinical settings:

- Childhood trauma: Individuals who have experienced traumatic events in childhood, such as abuse or neglect, may repress these memories. This can manifest as unexplained anxieties, fears, or relationship difficulties in adulthood.
- War veterans: Soldiers returning from combat may repress memories of traumatic events experienced in war. This repression can be a factor in post-traumatic stress disorder (PTSD), where the person might have flashbacks or nightmares when the repressed memories resurface.
- Survivors of accidents or disasters: People who have survived accidents, natural disasters, or other traumatic events may unconsciously repress the memory of these events to cope with the intense fear and distress they experienced.
- Victims of assault: Victims of physical or sexual assault may repress memories of the assault as a defense against the intense pain and trauma associated with the experience.

In all these cases, repression serves as a protective mechanism to shield the individual from psychological pain. However, the repressed memories can continue to affect emotional health and behavior and do so with no or limited awareness.

Projections

A psychological *projection* is a defense mechanism in which you attribute unwanted or unacceptable thoughts, feelings, or motives to another person. This mechanism can serve to reduce anxiety or guilt by allowing the individual to "see" their undesirable qualities in others instead of acknowledging them in themselves. Here are some examples of unconscious projections.

- Attributing anger to others: A person who is frequently angry or hostile might consistently perceive others as angry or looking for a fight.
- Jealousy projection: Interpreting others' actions as driven by being envious of them, even when there is little evidence for it.
- Projecting insecurities: Criticizing others for incompetence or lack of knowledge to avoid confronting your insecurities.
- Moral projection: Seeing others as dishonest or unethical may reflect your ethical dilemmas.
- Projecting relationship fears: Fearing betrayal or abandonment by accusing a partner of being unfaithful or planning to leave, even without accurately indicating such intentions.
- Projecting sexual or romantic feelings: Believing that another person is attracted to you and interpreting neutral actions as flirtatious.

In each of the above cases, projections externalize internal conflicts, emotions, or desires that a person finds challenging to confront.

Spiritual Bypass

Spiritual bypassing is a term used in psychology and spiritual circles to describe a tendency to use spiritual beliefs, practices, or experiences to avoid facing unresolved emotional issues, psychological wounds, and unfinished developmental tasks. Psychologist John Welwood first introduced this concept in the 1980s (142). It highlights how you might, consciously or unconsciously, sidestep personal growth challenges or painful feelings by immersing yourself in spiritual practices or beliefs. Critical characteristics of spiritual bypassing include:

- Using spirituality to maintain an exclusively positive outlook, neglecting or denying the existence of problems or painful emotions.
- Viewing emotions like anger or sadness as signs of spiritual failure rather than as natural, informative aspects of human experience.
- Attributing difficulties to external or cosmic causes without acknowledging one's role in their challenges or relationships.
- Holding unrealistic standards for oneself and others based on spiritual ideals can lead to disappointment and a sense of failure when those standards are unmet.
- Superficially applying spiritual concepts or quotes to complex life issues without engaging in deeper inquiry or emotional work.

Addressing spiritual bypass involves integrating spiritual practice with psychological understanding and emotional work, recognizing that genuine growth entails confronting and working through personal and emotional challenges, not merely transcending them.

Infatuation

Infatuation can be considered a psychological state that mainly operates unconsciously. For instance, individuals may not be fully aware of why they are drawn to someone, as it tends to be driven by intense emotions, desires, and idealizations. This can be explained by the *halo effect* (a cognitive bias we discuss in the next section), where positive impressions in one area unduly influence overall perception and can make people unconsciously overvalue certain traits in the person of their infatuation. In clinical psychology, infatuation is often associated with projection.

Infatuation may also be linked to unconscious emotional needs, such as seeking validation or projecting personal ideals onto another person or situation without deeper awareness. Infatuation triggers high dopamine levels, a neurotransmitter associated with pleasure and reward. Freud considered early forms of romantic attraction, including infatuation, to be rooted in unconscious desires, particularly those related to sexuality and early childhood experiences. He argued that infatuation often expresses unconscious needs, such as the desire for security or unresolved conflicts from earlier relationships (e.g., with parents).

Helen Fisher (143), a biological anthropologist, has extensively researched the neurobiological basis of romantic love, including infatuation. In her work, she describes infatuation as part of the early stages of romantic love, driven by heightened levels of dopamine and norepinephrine in the brain. Fisher's research shows that these neurotransmitters are associated with excitement, reward, and obsession—key elements of infatuation. Fisher views infatuation as a natural part of the brain's evolutionary drive toward pair bonding and reproduction.

Conscious Barriers to Consciousness

Many unconscious barriers become "consciously visible" over time, mainly when a therapist sheds light on these phenomena' psychological mechanisms. However, it is paradoxical that our psyche persists in employing filters that diminish our capacity to recognize and understand our thoughts and

Figure 60: Conscious Barriers (AI)

behaviors. Conscious mechanisms that limit your consciousness include self-doubt, self-censorship, self-limiting beliefs, denial, self-deception, and distractions. This is not a complete list, but it is good to consider if you are on a mission to clear the way to higher consciousness.

Finally, I discuss cognitive tactics that limit our awareness, called *cognitive biases*. They are systematic patterns of deviation from norm or rationality in judgment, essentially like mental shortcuts that the brain uses to simplify information processing. They often lead to perceptual distortion, inaccurate judgment, illogical interpretation, and reduced consciousness.

Self-doubt

Self-doubt is a lack of confidence in your abilities, decisions, or thoughts. It can be insidious in various ways, affecting individuals personally and professionally. Self-doubt gradually erodes self-confidence. If you constantly question or self-censor your capabilities, you will struggle to trust your judgment or skills. It can lead to a cycle where doubt breeds more, undermining your self-assuredness. Self-doubt can lead to indecisiveness. Individuals who doubt themselves might find it challenging to make decisions, fearing they will make the wrong choice. This indecision can hinder progress in various aspects of life, from career advancement to personal relationships (144).

Amid the intensity of the retreat, self-doubt crept in like an unwelcome companion. I questioned my place in this ancient setting, my ability to integrate its wisdom with my scientific foundation, and even the path that had led me there. The quiet moments were the hardest, as the jungle's hum seemed to mirror the uncertainty within me. Was I worthy of this transformation? Could I truly reconcile the worlds of science and spirit? Yet, as the days passed, I began to see self-doubt not as a flaw but as an invitation—to lean into vulnerability, to trust the process, and to let go of the need for immediate answers. In doing so, I discovered a resilience I hadn't known I possessed.

Consumed with self-doubt, we often hesitate to take risks or seize opportunities out of fear of failure. This avoidance can mean missing out on

214

potentially beneficial experiences, whether a new job, a relationship, or a personal goal. Self-doubt can also contribute to feelings of *imposter syndrome*, a belief that you are not as competent as others perceive you to be (145). This can be particularly prevalent in professional environments, often leading to stress and anxiety.

Indeed, chronic self-doubt can contribute to mental health issues like anxiety and depression. Constantly questioning one's abilities can lead to feelings of inadequacy and hopelessness. While self-doubt can be paralyzing if excessive, a healthy amount can signify self-awareness and openness to learning and growth. It prompts you to question your actions and beliefs, potentially leading to more thoughtful and deliberate choices. Addressing self-doubt leads to building self-esteem, seeking support from others, and challenging negative thought patterns.

Self-censorship

While related to personal and psychological experiences, *self-censorship* and self-doubt differ fundamentally in their nature and manifestations.

Self-censorship is the act of deliberately restraining or altering your expression, behavior, or actions due to perceived social pressures, fear of criticism, or potential repercussions. It is more about controlling or modifying external behavior or expression rather than an internal sense of doubt or inadequacy. It impacts freedom of expression and authenticity in social or professional settings. This can occur even when you are confident in your abilities or beliefs, and it is straightforward to study on social platforms.

For instance, one study examined "last-minute" self-censorship on Facebook. This refers to content users type but ultimately decide not to post or content filtered after being written on Facebook. Data on self-censorship behavior was collected from 3.9 million users over 17 days, using features that described users and their interactions. Results indicated that 71% of users exhibited some level of last-minute self-censorship in that period and supported the theory that a user's "perceived audience" lies at the heart of the issue. Also, males censored more posts than females (146). By examining what users started to write but later deleted, the study provided insights into the psychological and social factors influencing online self-expression, including concerns about privacy, audience perception, and potential negative feedback.

In summary, self-censorship is a behavior motivated by external social pressures and the desire to avoid negative consequences or judgment.

Self-limiting Beliefs

Self-limiting beliefs are assumptions or perceptions about yourself and the world that restrain your potential and opportunities. These beliefs are often deeply ingrained and can significantly influence your behavior and life choices. Self-limiting beliefs can be subtle and persistent, frequently unrecognized while profoundly impacting your life. These beliefs create invisible barriers to growth and achievement. They convince you that you cannot reach a certain point, whether in your personal, academic, or professional life. Self-limiting beliefs often do not align with your actual abilities or circumstances. They are based more on fear and doubt than on facts. Negative experiences can reinforce these beliefs, creating a cycle where failure is attributed to inherent limitations rather than external factors or lack of experience (147).

Self-limiting beliefs can lead to overly cautious or avoidant behavior, preventing individuals from taking risks or trying new things. Continual self-limiting thoughts can erode self-esteem and confidence, making recognizing and celebrating strengths and achievements challenging. Overcoming self-limiting beliefs involves recognizing and challenging these thoughts (more awareness!), seeking evidence against them, and gradually building confidence through small achievements.

Denial

Psychological *denial* is a defense mechanism in which you refuse to acknowledge or accept reality or facts, thus protecting yourself from facing distressing aspects of life or behavior (148). Denial can be insidious, often working subconsciously to distort your perception of reality, leading to negative consequences. It usually arises to avoid confronting painful or difficult truths or emotions by selectively ignoring or reinterpreting information that conflicts with what you wish to believe.

While denial can provide immediate, short-term relief from distress, it often exacerbates problems in the long term by preventing resolution or coping. Denial can strain relationships and impair decision-making, as the person in denial may reject valid concerns or advice from others. Denial can range from mild (e.g., downplaying a problem) to severe (e.g., completely refusing to acknowledge a glaring issue). It can manifest in various forms, such as denial of fact, denial of responsibility, or denial of impact.

It took me decades to accept that denial was closing the doors to more plasticity. Waking up was painful, but ultimately, it was the only way to break free.

Self-deception

Self-deception occurs when you convince yourself of a false or distorted reality, often to feel better about yourself or your situation. This process involves ignoring or *over-rationalizing* away evidence that contradicts the preferred belief. Self-deception is distinct from outright lying, as it is often subconscious; individuals are not fully aware they are deceiving themselves. It involves refusing to accept facts or information that are inconvenient or unpleasant, creating justifications for one's behavior, decisions, or beliefs that are not entirely valid, and considering information supporting preconceptions while ignoring contradictory evidence (149).

The goal of self-deception is interpreting events in an overly favorable way to oneself. *Cognitive dissonance* is at the root of self-deception because facing a discrepancy between one's beliefs and reality is unpleasant. For example, a person who smokes might experience cognitive dissonance because they know smoking is harmful to their health. To reduce this dissonance, they might change their belief about the risks of smoking, seek out information that downplays the risks, or downplay the importance of their health in favor of the enjoyment they get from smoking.

Self-deception can serve as a temporary coping mechanism, helping individuals avoid emotional pain or anxiety. However, over time, it can lead to poor decision-making, strained relationships, and a lack of personal growth. Recognizing and confronting self-deception often requires introspection and sometimes the help of specific modalities that can raise consciousness by addressing underlying issues and cognitive biases.

Distractions

Distractions can significantly limit your ability to maintain consciousness and awareness every minute of the day. A study conducted by Dr. Gloria Mark at the University of California, Irvine, investigated how digital media affects attention spans over time. In 2004, the research found that individuals spent an average of two and a half minutes on a (150) task before switching. By 2012, this average decreased to 75 seconds, and more recent observations indicate it has further reduced to approximately 47 seconds. This trend highlights the increasing impact of digital distractions on our ability to maintain focus. So, distractions fragment your attention, pulling you away from your current activities or thoughts. This scattered focus makes it hard to stay consciously engaged with a task, leading to a superficial understanding or incomplete information processing. Constant distractions can impair your ability to encode and retrieve information effectively (151, 152).

When we are not fully present due to distractions, our ability to remember details or learn new information is compromised. Managing multiple sources of distraction puts a strain on our cognitive resources. This increased cognitive load can lead to mental fatigue, reducing our ability to think clearly and maintain a conscious awareness of our actions and decisions. Distractions can also prevent us from being mindful of our emotions. This lack of emotional awareness can lead to poor emotional regulation and difficulty recognizing and addressing our emotional needs. Continuous distractions can make engaging in mindfulness and self-reflection challenging, essential for maintaining a robust conscious connection with ourselves.

Finally, distractions compromise the ability to engage in deep, focused work. Deep work requires sustained attention and consciousness, which distractions directly undermine. Minimizing distractions involves creating an environment conducive to focus, setting clear priorities, and practicing mindfulness to enhance our ability to remain present and consciously engaged in activities. Techniques like meditation, time-blocking for specific tasks, and setting boundaries around technology use can help reduce the impact of distractions on consciousness.

Cognitive Biases

Cognitive biases act as filters that shape our perception, often operating unconsciously and limiting our ability to expand awareness and make fully informed decisions. Understanding these biases helps us rewire thought patterns and foster greater openness and self-awareness. As previously noted, *cognitive biases* are systematic patterns of deviation from normative or rational judgment (153). These biases can greatly influence decision-making and consciousness by distorting perceptions and interpretations of reality. I chose to address cognitive biases in this section because, although they often operate below conscious awareness, many can be easily identified in hindsight when provided with clear operational definitions.

Created by Buster Benson in 2016, the cognitive codex highlights the complexity and ubiquity of human cognitive biases, helping to reflect on how these biases impact their judgment and offer pathways to minimize their influence on decision-making (154). It organizes and categorizes over 100 cognitive biases that influence human thinking and decision-making, providing a brilliant educational tool on affect, perception, memory, and behavior.

The biases are grouped into four main categories: 1/ Too much information, 2/ Not enough meaning, 3/ Need to act fast, and 4/ What should we remember. These categories reflect how biases mainly emerge when the brain processes overwhelming information, assigns meaning, acts quickly under pressure, or selectively retains specific details. The wheel helps illustrate how these biases function and provides a comprehensive way to reflect on cognitive distortions that shape human judgment.

Below are the definitions of some of the cognitive biases that may contribute to lower consciousness and their effects:

- *Loss avoidance* is a deeply ingrained bias that causes us to prioritize avoiding losses over acquiring equivalent gains. This bias is mainly rooted in the role of the PRIMAL brain, wired for survival-based decision-making, often leading to irrational risk aversion. Daniel Kahneman and Amos Tversky, pioneers in behavioral economics, demonstrated that the pain of losing is felt twice as intensely as the pleasure of gaining (155). Their Prospect Theory challenges classical utility theory based on rationality by showing that people's decisions are often shaped by emotional and psychological biases rather than logical value calculations. Kahneman's research, which earned him the 2002 Nobel Prize in Economic Sciences, underscores how loss aversion skews our judgment, pushing us to avoid risks even when taking them might lead to greater rewards. This bias limits our openness to change, making us more likely to stay in our comfort zones, avoid new opportunities, and resist transformation. For example, the distress of losing $100 typically outweighs the joy of gaining $100, even though the objective value is the same. This imbalance in perception can fuel fear-based decision-making, reducing cognitive flexibility and flow. Recognizing loss aversion is the first step in overcoming its restrictive influence.
- *Confirmation Bias* leads us to seek and favor information that aligns with our existing beliefs, making it harder to challenge personal narratives or embrace new perspectives. This limits our cognitive flexibility and reinforces unconscious mental loops (156).
- *Anchoring Bias* causes us to rely too heavily on the first piece of information we receive, shaping our judgments and decisions without more profound reflection. This prevents critical reassessment and keeps us locked in fixed thought patterns.

- *Availability Heuristic* makes us overestimate the importance of recent or emotionally striking information. This can distort risk perception, reinforcing fear-based thinking rather than rational awareness.
- *The Dunning-Kruger Effect* causes individuals with low expertise to overestimate their abilities, blocking self-growth by reducing receptiveness to learning.
- *Hindsight Bias* makes past events seem more predictable than they were, leading to oversimplified cause-and-effect reasoning and reducing our ability to learn from experience.
- *Status Quo Bias* keeps us attached to familiarity, resisting change even when it could lead to personal or professional evolution. It fosters mental rigidity and stagnation in consciousness.
- *Self-serving bias* causes us to attribute successes to personal qualities but blame failures on external factors. This hinders deep self-reflection, making personal growth more difficult.
- *Groupthink* suppresses independent thinking in favor of group cohesion, often limiting innovation and expanding consciousness through diverse perspectives.
- *The Bandwagon Effect* pushes us to adopt beliefs simply because they are popular, reinforcing conformity over authenticity and blocking self-inquiry.
- *The Halo Effect* makes us generalize one positive trait (such as attractiveness or success) to other unrelated traits, affecting how we judge people and situations without deeper awareness *(157).* .
- *The Hawthorne Effect* (closely related to the Halo effect) shows that we tend to modify their behavior when they know they are being observed. This suggests that conscious self-awareness can change patterns of thought and action, a helpful insight into habit reprogramming.
- *Optimism & Pessimism Bias* cause us to overestimate positive or negative outcomes. While optimism bias may lead to overconfidence, pessimism bias can fuel avoidance and anxiety, distorting a balanced perspective on reality.

Recognizing these biases is the first step toward rewiring them. Mindfulness, neuroplasticity training, and intentional self-reflection can help break these unconscious patterns, allowing for a higher state of awareness and flow.

Now that I have presented some of the top causes of constricted consciousness, our focus can move to techniques or approaches that have been used, sometimes for thousands of years, to provide relief and healing from the toxic effects of stress, anxiety, depression, trauma, and addictions. While raising awareness has proven to open states from which people can use plasticity to heal, some approaches insist that this process cannot happen unless the body is cleared from the trails of the emotional marking left by SADAT conditions.

Therefore, before discussing practices that can raise consciousness, I will review why clearing emotional wounding from the body and mind is central to your mental health. While I have often resisted the difficult, if not embarrassing, nature of cathartic methods, they provide a unique and probably irreplaceable substitute for countless hours of therapy.

Cathartic Methods

In the late 19th and early 20th centuries, psychologists and neurologists, including Sigmund Freud, explored the concept of *catharsis* as a way of providing relief from stress, anxiety, and trauma. Catharsis is a psychological concept that refers to the process of releasing strong or repressed emotions. Originally derived from the Greek word "katharsis," meaning "purification" or "cleansing," the concept has been utilized in various contexts in psychology, but it is also used as an artistic expression. Freud initially believed that hysterical and neurotic symptoms were due to the repression of traumatic memories and that, therefore, catharsis could be therapeutic.

Screaming, purging, colonic cleansing, and fasting are the most common ways to achieve cathartic clearance of stress and trauma. These methods have been used historically and in various cultural contexts for thousands of years. They are based on releasing pent-up emotions, particularly negative ones, to achieve emotional cleansing.

Screaming

Screaming, especially, can activate the body's natural relaxation response following the intense emotional expression. Screaming can give individuals a sense of empowerment. It can be a way of reclaiming control over one's emotions and body, particularly for individuals who may feel powerless in other aspects of their lives. Intense emotional expressions can lead to the release of endorphins, the body's natural painkillers, which can induce a feeling of euphoria or well-being. Historically, many cultures have had rituals involving screaming, shouting, and other forms of loud vocal expression, often in the context of religious or spiritual ceremonies. These practices were also believed to expel evil spirits or negative energies.

In the 1970s, Arthur Janov's *Primal Therapy* became famous as a form of psychotherapy that encouraged patients to re-experience and express repressed feelings through screaming (158). It was based on the idea that neuroses were caused by repressed trauma. While primal therapy and its techniques are not widely accepted in mainstream psychology today, the value of expressing and releasing emotions is still recognized as an essential aspect of psychological health. However, contemporary therapy (especially Western approaches) focuses more on understanding and controlling emotions than attempting to clear their somatic marking.

Purging

Purging is commonly referred to as a self-induced process to expel food from the body to avoid weight gain and, as such, is mainly discussed as a symptom of bulimia. However, physical purging can also help expel negative energies or toxins from the body, alleviating stress and emotional turmoil. For instance, *emetics*, substances that induce purging or vomiting, have been employed in various traditional healing practices worldwide to cleanse or purify the body and mind (159). Indigenous and shamanic cultures, including Native American ceremonies, often employ natural emetics in ritual contexts to prepare for significant events or facilitate healing.

In traditional Ayurvedic medicine, a practice known as *Vamana* is used to eliminate excess mucus (Kapha), believed to be a toxin, from the body. This is done under strict supervision as part of a broader cleansing process known as *Panchakarma*. In some Indigenous and shamanic cultures, emetics are also commonly used in ritual contexts, as I reported in my account of the Peru retreat in Chapter ONE.

Thus, the act of vomiting may release not just physical but emotional and spiritual impurities, according to many traditions. According to *humoral theory*, an imbalance of these humors within the body leads to illness. Body humor, or humoral theory, is an ancient concept that was once a central part of understanding human health in Western and Eastern medical traditions. This theory, developed mainly by the ancient Greeks and Romans, posited that human health and temperament were governed by a balance of four bodily fluids or "humors." These humors were believed to be *blood, phlegm, yellow bile, and black bile*. Treatments, therefore, focused on restoring balance, often through diet, herbal remedies, bloodletting, and other practices aimed at adjusting the levels of these fluids. Historically, the concept of balancing *bodily humors* was central to Greek medicine.

Blood was associated with a sanguine temperament, characterized by optimism, enthusiasm, and social extroversion. It was thought to be dominant in the spring and during adolescence. Phlegm was linked with *a phlegmatic* temperament, which was seen as calm, reliable, and thoughtful. Phlegm was believed to be dominant in the winter and during old age. Yellow bile (Choler) corresponded to a choleric temperament characterized by ambition, leadership qualities, and irritability. It was thought to be dominant in the summer and adulthood. Finally, black bile (Melancholer) was associated with a melancholic temperament, characterized by thoughtfulness, introspection, and often sadness or depression. It was believed to be dominant in the fall and later adulthood.

Modern medicine recognizes that many biological, environmental, genetic, and psychological factors contribute to health and illness rather than an imbalance of these four fluids. Nonetheless, humoral theory played a significant role in medicine's history and the development of early medical thinking. From a modern medical perspective, though, induced vomiting is generally not recommended to relieve stress or emotional issues. The practice can have adverse physical consequences, including electrolyte imbalance, dehydration, and damage to the esophagus and stomach lining.

Colonic Cleansing

Colonic cleansing is a technique for clearing emotional energy commonly found in traditional, alternative, or holistic health practices rather than mainstream medicine. This practice also stems from believing in the connection between physical cleansing and emotional or spiritual healing. Renowned trauma experts such as Dr. Gabor Maté, Dr. Bessel van der Kolk, and Dr. Peter Levine (17, 160) emphasize the critical need to release residual energy from past traumas to prevent it from manifesting as physical illness. Dr. Maté highlights that trauma is not just an event but the residual impact it leaves on the body and mind, potentially leading to chronic diseases if unaddressed. In his seminal work "The Body Keeps the Score," Dr. van der Kolk illustrates how unresolved trauma imprints on the body, affecting mental and physical health (161). Dr. Levine's "*somatic experiencing*" approach focuses on discharging this pent-up energy through bodily sensations, facilitating healing, and preventing trauma-related illnesses. It is often used in alternative medicine for various purported benefits, including detoxification and emotional release. Specific detox diets or cleansing regimens used by shamans, for instance, may include specific foods or concoctions that induce detoxification and emotional cleansing, often resulting in diarrhea.

In *Ayurveda*, Panchakarma also includes a series of cleansing procedures to purify the body. *Virechana,* one of the procedures, involves purgation or induced diarrhea and is believed to release toxins and balance the body's energies. In various Indigenous and shamanic cultures, the use of specific herbal preparations, sometimes including psychoactive substances like Ayahuasca, is believed to cleanse the body and spirit via its purgative effect. Some practices in *traditional Chinese medicine* use herbal formulas designed to "clear heat" or "eliminate toxins" from the body, occasionally resulting in diarrhea. This is sometimes considered part of emotional healing in traditional Chinese medicine philosophy.

It is important to note that these practices and their underlying principles are not universally accepted in the medical community. Using physical purging, such as diarrhea, to clear emotional energy is largely based on ancient beliefs. Western doctors have mostly ignored the approach because inducing diarrhea can pose health risks, including dehydration and electrolyte imbalance, that can bring complications and liabilities for practitioners to handle.

Fasting

Some spiritual traditions involve fasting or other forms of self-denial to purify the body and mind and deepen spiritual awareness. Fasting can be considered a cathartic method in various spiritual, cultural, and psychological contexts. Fasting, particularly as part of spiritual or religious practice, often aims to cleanse both the body and the mind, which can lead to emotional release and spiritual renewal.

Abstaining from food (and sometimes drink) allows the body to cleanse itself of toxins. This physical detoxification can also impact mental and emotional health, as physical well-being is closely linked to psychological states. Also, fasting requires discipline and self-control, which can lead to a deepened sense of personal reflection and introspection. This introspective process can help individuals confront and release emotional burdens, unresolved conflicts, and repressed emotions.

Like some of the previous methods, fasting itself can be challenging and might bring up a range of emotions, from frustration and anger to exhilaration and a sense of achievement. I felt that emotional lift when I completed my fasting challenge in Peru. Processing these emotions can be cathartic. In many religious traditions, fasting is seen as a way to cleanse the soul, making it a spiritual catharsis (162). It is sometimes used as a method of penance to purge oneself of sins or impurities.

The spiritual discipline involved in fasting is believed to heighten one's spiritual awareness and connection to the divine, which can lead to profound personal transformations and a renewed sense of spiritual purpose. Fasting can increase mindfulness and awareness of the present moment. This heightened awareness can make meditation and prayer more profound, allowing deeper spiritual insights and emotional processing. Fasting can interrupt daily routines and allow one to break free from habitual behavior, including negative emotional cycles. This reset can be cathartic, offering a chance to start anew with a clearer mind and healthier habits.

For Christians, fasting during Lent is a form of *repentance* and a way to share in the suffering of Christ. For Muslims, fasting during Ramadan is intended to teach Muslims patience, modesty, and spirituality. For Buddhists, periodic fasting is used to develop control over one's attachments and desires. For many Indigenous people, like the Shipibo community, fasting or dietary restrictions, known as a "dieta," play a significant role in their spiritual and healing practices. The significance of fasting revolves around spiritual purification, preparation for ceremonial practices, and enhancing connections with the spiritual world. Fasting is not just about physical detoxification but also spiritual cleansing. Abstaining from certain foods, and sometimes from sexual activity and social interactions, is believed to purify the body and spirit. This purification is considered essential for individuals preparing to participate in Ayahuasca ceremonies.

Meanwhile, some studies suggest fasting can improve mood and mental well-being. These effects are attributed to changes in neurotransmitter levels, reduced inflammation, and enhanced brain function due to metabolic shifts during fasting (163).

In summary, screaming, purging, colonic cleansing, and fasting can serve as cathartic methods by promoting physical detoxification, facilitating emotional release, deepening spiritual experiences, enhancing mindfulness, and helping individuals reset habitual behaviors. Their effectiveness and the experience of catharsis can vary widely among individuals, depending on your intentions, the context, and psychological and physical state.

Natural Modalities to Raise Consciousness

There are many natural ways (non-substance-based approaches) to restore homeostasis and deepen our mind-body connection. Most originated thousands of years ago. These practices can help enhance one's understanding of oneself and the world. In recent decades, psychologists and neuroscientists have given them more attention.

Exercise

Figure 61: Exercise (AI)

Exercise has numerous benefits for physical and mental well-being and can significantly raise consciousness. It increases blood flow to the brain, enhancing cognitive functions such as memory, attention, and processing speed.

Exercise increases blood flow to the brain, enhancing cognitive functions such as memory, attention, and processing speed. This heightened cognitive functioning is critical to understanding and seeking higher consciousness. Physical activity helps reduce stress hormones like cortisol and adrenaline while increasing endorphins, the body's natural mood lifters.

Lower stress levels allow for clearer thinking and improved focus, contributing to heightened consciousness. Exercise, especially mindful forms like yoga or tai chi, strengthens the connection between mind and body. This awareness of bodily sensations, movements, and breath can foster a more profound sense of self-awareness and presence.

Regular exercise can also improve emotional resilience and awareness. As people become more attuned to the emotional shifts that occur before, during, and after exercise, they better understand their emotional landscape. Exercise contributes to better sleep, which is essential for cognitive function and overall mental health. Quality sleep enhances clarity of thought, attention, and the ability to be present, all of which are important for consciousness.

Group exercise or sports can increase social interaction and a sense of community. These social connections can lead to a greater understanding of others and oneself, contributing to social consciousness. Exercise is known to improve mood by increasing the production of neurotransmitters like serotonin and dopamine (164). A positive mood state allows for a more open and expansive view of oneself and the world. Finally, regular participation in exercise requires discipline and willpower, traits beneficial for self-control and intentional living, and aspects of heightened consciousness.

Figure 62: Breathwork (AI)

I discovered *Holotropic Breathwork* nearly 20 years ago as I was looking for natural ways to cope with my anxiety and stress. Not only did the practice help me find some relief, but it allowed me to meet *Dr. Stanislav Grof*, the creator of the practice. Introduced earlier, Dr. Grof is known for his pioneering work in the field of psychedelic therapy and transpersonal psychology. His career spans several decades and is marked by significant contributions to psychology and consciousness studies.

In the 1950s, Grof was introduced to LSD and began researching its psychotherapeutic uses. His work in this area was pioneering, as he explored the potential of psychedelic substances for treating mental health disorders and for understanding the human psyche. In the 1970s, after the legal clampdown on the use of LSD and other psychedelic substances, Grof, together with his wife, Christina, developed Holotropic Breathwork to reach altered states of consciousness like those achieved with psychedelic drugs through controlled breathing, evocative music, and bodywork. It also includes a "sitter, another person who can support the participant during the process, encouraging him or her to release as much toxic energy as may be necessary to clear past traumas and other challenging mental conditions.

Thus, Holotropic Breathwork aims to access altered states of consciousness, which can lead to deep emotional release, self-discovery, and spiritual experiences (165, 166). Here is how Holotropic Breathwork impacts consciousness. Participants breathe slightly faster than usual using what Grof calls circular breathing, where the in-breath and the out-breath are of equal length. The resulting hyperventilation can alter the body's balance of oxygen and carbon dioxide, which may lead to altered states of consciousness. The sessions are accompanied by music that can range from calming to intense. Music helps guide the emotional journey and intensify the experience.

The practice is typically conducted in a group setting with facilitators who provide guidance and emotional support. This setting helps create a safe space for exploring consciousness. After the active breathing session, which usually lasts three to four hours, participants engage in an integration process, traditionally involving drawing a *mandala* or sharing experiences. This helps in grounding the insights gained during the session. In spiritual practice, a mandala is often a symbolic representation of the universe, the self, or the spiritual journey. Its intricate geometric patterns and radial symmetry serve as a visual aid for meditation, helping practitioners focus their attention, quiet the mind, and deepen their understanding of the interconnectedness of all things.

Mandalas are frequently used in various spiritual traditions, such as Tibetan Buddhism and Hinduism, to guide individuals toward inner clarity and harmony. Creating or contemplating a mandala can be seen as a metaphor for bringing order and balance to the chaos of life, aligning the individual with a greater sense of purpose and spiritual insight.

Meanwhile, the altered state of consciousness achieved through Holotropic Breathwork can bring subconscious thoughts and emotions to the surface. This can include repressed memories, symbolic visions, and emotional catharsis. Many participants report profound emotional releases and a sense of healing from past traumas or emotional blockages (167). As a result, this process can lead to significant personal transformation and increased mental clarity.

I have witnessed dozens of individuals experiencing these spiritual or even mystical states after Holotropic Breathwork sessions, reporting feelings of oneness, transcendence, or deep inner peace. The intense nature of the experience can lead to a feeling of rejuvenation, with participants often reporting a renewed sense of purpose or perspective on life. Participants share experiences with a wide range of physical sensations, from tingling and temperature changes to more intense vibratory experiences, as part of the process. I do not doubt that Holotropic breathwork can enhance self-awareness and generate valuable insights into personal issues, patterns, or life challenges. Meanwhile, it is essential to note that Holotropic Breathwork is an intense practice and may not be suitable for everyone, particularly those with cardiovascular issues, severe psychiatric conditions, or who are pregnant. It should be cautiously approached and practiced under the guidance of trained facilitators.

In general, consciously controlling and focusing on breathing patterns can significantly enhance our ability to maintain a state of mindfulness. By directing attention to the breath, an essential yet often automatic function, breathwork can reduce scattered thoughts and bring an individual to the present moment. This focused attention enhances mindfulness and awareness of the current experience. Breathwork can also regulate the autonomic nervous system, which controls unconscious processes like heart rate and digestion. Controlled breathing can shift the body from stress (sympathetic activation) to relaxation (parasympathetic activation). As a result, controlled breathing exercises can effectively reduce stress and anxiety on demand.

Yoga

Figure 63: Yoga (AI)

Yoga can be considered a form of exercise, but it is important to note that it encompasses much more than physical activity. Yoga is an ancient practice with a rich history that spans thousands of years. Its origins trace back to the Indian subcontinent. The earliest references to practices resembling yoga are found in the Indus-Sarasvati civilization in Northern India, from 3000 to 1500 BCE (168).

Carved stones discovered at archaeological sites depict figures in positions resembling yoga postures, suggesting early forms of yoga. The Vedic texts, the oldest scriptures of Hinduism during this period, mention rituals and ceremonies that foreshadow later yoga practices, which focus on controlling the mind and senses.

Yoga was introduced to the West in the late 19th and early 20th centuries. Key figures like Swami Vivekananda and Paramahansa Yogananda played a significant role. The 20th century saw the emergence of influential yoga teachers like T. Krishnamacharya, B.K.S. Iyengar, Pattabhi Jois, and others who developed and popularized different styles of yoga. In contemporary times, yoga has become a global phenomenon, encompassing various styles and interpretations.

Yoga is a holistic practice that combines physical postures (asanas), breathing techniques (pranayama), meditation (dhyana), and ethical precepts. Its benefits extend beyond physical fitness to mental, emotional, and spiritual well-being. It can also significantly enhance our ability to be conscious and mindful. It focuses on the harmony between mind and body and can profoundly impact mental clarity, emotional stability, and physical health. Yoga emphasizes the connection between physical movements and breathing, fostering heightened body awareness. This increased bodily consciousness can lead to better control over bodily functions and a more attuned sense of one's physical presence and capabilities. Holding yoga poses (asanas) and maintaining balance requires concentration and mental focus. This discipline of the mind can transfer to other areas of life, improving overall attention and focus. Yoga reduces stress and anxiety levels by encouraging relaxation and lowering the levels of the stress hormone cortisol. A calmer mind is more capable of conscious thought and less prone to being overwhelmed by stimuli.

Yoga often involves mindfulness, which is the quality of being fully engaged in the *now*. This can increase awareness and appreciation of the now, reducing preoccupation with past or future concerns. Through controlled breathing (pranayama) and meditation, yoga helps regulate emotions. This emotional balance is conducive to a state of heightened consciousness and self-awareness. Regular yoga practice has been linked to improved cognitive functions like memory, executive function, and processing speed. A clear and sharp mind is better equipped for conscious decision-making and problem-solving.

Better sleep is also a direct benefit of yoga, and good sleep contributes to improved cognitive function and alertness. A well-rested mind is more capable of maintaining conscious awareness throughout the day. Yoga provides an opportunity for introspection and self-reflection, leading to deeper self-understanding and a more mindful approach to life. The philosophical aspects of yoga promote a sense of connectedness with oneself, others, and the world, enhancing empathy and conscious interactions. By improving physical health, mental clarity, and emotional balance, yoga contributes to overall well-being, making it easier to stay conscious and aware of various aspects of life. Incorporating yoga into daily life can lead to a more mindful, conscious state of being.

I discovered yoga in my mid-forties and have practiced the discipline almost daily ever since. This approach nurtures my body and calms my mind, enabling me to engage more deeply with my inner self and the world around me. I discovered yoga in my mid-forties and have practiced the discipline almost daily ever since. I see it as a gift to my body and mind—an invitation to slow down, to be in harmony with myself and the world around me. While I have fallen off the horse of daily practice a few times, I have regularly incorporated yoga routines each morning that sustain my level of consciousness. I always find joy and peace in working on moves that relax my body and prepare my mind for the day.

Nature bathing, especially *forest bathing*, or "*Shinrin-yoku*," a practice originating in Japan, is a mindful practice of absorbing the forest atmosphere through all the senses, not just walking in the woods (169). The therapeutic value of the ancient practice applies similarly to any form of contemplative immersion in nature.

Nature bathing encourages mindfulness and focuses on the sensory experiences of the natural world around us. This heightened awareness of the environment and one's reactions to it can lead to an increased state of consciousness. Many scientific studies have shown that spending time in nature can decrease cortisol levels, a stress hormone, promoting relaxation (170). A calmer, less stressed mind is more capable of reflective thought and awareness.

Figure 64: Nature Bathing (AI)

Meanwhile, exposure to natural environments has been linked to improved mood and reductions in feelings of anxiety and depression. This emotional balance is crucial for a clear and conscious state of mind. Nature bathing fosters a deeper connection with the natural world, enhancing one's sense of interconnectedness and empathy, key aspects of consciousness. Nature stimulates attention, helping to replenish mental resources depleted in urban settings. This restoration can improve cognitive functions related to consciousness.

As noted earlier, forest environments can significantly boost the immune system, partly due to inhaling phytoncides natural oils within the forest air (171). Enhanced immune function is associated with overall better health and well-being, which is why forest bathing has been explored as a therapeutic practice for mental health conditions such as depression and anxiety. Several studies have shown promising improvements in cognitive functions, such as concentration and clarity of thought.

Finally, spending time in nature has been found to increase self-reported happiness and emotional stability (172).

Meditation practice significantly enhances our ability to be conscious or maintain a state of mindfulness and awareness in several profound ways. Meditation can improve cognitive and emotional functioning by training the mind to focus and redirect thoughts (173, 174). Meditation cultivates mindfulness, which is the ability to be fully present, aware of where we are and what we are doing, and not overly reactive or overwhelmed by what is happening around us. This heightened awareness strengthens our connection with the present moment.

Figure 65: Meditation (AI)

Regular meditation practice can increase the strength and endurance of your attention. It helps sustain attention and recover attention more quickly when it wanders, leading to better focus in daily tasks. Meditation can help you better understand yourself, helping you grow into your best self. It encourages self-reflection and can help you recognize thought patterns or emotional responses, contributing to greater self-awareness.

Meditation reduces stress by lowering cortisol levels. This calmer state of mind enhances conscious awareness, as stress often clouds judgment and impairs cognitive functions (175). Some forms of meditation can lead to an improved self-image and a more positive outlook on life, contributing to a more conscious and deliberate approach to daily activities. Meditation has been shown to improve memory, cognitive flexibility, and executive functioning. This can translate into better decision-making and problem-solving skills. Loving-kindness or compassion meditation fosters a sense of interconnectedness and empathy, enhancing your ability to be attuned to your feelings and thoughts and those of others. Better sleep through meditation improves overall cognitive function, alertness, and consciousness during waking hours.

Meditation also enhances the mind-body connection, fostering greater awareness of the physical self, which can help identify and address physical ailments or tensions early on. Incorporating meditation into daily life can help individuals become more contemplative and aware, enhancing overall well-being and quality of life. The key is consistency and practice, as the benefits of meditation typically accrue over time. While I have never considered myself a good meditator, I have learned not to judge whether I do it right but rather to focus on the immense benefits of reaching a neutral state from which I can embrace the power of mindfulness.

Cognitive Behavioral Therapy (CBT)

Figure 66: CBT (AI)

Cognitive Behavioral Therapy (CBT) is a form of psychotherapy that focuses on identifying and changing negative and unhelpful thought patterns, beliefs, and behaviors. It is a widely used approach for treating a variety of mental health disorders, including anxiety, depression, phobias, and stress-related disorders. CBT works on the principle that our thoughts, feelings, and behaviors are interconnected and that changing negative thought patterns can lead to changes in feelings and behaviors (176).

CBT helps individuals identify the distorted or negative thought patterns that limit their consciousness and contribute to their emotional difficulties. This process is known as cognitive restructuring. Once these thoughts are identified, the therapist challenges them and replaces them with more realistic and positive thoughts.

This is done through various techniques like reality testing, examining evidence, and weighing the pros and cons. CBT incorporates behavioral techniques like exposure therapy (for phobias or anxiety disorders) and activity scheduling (for depression). These techniques help individuals gradually face and overcome their fears or engage in activities that boost their mood. By identifying and understanding their thought patterns, individuals become more aware of how their thoughts influence their emotions and behaviors. This heightened self-awareness is a key aspect of consciousness.

Many CBT approaches incorporate elements of mindfulness, teaching individuals to be present and fully engaged in the current moment. Mindfulness enhances consciousness by promoting a non-judgmental awareness of thoughts and feelings. Also, CBT helps individuals better understand and manage their emotions, leading to greater emotional intelligence and awareness. So, by changing unhelpful behaviors, individuals can better understand the cause-and-effect relationship between their actions and their mental state, leading to a more conscious approach to their behaviors, decisions, and problem-solving.

While the approach can be considered natural from the perspective that it does not involve taking prescription drugs or working with entheogens, CBT is not easy to implement, especially for anyone affected by any of the SADAT conditions. It consists in learning coping skills and strategies to manage stressful situations, regulate emotions, and improve communication. Also, it requires completing homework assignments to practice the skills learned in therapy sessions. Finally, finding a competent practitioner and covering the treatment cost can be challenging. A treatment plan typically lasts several months and costs thousands of dollars, while the period after which the benefits of treatment may recede could be less than a year (177).

Some readers may wonder why I did not promote the benefits of other *conventional talk psychotherapies* in this book section, especially therapies based on psychodynamic, humanistic, dialectical behavioral, and mindfulness frameworks. All these therapeutic frameworks have the potential to raise consciousness by fostering greater self-awareness, emotional regulation, and insight into one's thoughts, behaviors, and relationships. They aim to provide individuals with tools to understand themselves better and navigate their inner worlds more effectively, promoting mental well-being and personal growth. However, even when people are open and willing to engage in formal therapy, they face considerable challenges, such as finding solutions, support, or tools to help them achieve significant shifts. Therefore, it may result in the individual telling and re-telling their same old story, further entrenching their dysfunctional habits, beliefs, and patterns

First, finding therapists in most countries is an ongoing struggle for millions who could benefit from this approach. Second, many cannot afford it. Lastly, it is often based on lengthy, frequent one-on-one sessions requiring significant active participation and, as with CBT, significant *cognitive effort*. Therefore, most patients suffering from SADAT are not likely to participate, considering how painful and mentally demanding sharing intimate struggles can be. For all those reasons, I believe that only a limited number of people suffering from SADAT may benefit from talk therapy in the long run unless it is done while engaging with other consciousness and neuroplastic modalities described in this book. However, I would encourage people who like the one-on-one format to address SADAT conditions to consider engaging with a practice like

Soul-Centered Coaching instead, where clients are guided to find active solutions using somatic, symbolic, or shamanic approaches that can genuinely help them transform (178). A soul-centered approach integrates spirituality, psychology, and personal growth to foster deep healing and transformation. Unlike traditional methods focusing solely on cognition or behavior, this perspective recognizes the soul as the essence of human experience, guiding individuals toward purpose, authenticity, and inner harmony. Influenced by depth psychology (Jung), trauma healing (Van der Kolk, Levine), and mindfulness (Kornfield and Poland), it emphasizes practices like meditation, somatic release, and narrative reframing to process past wounds. Reconnecting with one's higher self and intuition promotes self-awareness, emotional resilience, and flow, leading to profound, sustainable healing beyond the mind (179, 180).

EMDR

Eye Movement Desensitization and Reprocessing (*EMDR*) is primarily known as a psychotherapy approach for treating trauma and related conditions, but it can also be viewed as a natural tool for raising consciousness (181-183). EMDR helps individuals process and integrate traumatic memories, reducing their emotional charge. By resolving trauma, individuals can achieve a clearer and more peaceful mental state, which is conducive to higher levels of consciousness.

Figure 67: EMDR

Usually, the process begins with a thorough assessment, where the therapist gathers the client's history and identifies specific traumatic memories or distressing experiences to target. Then, the therapist and client identify specific memories to target, focusing on the associated images, beliefs, emotions, and physical sensations. These elements form the basis of the desensitization process.

A key component of EMDR is bilateral stimulation, which typically involves the client following the therapist's finger movements with their eyes. Other forms, like auditory tones or tactile pulses, can also be used. This bilateral stimulation is believed to facilitate the brain's information-processing system. During the bilateral stimulation, the client focuses on the traumatic memory while simultaneously experiencing the stimulation. This process helps desensitize the emotional response to the memory and enables the brain to reprocess the memory, integrating it to reduce its emotional impact.

The therapist guides the client to replace negative beliefs associated with the traumatic memory with positive, adaptive beliefs. For instance, believing "I am powerless" might replace "I am in control now." The client is asked to focus on their body and identify any residual tension or discomfort related to the memory. Any remaining distress is addressed using further bilateral stimulation until the client feels calm and neutral.

Each session ends with a closure phase, where the therapist ensures the client is stable and safe, providing relaxation techniques if needed. The client is also instructed to handle any distressing thoughts or feelings that may arise between sessions. At the beginning of subsequent sessions, the therapist and client reevaluate the targeted memories and any new insights or changes. This helps guide the therapy process and determine the next steps.

To summarize, EMDR aids in identifying and processing suppressed emotions. This increased somatic awareness can deepen an individual's understanding of their mind-body connection, fostering a more integrated sense of self. By resolving past traumas, individuals can become more present and mindful, qualities associated with higher states of consciousness. For some, deep emotional and psychological healing through EMDR can open doors to spiritual insights and experiences, contributing to a broader and deeper understanding of existence.

Entheogen-based Consciousness-raising Modalities

This section explores the neurobiological and therapeutic potential of entheogens, substances used for centuries in various spiritual and healing traditions. While recent research highlights their capacity to facilitate profound personal insights, emotional healing, and neuroplasticity, it is essential to approach this topic with caution and respect for the complexity of these substances and their effects. Also, I need to emphasize that this discussion is presented solely for educational purposes. I do not prescribe or recommend entheogens in any form. The decision to engage with such substances is highly personal and should be made in consultation with qualified professionals, adhering to legal and ethical guidelines in your region.

Entheogens can have profound effects on the mind and body, and their use carries risks, particularly for individuals with certain medical or psychological conditions. As with any therapeutic tool, the potential benefits must be carefully weighed against these risks. This section aims to inform and inspire thoughtful exploration of entheogens' scientific, therapeutic, and cultural dimensions, not to advocate for their indiscriminate use.

Documentaries and Films on Entheogens

In recent years, there has been a notable increase in films and documentaries exploring the benefits and drawbacks of psychedelics. While some of these programs are invaluable in dispelling widespread misconceptions often perpetuated on social media, they often fall short of capturing the lived, felt experience of those who engage with these substances. This is particularly true when case studies and narratives are not framed through a neurospiritual lens, limiting their depth of insight. Nonetheless, the following list includes valuable educational content worth considering for a broader understanding of psychedelics from historical, cultural, biological, spiritual, and therapeutic perspectives.

Hamilton's Pharmacopeia
Hosted by Hamilton Morris, this series investigates the history, chemistry, and cultural impact of various psychoactive substances, offering an in-depth look at their use and significance (184).

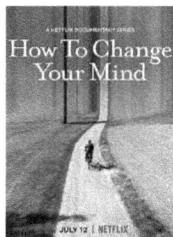

How to Change Your Mind
Based on Michael Pollan's book, this series explores the history and uses of psychedelics, including LSD, psilocybin, MDMA, and mescaline, highlighting their potential therapeutic benefits (185).

Fantastic Fungi
This visually stunning documentary delves into the world of fungi, particularly psilocybin mushrooms, and examines their role in nature, medicine, and human consciousness (186).

Psychedelia

This engaging film chronicles scientific research studies showcasing promising medical breakthroughs. It features first-person accounts from patients with end-of-life anxiety (187).

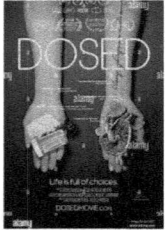

Dosed

This documentary follows a woman's journey as she turns to psychedelics to overcome her depression and addiction, shedding light on the therapeutic potential of these substances (188).

Ayahuasca: Vine of the Soul

This documentary follows individuals seeking spiritual awakening and healing through ayahuasca ceremonies in the Amazon, offering an intimate look at their transformative experiences (189).

The Last Shaman

The story is about the personal journey of a young man battling depression who travels to the Amazon to explore traditional healing methods, including ayahuasca ceremonies, in search of a cure (190).

DMT: The Spirit Molecule

While focusing on DMT, the active ingredient in ayahuasca, this film explores its effects on human consciousness through interviews with experts and participants (191).

Embrace of the Serpent

A dramatized account inspired by the real-life experiences of Amazonian tribes and their encounters with Western explorers, touching upon traditional plant medicines like ayahuasca (192).

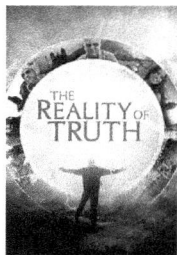

The Reality of Truth

This film investigates the relationship between spirituality, religion, and plant medicine, including ayahuasca. It features interviews with various thought leaders (193).

Trip of Compassion

This a documentary showcasing the transformative power of MDMA-assisted psychotherapy, offering a glimpse into its profound impact on treating trauma and PTSD (194).

While this list is exhaustive, these documentaries offer valuable perspectives on many of the substances discussed next.

Neurobiological and Therapeutic Effects of Entheogens

This section successively presents plant-based (PB), animal-based (AB), and synthesized-based (SB) entheogens. The word comes from Greek, meaning "generating the divine within," these substances are traditionally used in various cultural rituals to facilitate transcendental experiences, spiritual awakening, or deeper self-awareness. Entheogens are psychedelic substances used in religious, spiritual, or shamanic contexts to induce altered states of consciousness for spiritual, psychotherapeutic, or divinatory purposes. Each activates the mind through complex neurobiological and spiritual mechanisms under intense study.

Meanwhile, the word "psychedelic" is derived from the Greek words "psyche," meaning "mind" or "soul," and "delos," meaning "manifest" or "reveal." It was coined in 1956 by British psychiatrist Humphry Osmond during his correspondence with author Aldous Huxley. Osmond was seeking a term to describe substances like LSD and mescaline that can expand consciousness and reveal the inner workings of the mind. The term "psychedelic" thus literally means "mind-revealing" or "soul-manifesting," reflecting the profound alterations in perception and consciousness these substances can induce.

Figure 68: Entheogens (AI)

The use of entheogens is a growing anthropological and medical interest, particularly for their potential in treating SADAT conditions. However, their legal status varies by country and region. Again, the purpose of this discussion is not to recommend a specific substance but rather to present evidence supporting the value and purpose of psychedelic modalities that address SADAT conditions specifically.

When presenting the value of entheogens for raising consciousness or plasticity, it is essential to approach the topic with balance, acknowledging both the historical/cultural significance and the current scientific understanding, along with the legal and safety considerations.

Ayahuasca (PB)

Figure 69: Ayahuasca Vine (AI)

As discussed extensively in Chapter ONE, *Ayahuasca*, a psychoactive brew traditionally used in Indigenous rituals in the Amazon basin, is known for its potential to alter and raise consciousness significantly. Ayahuasca can enhance emotional processing by affecting brain regions involved in emotional regulation, such as the amygdala and insula, which could contribute to its potential therapeutic effects for conditions like depression and PTSD. The brew is typically made from the *Banisteriopsis caapi* vine and the *Psychotria viridis* shrub (Chacruna), the latter of which contains N-Dimethyltryptamine (DMT), a potent psychedelic compound. DMT acts primarily on the serotonin (5-HT) system, particularly the *5-HT2A receptor*, which regulates mood, cognition, and perception. The

240

Monoamine Oxidase Inhibitors (MAOIs) in Banisteriopsis caapi allow DMT to be active when ingested orally, which would not be the case otherwise.

Therapeutic benefits

Research has widely confirmed the therapeutic value of Ayahuasca in treating depression, anxiety, and PTSD (195-197). The altered state of consciousness induced by Ayahuasca can provide a unique perspective that might be beneficial in psychotherapy. In particular, Ayahuasca can induce a state of *ego dissolution or ego suspension* – a temporary loss of the sense of self. This experience can lead to an experience of oneness with the world and a breakdown of perceived barriers between the self and others, which can be highly transformative. Some studies also suggest that Ayahuasca can increase emotional empathy, enhancing the ability to understand and share the feelings of others (198).

Effect on consciousness

The primary psychoactive component, DMT, induces profound changes in consciousness, often including altered perceptions of time and space, vivid visualizations, and intense emotional experiences. These altered states can lead to a feeling of transcending reality. It can bring subconscious thoughts and repressed memories to the forefront, allowing for therapeutic emotional release and insight. My account of Aya sessions I completed in Peru provides evidence of how the brew changed my consciousness and helped me heal from severe anxiety and bouts of depression. I have also witnessed similar effects on hundreds of other people. Those who experience Ayahuasca frequently systematically describe feelings of interconnectedness, encounters with spiritual entities, or profound insights into the nature of the self and the universe. Thus, the reflective nature of the experience can enhance self-awareness and help users reevaluate their psychological struggles, life challenges, and more profound existential questions.

Potential Risks

Ayahuasca carries psychological and physiological risks, including intense emotional distress, reactivation of past traumas, and potential psychotic episodes, especially for individuals with underlying mental health conditions. Physiologically, it can cause nausea, excessive and violent vomiting, increased heart rate, and dangerous interactions with medications (such as SSRIs).

Unregulated ceremonies also pose risks due to improper dosing, lack of medical oversight, and unethical practices by facilitators. Those considering Ayahuasca should thoroughly research, ensure medical and psychological preparedness, and seek experienced, ethical practitioners to minimize harm while engaging in this powerful psychedelic experience.

Legal Status

Ayahuasca is illegal in many countries and is classified as a controlled substance due to its DMT content. It is legal in Peru, Brazil, Columbia, and Ecuador. Many countries, like the USA, Canada, and the Netherlands, tolerate the use of the brew for religious ceremonies, but DMT is typically a controlled substance in all of them.

Psilocybin (PB-Magic Mushrooms)

Figure 70: Magic Mushrooms (AI)

Psilocybin, a naturally occurring psychedelic compound, is primarily found in a variety of mushrooms commonly known as "magic mushrooms." This compound has gained significant attention in therapeutic mental health treatment and the exploration of human consciousness in the last few years (199). Psilocybin and its active metabolite, psilocin, primarily work by activating serotonin 5-HT2A receptors (the same receptors activated by Ayahuasca). These receptors are widely distributed in the brain, especially in regions that influence perception, cognition, and mood. Psilocybin affects areas of the brain involved in emotional processing, such as the amygdala. This can lead to an enhanced emotional response to stimuli and may contribute to the therapeutic potential of psilocybin for mood disorders.

Therapeutic benefits

Psilocybin has emerged as a promising tool in modern psychiatry. Recent research has begun to unravel its potential benefits, particularly for conditions often resistant to traditional forms of treatment. Studies have shown that psilocybin can produce rapid and sustained antidepressant effects, offering hope for individuals suffering from major depressive disorders, including treatment-resistant depression. Additionally, research indicates its effectiveness in alleviating end-of-life anxiety for terminally ill patients, as well as potential benefits in treating conditions such as Post-Traumatic Stress Disorder (PTSD) and substance addiction. The therapeutic action of psilocybin is linked to its ability to disrupt negative thought patterns, promote neural connectivity (plasticity!), and provide patients with new, insightful perspectives on their lives and conditions (200-202).

Effect on consciousness

Psilocybin can profoundly alter perceptions, emotions, and cognitive processes. Users often report experiences that include enhanced introspection, altered sense of time and space, and mystical or spiritual encounters. These experiences can lead to significant personal insights and reevaluating one's understanding of oneself and the world. As for Ayahuasca, the concept of ego dissolution or *ego suspension,* as I prefer to say, is a notable aspect of the psilocybin experience. This phenomenon can lead to greater feelings of connectedness with others, nature, and the universe. By lowering the activity of the brain's default mode network, psilocybin offers a unique window into exploring the neurobiological underpinnings of the human experience of self and reality.

Potential risks

While psilocybin is known to have a low risk of physical harm, it can pose significant psychological risks, particularly for individuals with a personal or family history of psychotic disorders like schizophrenia. When not used in a therapeutic session supported by an expert clinician or guide, Psilocybin, as with other substances that alter one's perception, can cause distressing psychological experiences, commonly known as "bad trips," which can include intense anxiety, fear, and paranoia. However, when ingested in an appropriate setting with a therapist present, the odds of a "bad trip" are reduced considerably because the therapist can immediately help the individual reframe their experience to something positive and valuable.

Legal status

Under federal law, psilocybin is illegal to possess, sell, or cultivate. This applies to the mushrooms containing the compound and the isolated substance itself. In recent years, however, the Food and Drug Administration (FDA) has granted "Breakthrough Therapy" designation for psilocybin therapy for treatment-resistant depression, which could lead to changes in legal status if clinical trials prove successful.

Under state law, some cities and states have begun to change their stance. For example, Denver, Colorado; Oakland and Santa Cruz, California; Ann Arbor, Michigan; and Washington, D.C., have decriminalized the possession and use of psilocybin mushrooms to varying degrees. However, decriminalization does not equate to legalization; it typically means that law enforcement treats it as a low priority. It is legal to consume psilocybin in Jamaica, Brazil, the Netherlands, and Australia.

Peyote (PB-Mescaline)

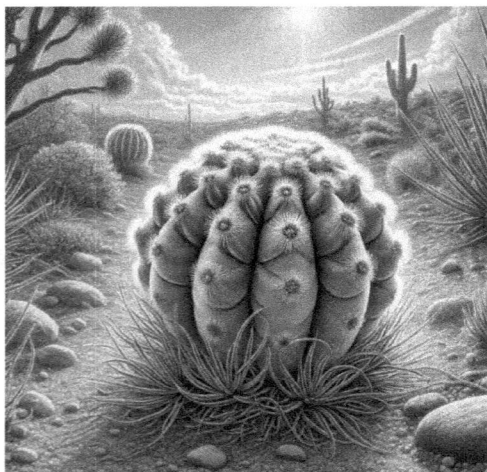

Figure 71: Peyote (AI)

Peyote is a small, spineless cactus native to Southern Texas and Mexico's deserts. It contains the psychoactive compound mescaline, which makes it the center of various spiritual, therapeutic, and consciousness exploration practices. Indigenous peoples have used peyote for centuries in spiritual and religious rituals. The Native American Church, which integrates Indigenous beliefs with Christian elements, uses peyote in its ceremonies as a sacrament.

Mescaline is found in peyote (Lophophora williamsii) and other cacti like San Pedro (*Echinopsis pachanoi*). It has distinct neurobiological effects on the brain, primarily through its interaction with serotonin receptors. Mescaline is an agonist at serotonin (5-HT) receptors, especially the 5-HT2A receptor. An agonist in pharmacology is a substance that binds to a specific receptor in the body and activates it, mimicking the action of a naturally occurring substance. Agonists can be drugs, hormones, or neurotransmitters that produce a biological response by interacting with receptor sites. As discussed for both Ayahuasca and psilocybin, the 5-HT2A receptor is crucial in regulating mood, cognition, and perception, and its activation is a key driver of many psychedelic experiences. Activation of the 5-HT2A receptors leads to altered sensory perceptions, visual hallucinations, and changes in cognitive processing.

Therapeutic benefits

While formal research on peyote as a therapeutic agent is limited compared to other psychedelics like LSD or psilocybin, studies suggest potential benefits in treating a range of psychological issues, including substance abuse and mental health disorders. Peyote users often report profound psychological insights and emotional catharsis, which can be therapeutic, especially in guided and controlled settings (203).

Effect on consciousness

The use of peyote is associated with altered states of consciousness and sharp changes in perception, thought, and emotion. Users often report intensified colors, geometric patterns, and altered time perception. Mescaline can induce a range of emotional responses, from euphoria and a sense of well-being to introspection and profound spiritual experiences.

These effects can increase empathy and a feeling of connectedness. The disruption of the default mode network (DMN) is closely related to the experience of ego dissolution often reported by users of psychedelics. This refers to losing the sense of self or the boundaries between the self and the external world. It can lead to feelings of unity and interconnectedness with the universe, often described as a mystical or profound spiritual experience.

Potential risks

While many experiences with peyote are positive, as with some other psychedelic substances, it can also cause psychological distress, including anxiety, fear, and paranoia, particularly at high doses or in uncomfortable settings. Physical effects may include nausea, increased heart rate, and changes in blood pressure. The intensity of the experience can be overwhelming for some.

Legal Status

Peyote is a Schedule I controlled substance in the United States, though there are exemptions for its religious use by members of the Native American

Church. Its legal status varies in other countries.

Wachuma (PB-San Pedro cactus)

Figure 72: Wachuma (AI)

Wachuma, also known as **San Pedro**, is a psychoactive cactus native to the Andes mountains in South America. Its use dates back thousands of years, primarily for spiritual and healing purposes in Andean traditional medicine. Like peyote, this cactus also contains mescaline, a potent psychedelic compound found in the peyote cactus. Indigenous cultures traditionally use Wachuma for healing rituals and spiritual ceremonies. It is revered for its perceived ability to heal the body and mind and connect individuals to the spiritual world.

San Petro is traditionally used in Peru as a complement to Ayahuasca ceremonies. It is often employed during the integration phase to help individuals ground and process the intense emotional and spiritual revelations experienced during Ayahuasca sessions. Unlike Ayahuasca, known for its intense and sometimes challenging visions, San Pedro offers a gentler, heart-centered experience. It helps individuals reflect on their Ayahuasca journeys with clarity and compassion, fostering a sense of connection to nature, self, and others. This grounding quality makes it ideal for integrating lessons and aligning them with everyday life.

Over the past 10 years, I have worked with San Pedro multiple times and deeply value its subtle yet profound properties. Its ability to provide a nurturing space for introspection has consistently supported my personal growth and healing, complementing the transformational work initiated through Ayahuasca.

Again, like peyote, since mescaline is the primary psychoactive component in San Pedro, it acts as an agonist at serotonin (5-HT) receptors, particularly the 5-HT2A subtype. As mentioned for Peyote, activating 5-HT2A receptors by mescaline leads to altered sensory perceptions, including visual hallucinations, enhanced colors, and geometric patterns. Users may experience changes in the perception of time and space and altered cognitive processing.

Therapeutic Benefits

In contemporary therapeutic settings, Wachuma can help treat a range of psychological issues, including depression, anxiety, and substance abuse disorders. Users often report profound insights and emotional catharsis, which can contribute to psychological healing. While scientific research on Wachuma is less extensive than other psychedelics like LSD or psilocybin, there is growing interest in its potential integration into modern psychotherapeutic practices, particularly for its empathogenic and introspective qualities (204). "Empathogenic" refers to a class of psychoactive substances that produce experiences of emotional communion, oneness, relatedness, and emotional openness. The term is derived from "empathy" and "genic," which means creating or generating. Empathogens are known for enhancing feelings of empathy and emotional connection with others.

Effect on consciousness

Mescaline, the active compound in Wachuma, induces altered states of consciousness characterized by visual hallucinations, altered perception of time and space, and profound introspective and mystical experiences. Users often describe experiences of deep spiritual significance while under the influence of Wachuma, including feelings of interconnectedness with nature and the universe, ego dissolution, and transcendence. The introspective nature of the Wachuma experience can foster enhanced self-awareness and facilitate a deeper understanding of personal and existential issues.

Potential risks

While generally considered physically safe, Wachuma can cause nausea and intense psychological experiences, which can be challenging or distressing for some individuals. As with other psychedelics, the set (mindset) and the setting (environment) are crucial for a safe and beneficial Wachuma experience. Proper guidance, particularly by experienced practitioners, is essential due to the intensity of the psychedelic experience.

Legal Status

In most countries, including the United States, mescaline is classified as a Schedule I controlled substance, making Wachuma illegal to use, possess, or cultivate.

Iboga (PB-Ibogaine)

Iboga, a plant found in West Africa, particularly in Gabon, Cameroon, and Congo, has been gaining attention for its unique psychoactive properties. The root bark of the iboga tree contains ibogaine, a potent compound used traditionally in spiritual ceremonies by the Bwiti religion in West Africa. Ibogaine affects several neurotransmitter systems, including serotonin, dopamine, and *glutamate* pathways.

Figure 73: Iboga (AI)

Ibogaine acts as a serotonin reuptake inhibitor that blocks the reabsorption (reuptake) of neurotransmitters into the presynaptic neuron, allowing these chemicals to remain in the synaptic cleft (space between neurons) and continue to exert their effects on postsynaptic neurons for a more extended period. This mechanism enhances the action of neurotransmitters in the brain, which can have various therapeutic effects like altering mood and perception and significantly modulating dopamine pathways, which are crucial in reward and addiction processes.

Ibogaine is an antagonist at NMDA (N-methyl-D-aspartate) receptors involved in learning and memory. This antagonism can disrupt addictive patterns and has been proposed as a mechanism for reducing withdrawal symptoms and cravings in substance use disorders. Ibogaine also acts on the Ona-opioid receptor system. This action is thought to contribute to its anti-addictive properties and the intense, dream-like states often reported during ibogaine experiences.

Therapeutic Benefits

The potential of ibogaine in addiction treatment, particularly for opioid addiction, is linked to its hypothesized ability to 'reset' specific neural pathways altered by addiction (205). This effect might involve mechanisms of neuroplasticity, allowing the brain to reorganize and form new connections not dominated by addictive behaviors. Users often report intense introspective

248

experiences that will enable them to confront past traumas and harmful patterns of behavior, leading to psychological healing and behavior change.

Effect on Consciousness

Iboga induces a "waking dream state" characterized by visions, introspection, and confronting personal issues. This state can last for several hours or even days, providing a profound journey into the psyche. In traditional Bwiti practices, iboga is used for spiritual growth, healing, and rites of passage. It is a way to connect with ancestors and gain deep spiritual insights. Similar to other psychedelics, iboga can lead to ego dissolution, providing a unique perspective on self and life, often leading to transformative experiences.

Potential Risks

Ibogaine therapy can be risky, particularly for individuals with heart issues, as it can affect heart rhythm. Treatment must be conducted under medical supervision.

Legal Status

Ibogaine is classified as a Schedule I controlled substance in the United States and is illegal in several other countries. However, some countries, like Mexico and Canada, allow it.

Salvia (PB)

Salvia divinorum, often referred to as Salvia, is a psychoactive plant native to Mexico, known for its unique and potent effects on the brain and consciousness. The primary active ingredient in Salvia is salvinorin A, a compound that differs chemically from other well-known psychedelics and has distinct neurobiological effects. Salvinorin A is a potent kappa-opioid receptor (KOR) agonist. This action is unique among psychoactive

Figure 74: Salvia (AI)

249

substances and is responsible for Salvia's distinctive effects. KOR activation is linked to the modulation of perception, consciousness, and mood. Unlike many classic psychedelics like LSD or psilocybin, salvinorin A does not act on the serotonin system. This lack of action on serotonin receptors contributes to its unique psychoactive profile.

Therapeutic Benefits

Research into the therapeutic benefits of Salvia and salvinorin A is still in the early stages (206, 207). There is interest in its potential for treating addiction, particularly given its effects on the kappa-opioid system, which plays a role in substance abuse. However, clinical applications are still largely theoretical at this stage.

Effects on Consciousness

The psychoactive effects of Salvia are rapid in onset but short-lived, typically lasting only a few minutes when inhaled. Salvia induces a profound alteration of consciousness, often described as a dissociative experience. Users may report out-of-body experiences, traveling to other dimensions, feelings of merging with objects, or profound shifts in perception.

Potential Risks

While Salvia is generally considered to have a low toxicity profile, its potent and disorienting psychoactive effects can be psychologically challenging and may pose risks, particularly in unsupervised settings. Also, KOR activation can induce dysphoric reactions, including feelings of unease or discomfort, which contrasts with the euphoric effects of serotonin-acting psychedelics.

Legal Status

Salvia's legal status varies significantly by region. In some countries and U.S. states, it is legal, while in others, it is a controlled substance.

Cannabis (PB)

Cannabis can be considered an entheogen by some definitions because it has a long history of use in spiritual and religious practices in many cultures worldwide. Its psychoactive properties can alter perception, mood, and consciousness, which, for some users, can lead to experiences that are interpreted as spiritual or enlightening. Because of these effects and its use in spiritual practices, cannabis fits into the broader category of substances considered entheogens.

Figure 75: Cannabis (AI)

Cannabis affects the endocannabinoid system, a complex network in the body that helps maintain homeostasis by regulating mood, appetite, pain sensation, and memory. THC, a compound in cannabis, binds to cannabinoid receptors in the brain and alters these functions, potentially leading to euphoria, altered senses, and decreased anxiety. CBD, another cannabis compound, may counteract the cognitive impairment and memory problems induced by THC.

This means that using a cannabis product with both THC and CBD might result in a more balanced experience, potentially reducing the intensity of the high and making the overall effect more tolerable for some users. CBD is also being studied for therapeutic benefits like reducing inflammation and anxiety without intoxicating effects. This interaction with the endocannabinoid system underlies both the psychoactive and potential therapeutic effects of cannabis.

Therapeutic Benefits

Therapeutically, cannabis has been studied for its potential to alleviate symptoms of stress, anxiety, depression, addiction, and post-traumatic stress disorder (PTSD). It may help reduce anxiety levels, improve sleep, and manage pain, contributing to its therapeutic use. The debate about such benefits continues to rage among medical professionals (208).

Effects on Consciousness

Cannabis may expand consciousness by altering perception, thought processes, and emotional states, potentially facilitating healing by enhancing introspection and emotional release. This expanded state can offer new perspectives on personal issues, fostering a sense of peace and connectedness. In therapeutic settings, guided use can help individuals confront traumas and resolve deep-seated emotional conflicts, contributing to mental and emotional healing.

Potential Risks

Cannabis use also carries risks, including the potential for addiction, cognitive impairment with long-term use, and worsening of symptoms in some individuals.

Legal Status

The legal status of cannabis varies widely, with some regions allowing medical and recreational use and others imposing strict prohibitions. Legal limitations often dictate the conditions for which it can be prescribed, the forms in which it can be used, and the amount that can be possessed or cultivated.

Bufo and 5-MeO-DMT (AB & S)

Figure 76: Bufo Alvarius (AI)

Bufo alvarius, also known as the Colorado River toad or the Sonoran Desert toad, secretes a venom that contains 5-MeO-DMT, a potent psychedelic compound. 5-MeO-DMT can also be synthesized in a laboratory setting. The neurobiology of 5-MeO-DMT is complex and involves several key aspects of brain function. Like other classic psychedelics, 5-MeO-DMT primarily exerts its effects through the serotonin (5-HT) system. It acts as an agonist, particularly at the 5-HT1A and 5-HT2A receptors. The 5-HT2A receptor is crucial in regulating mood, cognition, and perception and is a key mediator of the psychedelic experience.

Therapeutic Benefits

While the therapeutic use of 5-MeO-DMT in Western medicine is relatively new and still under research, various Indigenous groups have used bufo alvarius venom for spiritual and healing purposes for hundreds of years (209). 5-MeO-DMT has gained attention for its intense psychoactive effects and potential therapeutic benefits. It is known for its rapid onset and intensely profound effects, often described as transformative or life-changing.

While the experience is short-lived, typically about 20-30 minutes, it can have long-lasting psychological impacts. Preliminary research and anecdotal reports suggest potential therapeutic benefits for various mental health conditions, including depression, anxiety, and PTSD. The profound experiences induced by 5-MeO-DMT can lead to significant personal insights, emotional release, and a reevaluation of life perspectives.

Effects on Consciousness

One of the most notable effects of 5-MeO-DMT is the experience of nearly "complete" ego dissolution, where users consistently report losing their sense of individual identity and merging with a greater consciousness of the universe. That is why many users report profound spiritual and mystical experiences, including feelings of unity, interconnectedness, and transcendence. These experiences can lead to significant shifts in worldview and self-awareness. As detailed in Chapter 1 of this book, my own encounter with bufo during my neurospiritual journey left an indelible mark on my path toward heightened awareness and healing. The intense experience was both humbling and enlightening, creating a gateway to profound self-awareness and emotional release.

Potential Risks

Given its potency, 5-MeO-DMT should be approached with extreme caution. The experience can be overwhelming and may not be suitable for individuals with certain psychological conditions or a history of psychosis.

Legal Status

5-MeO-DMT is a Schedule I controlled substance in the United States and is illegal in many other countries. However, its legal status can vary, especially when using bufo alvarius venom in traditional or religious contexts. The increasing popularity of bufo alvarius for psychedelic use raises concerns about the ethical treatment of these toads and their impact on their populations in the wild.

LSD (S-Lysergic Acid Diethylamide)

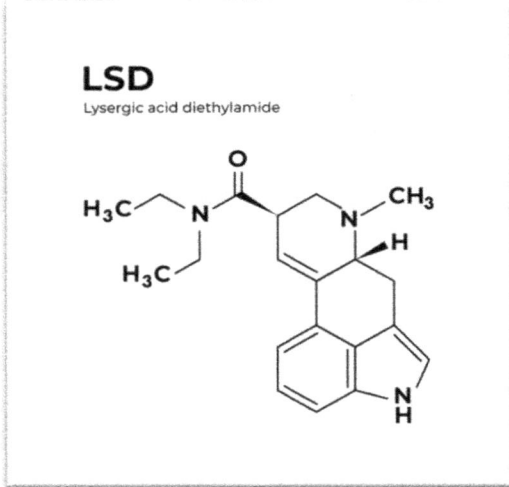

Figure 77: LSD Molecule (Istock)

Lysergic Acid Diethylamide (LSD) is a powerful psychedelic substance known for its profound impact on human perception, mood, and thought. Discovered in 1938 by Swiss chemist Albert Hofmann, LSD has since traversed a complex path, from psychiatric research to countercultural emblem, and now, a subject of renewed interest in the therapeutic and consciousness exploration realms. Its neurobiological impact is complex, involving various neurotransmitter systems, particularly the serotonin system, and it affects several brain areas. LSD primarily acts as a potent agonist at serotonin (5-HT) receptors, especially the 5-HT2A receptor. The activation of these receptors in various brain regions is central to the psychedelic effects of LSD.

Note that while LSD is presented here as a synthesized-based psychedelic, Morning Glory and Hawaiian Baby Woodrose seeds contain lysergic acid amide (LSA), a natural analog of LSD. While LSA and LSD share some similarities in their effects due to their chemical relation, there are notable differences in their psychoactive profiles, intensity, and overall experience. Both LSA and LSD can induce psychedelic experiences, including altered visual and auditory perceptions, enhanced emotional states, and changes in the perception of time and space.

However, LSD is significantly more potent than LSA. The psychedelic effects of LSD are typically more intense and profound. LSA experiences are often described as milder and less visually intense. LSD typically has a longer duration of action, with effects lasting up to 12 hours or more, whereas the effects of LSA tend to be shorter. LSA's visual effects are generally less pronounced and more subtle. LSD is often reported to induce a state of clear-headedness or lucidity during the peak of its impact, whereas LSA might lead to a more soothing or dreamy state.

Therapeutic Benefits

Initially, LSD was extensively studied for its potential therapeutic applications, particularly in psychiatry. In the 1950s and 1960s, it was investigated as a tool for psychotherapy, known as "psycholytic" and "psychedelic" therapy. Researchers explored its use in treating a variety of conditions, including depression, anxiety, and alcohol addiction. Its ability to induce profound psychological experiences aids therapeutic breakthroughs (210).

After decades of legal and social stigmatization, recent years have seen a resurgence in scientific interest in LSD. Contemporary studies are examining its potential in psychotherapy, especially for conditions like anxiety associated with life-threatening diseases, substance abuse disorders, and major depressive disorder. Many users report experiences that are often described as spiritual or mystical. These experiences can include a sense of interconnectedness with the universe, deep existential insights, and feelings of transcendence.

Effects on Consciousness

LSD has become a valuable tool in the scientific study of consciousness (211-213). By altering conscious experience so significantly, researchers can explore the relationship between brain activity and subjective experience, including the construction of reality and the nature of the self. LSD induces a unique pattern of brain connectivity. LSD affects brain regions involved in emotion and sensory processing, producing heightened emotional responses and vivid sensory experiences. LSD use can result in altered cognitive processing, including changes in thought patterns, time perception, and a heightened state of suggestibility. LSD is renowned for inducing dramatically altered states of consciousness that produce profound changes in sensory perception, emotional shifts, cognitive flexibility, and a sense of ego dissolution.

Anecdotal reports and some research suggest that LSD can enhance creativity, divergent thinking, and problem-solving abilities, making it of interest not only in therapeutic settings but also in fields that value cognitive flexibility and innovation.

Finally, LSD can decrease connectivity within the default mode network (DMN), disrupting self-referential thought processes and increasing global connectivity across different brain networks. This alteration is thought to contribute to the experiences of ego dissolution and altered consciousness.

Potential Risks

LSD is generally considered physically safe and non-addictive, but it can cause significant psychological effects. While many find the LSD experience enlightening, as with other potent psychedelics, there is a risk of negative experiences, particularly in unsupervised settings or among individuals with a predisposition to mental health disorders.

Legal Status

LSD is classified as a Schedule I substance under the Controlled Substances Act in the United States and is similarly controlled in many other countries, posing significant legal and accessibility barriers to its use and research.

Ketamine (S)

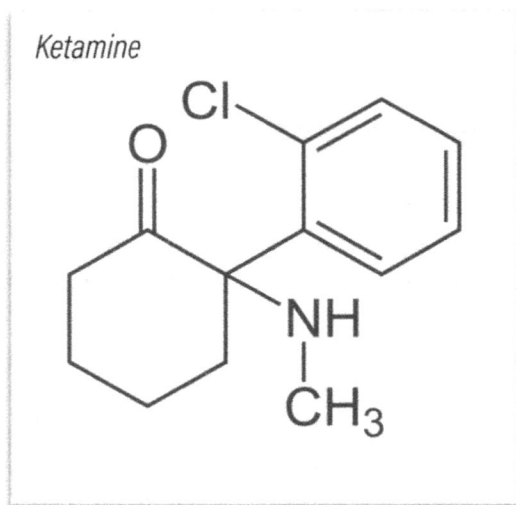

Figure 78: Ketamine Molecule (Istock)

Ketamine, initially developed as an anesthetic, has gained attention in recent years for its potential therapeutic effects, particularly in treating depression and other mental health disorders. Ketamine primarily acts as an antagonist at the N-methyl-D-aspartate (NMDA) receptors, which are involved in glutamate signaling. This action contributes to its anesthetic properties and its antidepressant effects. While its primary action is on NMDA receptors, ketamine also affects other neurotransmitter systems, including dopamine and serotonin pathways, contributing to its complex pharmacological profile.

Therapeutic Benefits

Ketamine has been found to produce rapid antidepressant effects in individuals with treatment-resistant depression, often within hours of administration (214). It can rapidly reduce suicidal ideation, which is a significant benefit in acute psychiatric emergencies. Ketamine is also used in managing certain chronic pain conditions due to its analgesic properties.

Ketamine has been found to have rapid antidepressant effects, particularly in treatment-resistant depression (215). Unlike traditional antidepressants that can take weeks to work, ketamine can produce relief within hours. It acts on the NMDA receptors in the brain, which is a different mechanism compared to conventional antidepressants. This action can result in rapid changes in mood and thought patterns.

There is ongoing research into its effectiveness in treating conditions like PTSD, anxiety, and chronic pain. Ketamine can be administered intravenously, intramuscularly, orally, or nasally. The nasal spray form, esketamine (*Spravato*), has been approved by the FDA for treatment-resistant depression in conjunction with an oral antidepressant. For depression, ketamine is often administered in a controlled clinical setting, under medical supervision.

Effects on Consciousness

At sub-anesthetic doses, ketamine can produce a *dissociative state* characterized by altered perceptions of time and space and vivid imagery (216, 217). This state can lead to introspection and significant shifts in perspective, which many users describe as spiritually meaningful. The altered state induced by ketamine can also enhance the effectiveness of concurrent psychotherapy, providing a different psychological vantage point for the user and the therapist. This is referred to as psychedelic-assisted therapy (PAT) among medical professionals.

Potential Risks

Ketamine has the potential for abuse and is a controlled substance. Recreational use can lead to harmful psychological and physical effects. Common side effects include dissociation, blood pressure and heart rate changes, dizziness, nausea, and perceptual disturbances, although these effects are typically transient.

Legal Status

In the therapeutic context, ketamine is used legally under prescription and medical supervision. Its use is regulated due to its potential for abuse.

MDMA (S)

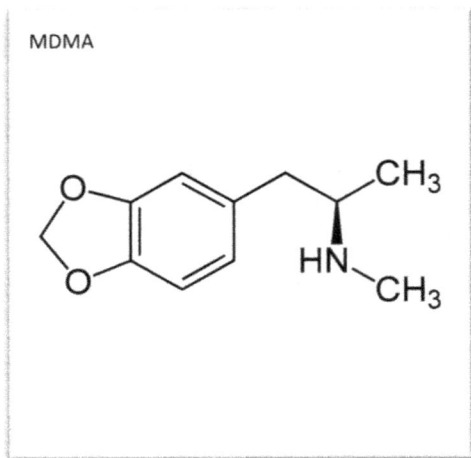

MDMA

CH_3

HN—CH_3

Figure 79: MDMA Molecule (IStock)

MDMA, short for 3,4-methylenedioxymethamphetamine, is a synthetic psychoactive drug known for its unique combination of stimulant and empathogenic properties. Commonly associated with the party scene and known as "Ecstasy" or "Molly" in its street forms, MDMA's potential in therapeutic settings and its impact on consciousness are areas of growing scientific interest. MDMA's most pronounced effect is on the serotonin system. It causes a massive release of serotonin, along with lesser releases of dopamine and norepinephrine. Serotonin affects mood, appetite, sleep, and other functions. Its release under the influence of MDMA leads to an elevated mood, emotional closeness, and empathy often reported by users.

Therapeutic Benefits

One of the most significant therapeutic applications of MDMA is in the treatment of Post-Traumatic Stress Disorder (PTSD). Clinical trials have shown that MDMA-assisted psychotherapy can be highly effective for individuals who have not responded to traditional treatments (218, 219). MDMA's ability to induce empathy, openness, and reduced fear responses is particularly beneficial in a psychotherapeutic setting. It facilitates emotional engagement and the processing of traumatic memories, allowing patients to address complex emotional content more effectively. MDMA's potent empathogenic effects — its ability to enhance feelings of empathy and connectedness — make it unique among psychoactive substances. This can lead to deeper introspection and an improved understanding of interpersonal dynamics.

Effects on Consciousness

While MDMA does not induce hallucinations like classical psychedelics, it significantly alters emotional and perceptual experiences. Users often report heightened sensations, emotional warmth, and a sense of well-being and connectedness. Unlike the ego dissolution commonly reported with substances like LSD or psilocybin, MDMA is often described as an "ego softener," where users maintain their sense of self but experience a reduction in defensive behaviors, fostering a sense of openness and emotional connection.

Potential Risks

Therapeutic use of MDMA is generally considered safe when administered in a controlled, clinical setting. However, recreational use can pose risks, including hyperthermia, dehydration, and serotonin syndrome. Also, following the initial surge, the brain experiences a significant depletion of serotonin, contributing to the comedown effects after MDMA use, such as depression, irritability, and fatigue. This depletion can affect the brain's serotonin levels for days to weeks after use.

Legal Status

MDMA is classified as a Schedule I controlled substance in the United States and is illegal in most countries. However, its status as a potential therapeutic agent is changing, with ongoing Phase 3 clinical trials for PTSD treatment.

- *Open Your Mind* teaches you how expanding your consciousness can break through limiting beliefs, enhancing self-awareness and mental flexibility for deeper personal growth.
- Barriers to Consciousness represent coping mechanisms that may restrict or filter information flowing to your consciousness.

Understanding Unconscious barriers

- Ego resistance drives low consciousness as a defense mechanism to protect our self-image.
- Repression serves as a protective mechanism to shield the individual from psychological pain.
- Projections are a defense mechanism in which individuals attribute their unwanted or unacceptable thoughts, feelings, or motives to another person.
- Spiritual bypassing is a tendency to use spiritual beliefs, practices, or experiences to avoid facing unresolved emotional issues and psychological wounds.
- Infatuation affects individuals who may not be fully aware of why they are drawn to someone or something, as it tends to be driven by intense emotions, desires, and idealizations. It often expresses unconscious needs, such as the desire for security or unresolved conflicts from earlier relationships

Understanding Conscious barriers

- Self-doubt is a lack of confidence in one's abilities, decisions, or thoughts. It can be insidious and affect individuals personally and professionally.
- Self-censorship is deliberately altering one's expression, behavior, or actions due to the fear of potential repercussion or judgment.
- Self-limiting beliefs are assumptions or perceptions about oneself and the world that restrain one's potential and opportunities.
- Denial is a defense mechanism in which a person refuses to acknowledge or accept reality or facts, thus protecting themselves from facing distressing aspects of life or their behavior.
- Self-deception is a psychological process where an individual convinces themselves of a false or distorted reality, often to feel better about themselves or their situation.
- Distractions compromise the ability to engage in deep, focused work.

- Cognitive biases can also significantly impact decision-making and consciousness by skewing our perceptions and interpretations of reality. Loss aversion and confirmation bias are two examples of mental states during which the understanding of a situation is skewed by either the fear of losing money or avoiding cognitive dissonance.

- Screaming, purging, and colonic cleansing are the most common ways to achieve cathartic clearance of stress and trauma.

Many natural consciousness-raising modalities can enhance self-awareness and present-moment focus, such as:

- Exercise has numerous benefits for physical and mental well-being and can also significantly raise consciousness.
- Holotropic Breathwork is a technique designed to reach altered states of consciousness like those achieved with psychedelic drugs through controlled breathing, evocative music, and bodywork.
- Yoga is a holistic practice that nurtures the body and calms the mind, enabling practitioners to engage more deeply with their inner selves and the world around them.
- Nature bathing is a mindful practice of being in nature, absorbing its atmosphere through all the senses, not just walking, hiking, or running trails.
- Meditation is the ability to be fully present, aware of where we are and what we are doing, and not overly reactive or overwhelmed by what is happening around us.
- Cognitive Behavioral Therapy (CBT) is a form of psychotherapy that focuses on identifying and changing negative and unhelpful thought patterns, beliefs, and behaviors.
- EMDR aids in identifying and processing suppressed emotions. This increased somatic awareness can deepen an individual's understanding of their mind-body connection
- Entheogens are plant-based, animal-based, or synthesized substances that alter and possibly raise consciousness. Used in religious, spiritual, or shamanic contexts, they induce altered states of consciousness for psychotherapeutic or divinatory purposes.
- These substances may help individuals access more profound levels of self-awareness, emotional healing, and connection to the spiritual realm. While growing research supports their therapeutic benefits for mental health conditions like PTSD, depression, and addiction, their legal status varies

widely. In some countries and U.S. states, entheogens are decriminalized or allowed for therapeutic use, but in many places, they remain illegal.

- Safe use is critical and typically requires a controlled, guided environment, such as psychedelic-assisted therapy, where trained facilitators help individuals navigate the experience, ensuring psychological and physical well-being. Without proper preparation and guidance, entheogens can lead to challenging or unsafe experiences, underscoring the importance of responsible usage and legal considerations.

The most prevalent entheogens are:

- Ayahuasca (PB) is a psychoactive brew traditionally used in Indigenous rituals in the Amazon basin known for its potential to alter and raise consciousness significantly.
- Psilocybin (PB), commonly known as "magic mushrooms," is a naturally occurring psychedelic compound that can disrupt negative thought patterns and promote neural connectivity.
- Peyote (PB) contains a psychoactive compound called mescaline, which is associated with altered states of consciousness and sharp changes in perception, thought, and emotion.
- Wachuma (PB), also known as San Pedro, is a psychoactive cactus that can provide profound insights and emotional catharsis.
- Iboga (PB), a plant found in West Africa, has a unique psychoactive ability to 'reset' specific neural pathways altered by addiction.
- Salvia (PB) is a psychoactive plant native to Mexico, known for its unique and potent effects on the brain and consciousness.
- Cannabis (PB) is a psychoactive plant that affects the endocannabinoid system, a complex network in the body that helps maintain homeostasis.
- Bufo alvarius (AB & S) contains 5-MeO-DMT, a potent psychedelic compound that can affect mood, cognition, and perception.
- LSD (S) is a powerful psychedelic substance known for its profound impact on human perception, mood, and thought.
- Ketamine (S), initially developed as an anesthetic, has psychoactive effects that are known to treat depression and other mental health disorders.
- MDMA (S) is a synthetic psychoactive drug known for its unique combination of stimulant and empathogenic properties.

As you open your mind and expand your awareness, you unlock the ability to see yourself and your experiences from a new perspective. This heightened consciousness lays the foundation for the next essential step—*OPEN*

Your Self. In the following chapter, we will explore how narratives you hold about your Self shape your life and how you can reframe these stories through awareness and intentional change to foster healing, growth, and a more profound sense of purpose.

CHAPTER FOUR

OPEN Your Self

"There is no greater agony than bearing an untold story inside you."

--**Maya Angelou**, Poet and Civil Rights Activist

Now that I have identified multiple modalities to raise consciousness and made the case that expanding your awareness brings therapeutic value to many SADAT conditions, I can move our discussion to the benefit of reclaiming your story. As we open ourselves to new ways of thinking and being, an essential part of the journey is turning inward to understand the stories we tell ourselves about who we are. These stories, shaped by our experiences, habits, and even our personality traits, are deeply embedded in the neural pathways of our brains.

To truly *OPEN your Self,* you must first examine these conscious and unconscious narratives and how they influence your thoughts, emotions, and behaviors. Note that writing "Self" with a capital "S" often signifies a distinction between the everyday, individual identity (the lowercase "self") and a more profound or more universal aspect of identity (the capitalized "Self"). This distinction is widely used in psychology, philosophy, spirituality, and literary theory to differentiate between two dimensions of human experience.

Narratives and their effect on the brain have been a significant focus of my professional life for over two decades. In this chapter, I will first discuss the effect stories have on our brains, especially the phenomenon known as *narrative transportation.* Then, I will unpack how your personal stories affect our well-being.

The root of "story" lies in the Latin word "historia," which means "a history, an account, or a narrative." This Latin term was also derived from the Ancient Greek word for "historía," which meant "inquiry" or "knowledge from inquiry," and also "judge." The meaning expanded to include factual and fictional narratives, written, spoken, or depicted. Today, "story" encompasses many narratives, from literary works to everyday anecdotes.

I define *a personal story* as a narrative about someone's life experiences. It is a recounting of events, emotions, thoughts, and insights that are unique to a person's journey. Personal stories are inherently subjective, reflecting the individual's perspective, feelings, and interpretations. They are based on real-life experiences from the storyteller's life and can include many experiences, from everyday occurrences to significant life events.

Personal stories often incorporate an emotional aspect, conveying what happened and how it affects a storyteller. Like most stories, personal stories typically have a narrative structure, including a beginning (setting the scene), a middle (describing the events or experiences), and an end (conclusion or reflection). However, personal stories often involve reflection on the part of the storyteller that can provide insights into the individual's values, beliefs, and learnings from their experiences.

The purpose of sharing a personal story can vary. It might be to entertain, educate, connect with others, inspire, or heal. Personal stories can build empathy and understanding while fostering connections between people. Therefore, personal stories play a significant role in shaping and expressing your identity. They help you make sense of your experiences and communicate who you are to others and yourself.

In this book, reclaiming your story means integrating new perspectives that may change interpretations, feelings, and beliefs embedded in your story. It is your ability to free your nervous system from the agony of worries, stress, regrets, shame, anger, and so much more.

The Effect of Stories on Your Brain

In *The Persuasion Code*, in collaboration with Patrick Renvoise, I present a model to help marketers create compelling stories in corporate videos, commercials, and customer testimonials. The book explores the science of persuasion from a neuromarketing perspective, which combines neuroscience with marketing and sales principles. It emphasizes the role of the dual-brain model (PRIMAL and RATIONAL) in processing all messages, including stories. The book demonstrates that successful narratives must start by addressing a customer's pain points, as the reptilian instinct of our PRIMAL brain is to avoid discomfort or threats at all costs.

Indeed, my research on stories demonstrates that successful narratives must activate the PRIMAL brain first, bypassing the RATIONAL brain initially while triggering emotions and tapping into fundamental human needs and desires. Additionally, captivating stories help us experience the world through the eyes of their creators, their characters, or our own, building empathy (or self-compassion) and emotional connections. When we see, hear, write, or share a story, our neurons fire in a way that mirrors (or amplifies) the characters' experiences, causing us to feel their joy, fear, or hope. This creates a bond between the storyteller and the audience (which could be yourself), making the message more impactful and persuasive.

Meanwhile, my research also confirms that compelling stories make information automatically memorable. Facts and figures are easily forgotten, but good stories embed information in a context that makes them more relatable and more accessible to recall from our memory banks. The structure of a story, with its beginning, middle, and end, provides a scaffolding for our brains to organize and retain information. Also, the PRIMAL brain is more likely to remember the beginning and end of a story. This is called the *primacy and recency effect*. Stories also allow us to visualize abstract concepts (remember that our visual system is the most dominant sense!), making them easier to understand and act upon. Finally, stories prime our brains for specific actions. They nudge us towards similar behavior (or reinforce existing behaviors) by exposing us to characters who behave in a certain way. This technique is often used in marketing and advertising to make viewers associate a product or service with positive outcomes.

I discovered that using contrast at critical points in a story (before/after, with/without, us/them) enables the PRIMAL brain to make quick, instinctual decisions or conclusions about a story. Story elements must be tangible and easily understandable to minimize cognitive effort to appeal to the PRIMAL brain since it cannot process complex, abstract information. Finally, visual elements are crucial in grabbing attention and accelerating understanding in the PRIMAL brain, which relies on the visual sense as the most dominant channel through which it can assess the importance and urgency of a decision. Using or describing compelling images can significantly enhance the stickiness of a story. So can emotional appeals rather than facts. Creating an emotional connection can be more influential than relying solely on rational arguments.

So, stories create emotional experiences and make complex concepts more relatable. When presented or processed with intention, they shape who we are and what we think, explaining most of our choices and values. Even today, studying media effects remains complicated and often confusing, riddled with assumptions and anecdotal evidence, not science. I continue to find that conventional methods investigating the impact of stories, especially ads, are underfunded and lacking.

For instance, surveys and focus groups collect self-reported information that is highly biased, inconsistent, and unreliable. This is why I decided to become a neuroscientist and promote research methods that could gather feedback on stories in real-time. I also became one of the founders of neuromarketing (as did my business partner Patrick Renvoise), authoring multiple international bestsellers on the topic and delivering thousands of lectures worldwide.

Meanwhile, one of the easiest ways to experience how stories impact your life is to take inventory of them at the end of each day (or week). Reflect on the extent to which they occupy your mind and how much you consider them when approaching a decision. To what extent do stories color your moods, even explain your temper? How is your personal story evolving? Does it limit you? Does it guide you?

As a result, you will likely realize that you process many stories without reflecting on their significance. Blame it on the noise of our media consumption or our lack of self-awareness. However, more importantly, you will notice how crucial stories are to your brain and how much energy is consumed by attending to them, processing them, and storing their meaning. Moreover, if you are meta-conscious, you will realize how some stories lock you in projections that do not serve you and keep you in the vicious loop of rumination and sometimes even anger. Finally, you will see how much stress or anxiety may be related to the energy you spend rehashing the same stories.

The Default Mode Network (DMN) In Self-referential Activity

A significant portion of our thinking about ourselves occurs when the brain is in a *default mode network (DMN)*, which is active during periods of rest and introspection (220). The DMN (see section on emotions, memory and cognition in Chapter Two) is associated with *self-referential thought* processes that manifest in various ways, including thinking about one's character, actions, or life experiences. They can be generated from a continuous stream of thoughts relating to oneself or one's actions or from evaluating one's abilities, actions, and behaviors against personal or societal standards. Recalling past experiences that are personally significant is also considered self-referential, and so is considering how future events or actions will affect oneself.

Studies suggest that the DMN is active for a considerable portion of our waking hours, indicating a substantial amount of time spent in self-referential thinking. Research using experience sampling methods, where individuals report their thoughts randomly, suggests that self-thoughts occur frequently throughout the day. Estimates suggest that people might think about themselves or their lives for a few minutes every hour, but this can vary greatly.

An *f*MRI study estimates that the average person may have between 6,200 per day (221). Though the researchers did not identify the specific content of subjects' thoughts, their method allowed them to count each. According to the study's author:

"What we call 'thought worms' are adjacent points in a simplified representation of activity patterns in the brain," as *"The brain occupies a different point in this 'state space' at every moment. When a person moves onto a new thought, they create a new thought worm that we can detect with our methods."*

Part of the challenge in estimating thought frequency lies in defining what constitutes a single thought. Thoughts can be fleeting or sustained, clear or vague, making them difficult to quantify. Additionally, people may not always be conscious of their thoughts, and self-reporting is unreliable. Meanwhile, individuals with certain personality traits (e.g., high neuroticism) or mental health conditions (like anxiety or depression) may experience a higher frequency of self-referential thoughts. Conversely, people who are more extroverted or engaged in highly demanding or external-focused tasks may have fewer self-referential thoughts.

Cultural background can also impact the frequency of self-referential thoughts. People from individualistic cultures, which emphasize personal achievement and individual identity, might have more self-focused thoughts than those from collectivist cultures like China. Self-referential thoughts can range from reflections on one's feelings, abilities, and goals to rumination about past events or planning for the future. The content and nature of these thoughts can vary widely based on the individual's current situation and life experiences. From a neuroscience standpoint, thoughts result from complex neural processes involving various brain regions. Measuring these processes accurately to count thoughts is currently beyond our scientific capability.

Why Do Stories Captivate Your Brain?

Stories often evoke emotional responses, which are crucial for memory and attention. Emotional content activates the amygdala, which also enhances the encoding and consolidation of memories in the hippocampus. Emotionally charged information is more likely to be remembered than neutral information. When you listen to a story, a phenomenon known as ***neural coupling*** occurs, where your brain's activity mirrors that of the storyteller; this synchronization allows for better understanding and retention of the story (222). It engages various brain regions, including those responsible for processing language, emotions, and social cues.

Stories often require us to understand the perspectives and motivations of characters, a task handled by our DMN. According to ***the theory of mind*** (TOM is a good acronym to remember it!) – the ability to attribute mental states to oneself and others, engaging in this mentalizing process makes stories more captivating and memorable since the human brain can recognize patterns and predict outcomes (57). Furthermore, stories often follow a recognizable

Figure 80: Theory of Mind (AI)

structure (such as a beginning, middle, and end) and include schemes of conflict and resolution. This structure engages our brain's desire for pattern recognition and makes us anticipate what will happen next, keeping our attention.

Meanwhile, stories require active cognitive engagement, such as visualizing scenes, understanding character development, and following complex narratives (223). This engagement stimulates areas of the brain associated with complex cognitive functions, such as the prefrontal cortex. Stories that are relatable or personally relevant are more likely to be remembered. Also, since novelty is a powerful attention grabber that activates the brain's reward system, it plays a crucial role in motivation and learning. Stories often provide a form of escapism and introduce novel experiences or perspectives that connect with our experiences and knowledge, recruiting the prefrontal cortex to think about the self.

Many researchers have explored how narratives affect brain function, emotional processing, and cognitive development. For instance, Uri Hasson is a neuroscientist known for his pioneering work in neurocinematics. His research demonstrated how the brains of a storyteller and a listener could synchronize during storytelling, revealing the profound impact of stories on neural processes. Hasson and his colleagues used functional magnetic resonance imaging (fMRI) to demonstrate that viewing a movie produces a remarkable synchronization of brain activity across viewers (224). Their study highlights how similar patterns of neural activity can emerge across individuals when exposed to the same dynamic stimuli, suggesting a neural mechanism for shared experiences and understanding during communication.

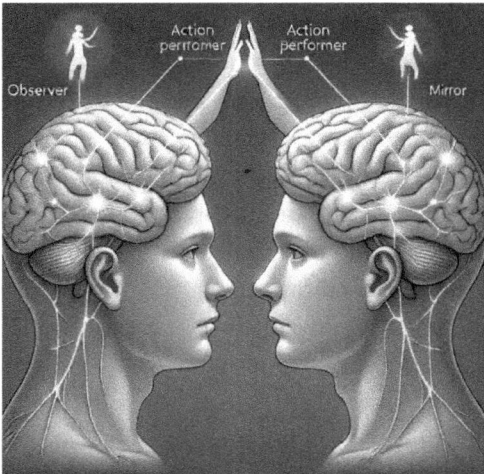

Figure 81: Mirror Neurons (AI)

Meanwhile, Keith Oatley, an Anglo-Canadian novelist and Professor Emeritus of Cognitive Psychology, has also extensively studied the effects of fiction on the human mind. His research indicates that engagement with fictional stories enhances our empathy and social cognition, improving our ability to understand others in the real world (225). Finally, Vittorio Gallese, a neuroscientist known for his work on *mirror neurons*, explored how our neurons play a role in understanding stories. His research suggests that when we engage with stories, our brains simulate (or mirror) the actions, emotions, and sensations narrated, contributing to our understanding and emotional involvement (226).

Like me, these researchers have employed various methods, including neuroimaging techniques like *f*MRI, physiological measures, psychological experiments, and literary analysis, to understand the profound effects of storytelling on the human brain. Our collective work highlights the role of narratives in shaping our emotions, social understanding, and cognitive functions. In essence, stories hold our attention and stay in our brains because they engage both the PRIMAL and RATIONAL brain systems, from emotional responses to cognitive processes. A symphony of neurobiological mechanisms makes stories powerful communication, learning, and entertainment tools.

Narrative Transportation

From psychological and neuroscientific perspectives, narrative transportation refers to the phenomenon where an individual becomes deeply engaged or absorbed in a story as if they were part of it. When we see or hear about an action in a story, the same neurons fire as if we were taking action, creating a deeper connection to the story. Imagine you are in a room, and someone begins describing the smell and taste of a freshly cut lemon. They describe how its sharp, tangy scent fills the air with hints of sweetness and a touch of bitterness. Then, they mention taking a slice, the juice beading on the surface, and squeezing it so the sour juice touches your tongue. You might notice your mouth watering even though no lemon touches your tongue. This reaction shows how mere suggestions can trigger bodily reactions like salivation as if you were smelling or tasting it.

This phenomenon stems from how the brain processes sensory information. Visualizing or imagining something vividly activates areas linked to those senses so that the brain may stimulate physical responses like salivation or a taste memory based on suggestion alone. This activity involves cognitive, emotional, and neural processes that lead a person to experience a story as if they were inhabiting the actors or characters of the narrative.

Richard Gerrig is a cognitive psychologist and professor at Stony Brook University, renowned for his contributions to the concept of narrative transportation in the early 1990s. His book introduced the concept of being 'transported' by a story, even though, at the time, he did not present evidence from brain studies proving many of his working assumptions (227). Instead, he explored how readers can become emotionally involved and lose themselves in a narrative. However, we now understand that listening to or watching a story can activate mirror neurons responsible for empathy and understanding others' actions. This phenomenon is a central paradigm that helps media psychologists like me decode the impact commercials and movies have on our brains.

Remarkably, narrative transportation causes a temporary suspension of disbelief, where the audience accepts the story world as authentic for the duration of the engagement. It often includes identifying with characters, feeling empathy for them, and experiencing emotions similar to those depicted in the story. During narrative transportation, individuals may lose awareness of their physical surroundings and the passage of time, feeling as if they are "transported" into the narrative world. This deep engagement can also influence a person's beliefs and attitudes, especially if the narrative is persuasive or relatable. For example, a person deeply engaged in a novel about overcoming adversity may feel more optimistic or inspired in their own life.

As discussed earlier, engaging with a story activates various brain networks associated with language processing, sensory perception, and emotional understanding. These include temporal and frontal lobes and the default mode network. Brain regions involved in emotional processing, like the amygdala, are highly involved, especially when the narrative elicits potent affective responses. Also, stories stimulate the brain's regions responsible for visual imagery and simulation, like the visual cortex and prefrontal areas, allowing the individual to vividly imagine the story's events.

Meanwhile, there are many situations during which narrative transportation may affect your experience of a story. While reading a gripping novel, you may feel so absorbed in the narrative that you are oblivious to your surroundings, experiencing the emotions, sights, and sounds described in the book as if you were living through this for real. Likewise, during a fascinating film, you might feel emotions in line with the characters, and for a while after the movie, you might continue to reflect on its themes or feel as though you were still part of the narrative.

A skillful storyteller can transport listeners to different times or places, making them feel like they are witnessing the events firsthand. Narrative transportation is a testament to the power of stories to deeply engage human cognition and emotion, illustrating the complex interplay between psychological engagement and neurobiological processes in our experience of narratives. Researchers in narrative transportation have contributed significantly to understanding this phenomenon.

For example, Melanie Green and Timothy Brock are particularly notable researchers for developing the *Transportation-Imagery Model*, a foundational text in narrative transportation research (228). They explore how being transported into a narrative world can change beliefs and attitudes, emphasizing the role of imagery and emotions in this process. Also, neuroeconomist Paul Zak has extensively researched the neurobiological effects of narrative transportation (229). His studies involve measuring changes in brain chemistry, like oxytocin levels, in response to storytelling, providing insight into the physiological basis of how stories can engage and influence us. Indeed, oxytocin, a neuropeptide and hormone often referred to as the "love hormone" or "bonding hormone" is released in various social interactions, including childbirth, breastfeeding, and interpersonal relationships. Research also suggests that oxytocin enhances narrative transportation—contributing to how absorbed or emotionally engaged we become with a good story and temporarily disconnected from our immediate surroundings. I was fortunate to collaborate with Paul on several studies assessing the neurobiological effect of public service announcements and the role of oxytocin in that process (230).

These researchers and others have used a range of methodologies, from psychological experiments and surveys to neuroimaging and physiological measurements, to investigate how narratives capture our attention, evoke emotions, and even change our beliefs and attitudes. Their work collectively underscores the power of storytelling in human cognition and social interaction.

The Therapeutic Use of Stories

Many forms of therapy focus on individuals' stories to achieve relief and healing. Stories help individuals reframe and reinterpret their experiences, viewing themselves as separate from their problems. By reshaping their narratives, people can find more empowering and constructive ways to perceive their challenges. Sharing personal stories, especially those involving trauma or difficult experiences, can provide a sense of catharsis. Expressing emotions tied to these stories can be a significant step in the healing process, helping to alleviate symptoms of stress, anxiety, and depression.

Stories allow individuals to make sense of chaotic or painful experiences. By constructing and reflecting on personal narratives, people can find meaning in their experiences, which is particularly helpful in addressing feelings of confusion, hopelessness, or despair often associated with depression and trauma. Therapeutic storytelling usually involves identifying with characters or scenarios that mirror an individual's experiences. This process can enhance self-compassion and empathy, aiding in the understanding and processing of one's own emotions and experiences.

For trauma and certain types of anxiety disorders, controlled exposure to traumatic or anxiety-provoking events through storytelling can even be therapeutic. It allows individuals to confront and process their fears safely, gradually reducing their emotional response. In *Remapping Your Mind*, Mehl-

Madrona and Mainguy delved into the power of storytelling to self-transform since it shapes our perceptions, beliefs, and behaviors (12). Drawing on principles from neuroscience, psychology, and Indigenous wisdom traditions, they examined how narrative techniques could influence brain function and promote self-healing. The authors explored multiple storytelling methods, including narrative therapy, guided imagery, and ritual practices, as tools for rewiring the brain and creating positive change.

Throughout their book, Dr. Lewis Mehl-Madrona and Barbara Mainguy offer practical exercises, case studies, and insights from their own experiences to illustrate how storytelling can be a therapeutic tool for personal growth and empowerment. They discuss the role of storytelling in healing trauma, overcoming limiting beliefs, and cultivating resilience. Overall, *Remapping Your Mind* provides readers with a holistic framework for harnessing the power of storytelling to reshape our lives, develop resilience, and unlock our full potential.

Stories, especially those that highlight resilience and overcoming adversity, undeniably provide potent models for coping strategies and foster a sense of hope and strength. Sharing personal stories can also enhance feelings of connection and support, particularly in group therapy settings. Realizing that others have had similar experiences can reduce feelings of isolation and stigma, which is especially beneficial in dealing with addiction and mental health issues. Also, certain types of storytelling, such as guided imagery or visualization, can promote relaxation and mindfulness. This can be particularly helpful in managing stress and anxiety.

So, through storytelling, individuals can reconstruct their self-identity, which, for people suffering from addiction and trauma especially, can help rebuild a stronger sense of self. In cognitive-behavioral therapy (CBT), stories and narratives are typically used to identify and challenge negative thought patterns. By examining, altering, or reframing the stories they tell themselves, individuals can develop healthier thinking patterns, reducing symptoms of anxiety and depression. Therapeutic storytelling is a multifaceted tool that can address various aspects of mental health issues. It allows individuals to express, process, understand, and reframe their experiences, fostering healing and resilience.

How Stories Explain the Success of Alcoholics Anonymous Success

Personal stories play a central role in the success of Alcoholics Anonymous (AA), a global organization dedicated to helping individuals recover from alcoholism through a well-established program that includes regular meetings and a supportive community (231). Personal stories allow members to share their experiences with alcoholism, including struggles and triumphs. This sharing fosters a sense of empathy and understanding among members as they realize they are not alone in their challenges.

Sharing personal stories in a confidential setting helps create an atmosphere of trust and safety. Members are more likely to open up and be honest about their struggles when they hear others doing the same, which is crucial for recovery. Hearing stories from members who have successfully managed their addiction also provides hope and inspiration to others through the narrative transportation effect. These stories act as powerful testimonials that recovery is possible, offering practical insights and strategies that new or struggling members can apply.

Figure 82: AA Meeting (AI)

Sharing personal stories encourages individuals to reflect on their own experiences with alcoholism. This reflection can lead to greater self-awareness and understanding of the personal triggers and consequences of their addiction. Telling one's dramatic story can be cathartic, helping individuals to process emotions and experiences related to their predicament. This emotional release is essential to healing and recovery since regularly sharing personal progress and setbacks reinforces an individual's commitment to sobriety. It serves as a reminder of where they have been and where they are heading, strengthening their resolve to stay sober.

Personal stories also help build a network of support. Members who share their stories often find that others offer support, advice, and encouragement based on their narratives' teachings. This social support is a critical factor in recovery and helps create a sense of accountability. Hearing others share similar struggles normalizes the experiences associated with recovery from alcoholism. It helps members understand that their difficulties are common and manageable. It improves the sense of agency and power that comes from a boost in self-efficacy. The therapeutic use of "second stories" is also worth noting. A second story is a protocol allowing someone to share their analysis and understanding of the first story to gain therapeutic value (232).

In summary, personal stories in Alcoholics Anonymous play a multifaceted role. They facilitate empathy, trust, learning, and support, offering both emotional and practical benefits to individuals recovering from alcoholism. The power of these stories lies in their ability to connect people, inspire change, and reinforce the principles of the AA program.

The Role of Stories in Psychedelic-Assisted Therapy (PAT)

In psychedelic-assisted therapy (PAT), personal stories can play a significant role in both the therapeutic process and the patient's journey toward healing and self-discovery. Before a psychedelic session, PAT therapists are usually trained to encourage patients to share their personal stories. I used the term "usually" because I have not seen this practice systematically implemented since the legalization of PAT, especially for Ketamine treatments. Since the process to legalize PAT has been slow and erratic, it has allowed many actors with questionable credentials or training to deliver sessions with limited psychological preparation and guidance (233).

Discussing one's life history, struggles, and the specific issues one hopes to address through PAT is critical for the entire program's success. This sharing helps set the context for the psychedelic experience and guides the therapeutic focus. Since psychedelic experiences can heighten introspection and emotional openness, patients may revisit and narrate significant memories or events in this state, gaining new perspectives or insights. Their personal stories, under the influence of psychedelics, can become more vivid and emotionally charged, facilitating deeper self-reflection.

For individuals with trauma-related disorders, recounting personal stories during therapy can be crucial, even if it is painful to do so. Indeed, psychedelics can alter the usual emotional responses to these memories, allowing patients to re-examine traumatic events with less fear or avoidance and more compassion towards themselves. After the psychedelic experience, integration sessions are usually held where patients discuss and make sense of their experiences.

Sharing personal stories also helps build trust and rapport between the therapist and the patient. A solid therapeutic alliance is crucial for the success of psychedelic-assisted therapy, as it provides a safe and supportive environment for patients to explore their psyche.

According to many studies, the insights and altered perspectives gained from revisiting personal stories under psychedelics can lead to transformative psychological changes (234). Patients often report shifts in their understanding of themselves and their relationships, leading to improved mental health outcomes. Sharing personal stories, especially in an emotional state induced by psychedelics, can be cathartic. As I insisted in Chapter THREE, this emotional release is often essential to healing. By revisiting their life stories in a new light, patients can develop greater empathy and compassion for themselves, understanding their life experiences and choices with more kindness and less judgment.

In summary, in psychedelic-assisted therapy (PAT), personal stories are not just narratives of past experiences; they become dynamic tools for exploration, healing, and transformation. The treatment provides a sacred space where these stories can be revisited, reinterpreted, and integrated, leading to profound personal growth and change.

The Role of Sharing Stories in Ayahuasca Group Sessions

Ayahuasca ceremonies and *sharing circles*, often rooted in Indigenous traditions and increasingly incorporated into Western therapeutic contexts, emphasize personal stories. Participants usually start by sharing their intentions

Figure 83: Aya Sharing Circle (AI)

or reasons for participating in an Ayahuasca ceremony. This can include personal struggles, life questions, or emotional burdens. Sharing these stories helps to set an individual and collective context for the experience.

During the Ayahuasca experience, individuals often encounter intense and profound emotional and psychological states. These can manifest vivid recollections, life reviews, or symbolic representations of personal stories. This process allows for deep introspection and confrontation of past traumas or unresolved issues. Sharing personal stories in a supportive, empathetic environment can also be cathartic.

Since I have attended many Ayahuasca sessions, I can confirm that participants often report a sense of emotional release and healing through their experiences and the subsequent sharing of their stories. Sharing personal stories in this context fosters a sense of connection and empathy among participants. It helps build a community where individuals feel seen, heard, and understood, an essential aspect of the healing process. Personal stories can also guide the shaman or facilitator in tailoring the ceremony. Understanding each participant's background and issues can help provide more personalized guidance and support during the ceremony.

Combining the Ayahuasca experience with sharing personal stories can lead to enhanced self-understanding and shifts in consciousness. Participants often gain new insights into their life experiences and relationships. In many cases, the insights and revelations from personal stories contribute to long-term spiritual and psychological growth (235). These narratives are pivotal in transformative and healing experiences as they can manifest changes in behavior, improved mental health, and a more profound sense of life's purpose.

I have reviewed multiple perspectives on how personal stories influence the trajectory of our lives. Personal stories are more than just a factual account of what has happened. They set the emotional direction of what may occur in the future. This is why the story concept is central to our ability to heal and rewire our brains. However, so is the notion that we have relatively stable patterns of thoughts, feelings, and behaviors during our lifetime. So, let us explore personality traits and their importance in accessing and rewiring your personal story.

The Make-up of Your Self

Personality traits significantly influence our personal stories and how we interpret and narrate them to ourselves. For example, an optimist might view challenges as opportunities for growth, while a pessimist might see them as insurmountable obstacles. These perceptual biases become integral to narratives about our lives (236).

Personality traits can influence what we notice and remember about an event. For instance, someone who is very anxious might be more likely to remember negative details. In contrast, someone highly open to experience may focus on novel and complex aspects of the same event. An individual high in emotional stability might recount a stressful event more calmly, while someone high in emotional sensitivity might describe the same event more intensely. Extroverted individuals might seek out social interactions and adventures, leading to stories filled with interpersonal dynamics and external activities. In contrast, introverted individuals might have more reflective stories, focusing on internal thoughts and solitary experiences.

Meanwhile, a highly conscientious individual might approach problems methodically, leading to narratives about overcoming challenges through discipline and planning. On the other hand, someone high in agreeableness might focus on resolving conflicts through cooperation and empathy. These biases frame the stories we tell about our experiences. Moreover, the emotional coloring can significantly alter their tone and content.

Personality traits can even influence the way we narrate our stories. For example, a person with high verbal intelligence and extraversion might tell their story with more engaging details. At the same time, someone more reserved might offer a brief, straightforward account. Additionally, how we interact with others and form relationships is influenced by our personality traits. An amiable person might have many stories about helping others, while someone high in narcissism might focus their narratives on their achievements and importance. Finally, our personality traits can lead us toward specific life paths and experiences. For example, risk-takers might have stories filled with adventure and spontaneity, whereas risk-averse individuals might tell stories about stability and caution.

In summary, personality traits are crucial in shaping our personal stories. They influence how we perceive and respond to events, remember and interpret experiences, interact with others, and, ultimately, narrate our life journey. These stories reflect who we are, influenced by the unique combination of traits that define our personalities.

In this next section, I will examine how our DNA and birth story shape our character. Then, I will review the two most scientifically acclaimed personality systems in the world: Big Five and the Enneagram.

Your DNA Story

Figure 84: Your DNA Story (AI)

Our DNA plays a significant role in shaping our personality, operating through the complex interplay of genetics and environment (237). *Heritability*, the proportion of variation in a trait that can be attributed to genetic differences, is estimated to be around 40-60% for personality traits.

This suggests that while genetics has a significant influence, it is not the sole determinant of personality. Many genes influence personality traits, each contributing a small effect, however. This makes the genetic basis of personality complex and multifaceted.

Research in *behavioral genetics* has revealed that specific genes are associated with various aspects of personality, such as temperament, extraversion, neuroticism, and openness to experience. These genetic predispositions can influence baseline tendencies in behavior, mood, and thought patterns. Also, twin and family studies have been instrumental in understanding the genetic influence on personality. They show that an individual's environment can interact with genetic predispositions to shape personality (238). For example, a genetic tendency towards high reactivity may lead to different outcomes depending on whether a person grows up in a supportive or stressful environment. This is considered an *epigenetic mechanism,* which we discussed briefly in Chapter TWO

Epigenetic mechanisms involve changes in gene expression without altering the DNA sequence, yet they can influence personality and trigger dramatic health and behavioral changes (239). Environmental factors and experiences can drive these long-lasting or even heritable changes. They interact with genetic predispositions in complex ways, often modulating the expression of genes without modifying the DNA sequence itself. For example, acute and chronic stress can lead to alterations in gene expression, particularly in genes associated with the stress response, immune function, and mental health. Exposure to environmental toxins and pollutants, such as heavy metals, pesticides, and air pollutants, can also disrupt gene expression. These substances may interfere with transcription factors, epigenetic modifications, and other regulatory mechanisms, leading to adverse health effects.

Social interactions and psychological well-being can also impact gene expression. For instance, social isolation and depression can change the expression of genes involved in immune response and neuroplasticity. Also, light, including natural and artificial light, can influence gene expression, particularly concerning circadian rhythms and mood. Genes that control the biological clock are responsive to light, affecting sleep patterns, hormone release, and other physiological processes. Additionally, genetics influences the functioning of various neurotransmitter systems, such as serotonin, dopamine, and norepinephrine. Variations in genes related to these neurotransmitters can affect mood regulation, impulsiveness, reward sensitivity, and other traits relevant to personality.

In summary, genetic factors contribute to developing brain structures and functions that underlie personality traits. For instance, the structure and connectivity of the prefrontal cortex, which is involved in decision-making and self-regulation, are partly genetically determined. While genetics play a significant role, it is essential to recognize that DNA alone does not predetermine personality. However, the expression of these genetic predispositions is modulated by environmental factors, life experiences, and personal choices, highlighting the intricate nature of personality development.

Stan Grof Pre-natal Stages: How Birth Stories Matter

Figure 85: Grof Pre-natal Stages (AI)

Meanwhile, SADAT conditions may be caused by many events that could date as far back as your own *birth story*. Stanislav Grof, a psychiatrist who focused his career on studying altered states of consciousness and a pioneer in transpersonal psychology, proposed a model explaining how prenatal development and birth stages influence an individual's psyche and subsequent life experiences (240). I have attended many workshops with Dr. Grof, whose teachings have significantly impacted my life.

According to Grof, there are four fundamental *perinatal matrices* (BPMs), each corresponding to a different stage of the birth process and each with its psychological implications. It is important to note that Grof's theories, particularly those related to prenatal experiences, are considered speculative by the broader scientific community. Nonetheless, they have been very influential among psychotherapists and transpersonal psychologists. Let us review each stage and the specific psychological trajectory of our lives they may influence.

- **BPM I - The Amniotic Universe**: This stage corresponds to the prenatal experience within the womb. Grof theorized that a positive experience during this stage (such as a healthy, low-stress pregnancy) could lead to a sense of essential trust and connection in an individual. In contrast, a negative experience (such as maternal stress or illness) could lead to feelings of alienation or existential anxiety.

- **BPM II - Cosmic Engulfment and No Exit:** This stage is associated with the onset of childbirth. The contractions begin, but there is no movement through the birth canal. Grof suggested that experiences in this stage might be linked to feelings of being trapped, oppressed, or experiencing high levels of stress without a clear resolution.

- **BPM III - The Death-Rebirth Struggle:** This stage corresponds to the actual process of birth, where the infant struggles through the birth canal. Grof connected this stage to later experiences of struggle, resistance, and the fight for survival. He theorized that navigating this stage could impact one's approach to challenges and stress in life.

- **BPM IV - The Death-Rebirth Experience**: This final stage represents birth itself, the emergence from the birth canal into the world. Grof associated this stage with the potential for transformative experiences, a sense of rebirth, and the overcoming of obstacles. He believed that a positive experience at this stage could lead to an ability to transform challenging situations and embrace change.

According to Grof, these stages of experiences and emotional environment deeply imprint on an individual's subconscious and influence their psychological development, behaviors, and patterns in adulthood. This could manifest in various ways. For instance, traits such as resilience, anxiety, trust, or fear might be influenced by one's perinatal experiences as conceptualized by these matrices. Early experiences could influence relationship patterns, life choices, and reactions to stress or change. In Grof's therapeutic practice, revisiting and understanding these perinatal experiences, often through methods like Holotropic breathwork (presented in Chapter THREE), is a path to resolving deep-seated psychological issues.

The Big Five Personality Model

The *Big Five* personality traits model, or the *Five-Factor Model*, is a widely acclaimed framework for understanding human personality. It posits that five broad dimensions define human personality. You can remember the traits by thinking about the acronym OCEAN, represented by the first letter of each trait:

- *Openness*: This trait describes people who are curious, adventurous, creative, and always seeking new experiences. They tend to be more aware of their feelings than others. However, they are also more likely to engage in risk-seeking behaviors. It contrasts with people who exhibit stubbornness, skepticism, and less risk-seeking behavior.

- *Conscientiousness*: This trait describes people who are outstanding at self-discipline, organization, dependability, and goal-directed behavior. It contrasts with people who are disorganized but tend to be open and creative.

- *Extraversion:* This trait describes people who seek extensive stimulation from the external world. They interact with others and are often perceived as having plenty of energy and enthusiasm. However, they can also appear dominant and controlling. They contrast with people who exhibit introversion and have a low appetite for controlling others.

- *Agreeableness*: This trait describes people who seek harmony and peace. They are considerate and kind and continuously display an optimistic view of human nature. This trait may also make people eager to avoid conflict and have trouble making decisions or disrupting the status quo. It contrasts with confrontational people, who seek conflict or disturb the status quo.

- *Neuroticism:* This trait is characterized by moodiness, anxiety, irritability, and sadness. It describes people who tend to experience more negative emotions than positive ones. They can be intense and overly questioning but may point out problems others do not see. This contrasts with calm people who maintain emotional stability regardless of the circumstances.

As noted earlier, research in behavioral genetics has provided evidence that genetic factors contribute significantly to these personality traits, although the exact mechanisms are complex and not fully understood (241, 242).

For openness, genetic factors can influence brain systems that govern novelty-seeking and intellectual curiosity. For example, gene variations associated with dopamine neurotransmission have been linked to creativity and exploration (243). Genes may affect the development and functioning of brain regions involved in self-control and goal planning, such as the prefrontal cortex. Genetic influences on temperament in early life can also predispose individuals to higher levels of conscientious behavior.

Genetics can impact the functioning of neurotransmitter systems, such as those involving dopamine, which are associated with extraversion, reward sensitivity, and sociability. This can predispose individuals to seek out social interactions and experience positive emotions in social settings. Genetic factors play a role in developing brain regions involved in empathy and processing social information. Additionally, genes related to oxytocin and serotonin neurotransmission influence prosocial behaviors, impacting levels of agreeableness. Finally, some genes can influence the amygdala's reactivity and other parts of the brain's limbic system, which are involved in emotional processing. This can affect an individual's sensitivity to harmful stimuli and predisposition to anxiety and mood disorders.

Meanwhile, multiple genes influence personality traits, each contributing to a small effect. This makes the genetic basis of the Big Five model complex and multifaceted. Additionally, environmental factors influence the expression of genetic predispositions for the five traits. For example, a genetic disposition towards high neuroticism may manifest differently depending on an individual's life experiences.

Having explored the Big Five personality model and its profound impact on our behaviors, I must delve deeper into the nuanced interplay between these traits and our capacity for growth and adaptability. Specifically, conscientiousness is often celebrated as a hallmark of reliability and success. However, when taken to extremes, this trait can quietly undermine the flexibility and openness required for neuroplasticity. For much of my life, I took immense pride in this trait. I saw it as the cornerstone of my identity—my thorough, reliable, and disciplined ability set me apart. I viewed myself as someone who could tackle any challenge through sheer focus and determination. I also found myself quietly judging those who did not seem to share the same drive, dismissing them as careless or lacking ambition.

This rigid sense of conscientiousness, however, came with its traps. It was not until I began exploring the brain's capacity to change that I recognized how my unwavering commitment to structure and perfection was, in many ways, a hindrance. Neuroplasticity thrives on flexibility, letting go of rigid patterns, and embracing uncertainty. My high conscientiousness often led me to cling tightly to routines and expectations, leaving little room for spontaneity or creative exploration—the elements needed for the brain to rewire and grow.

The Enneagram

The *Enneagram* model is a typology system that describes human personality as nine distinct types. The model is popular in various domains, including psychology, spirituality, and business leadership (244). The Enneagram is often used for personal self-awareness and spiritual development and as a tool in counseling and therapy. The precise origins of the Enneagram are unclear and subject to debate and discord among scholars. It has roots in ancient traditions, including Sufism, Christianity, and possibly other spiritual and philosophical traditions. The model commonly used today was developed in the 20th century, with significant contributions by Oscar Ichazo and Claudio Naranjo (29).

Each of the nine Enneagram types has distinct characteristics, motivations, fears, and ways of interacting with the world. Each typology is typically given a descriptive name. This is one such architecture with a short description of each type.

- **Type 1: The Perfectionist** - Rational, idealistic, principled, purposeful, self-controlled, and often perfectionistic.

- **Type 2: The Helper** - Caring, interpersonal, generous, people-pleasing, and possessive.

- **Type 3: The Achiever** - Success-oriented, adaptive, excelling, driven, and image-conscious.

- **Type 4: The Individualist** - Sensitive, withdrawn, expressive, dramatic, self-absorbed, and temperamental.

- **Type 5: The Investigator** - Intense, cerebral, perceptive, innovative, secretive, and isolated.

- **Type 6: The Loyalist** - Committed, security-oriented, responsible, anxious, and suspicious.

- **Type 7: The Enthusiast** - Busy, fun-loving, spontaneous, versatile, distractible, and scattered.

- **Type 8: The Challenger** - Powerful, dominating, self-confident, decisive, willful, and confrontational.

- **Type 9: The Peacemaker** - Easygoing, self-effacing, receptive, reassuring, agreeable, and complacent.

The Enneagram types are often grouped into three triads based on shared core fears, reflecting deeper motivations and emotional responses. This grouping is not immediately apparent from the basic descriptions of each type, but it becomes more meaningful when considering their underlying motivations.

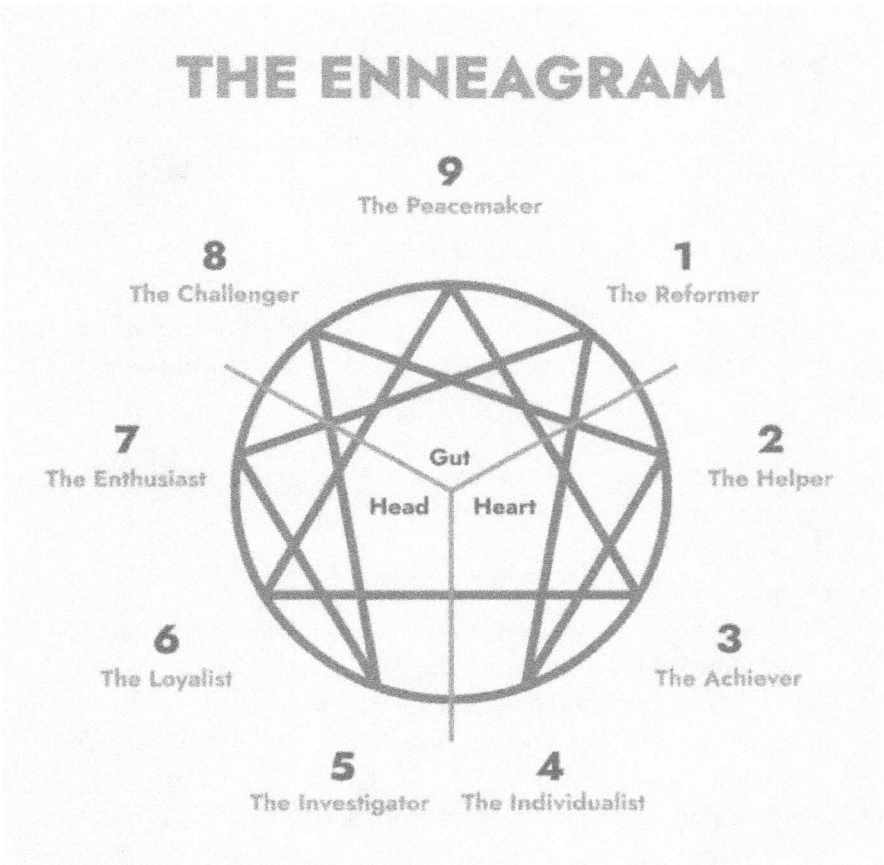

Figure 86: The Enneagram Model (Istock)

Types 8, 9, and 1 (The GUT triad)

These types are primarily concerned with control and resistance to control, relating to physicality and anger. The core fear is to feel powerless.

- **Type 8 (The Challenger):** Their core fear is being controlled or harmed by others, which drives them to assert control over their environment and maintain independence.
- **Type 9 (The Peacemaker):** They fear loss and separation, leading to a desire for peace and stability. This is often a reaction to a perceived

threat to their comfort and status quo, which they seek to maintain through avoidance of conflict.

- **Type 1 (The Perfectionist):** Their core fear is of being corrupt, evil, or defective. This fear drives their desire for order and integrity, often manifesting as a form of internal control over impulses they deem wrong.

The core fear relates to autonomy, control, and physical integrity in all these types. Each type copes with this fear differently: Type 8 by asserting control, Type 9 by avoiding conflict to maintain control, and Type 1 by self-controlling to prevent being 'bad.'

Types 2, 3, and 4 (The EMOTIONAL Triad)

These types are primarily concerned with identity and recognition, dealing with issues of shame and worthiness. Their core fear is not being seen and loved.

- **Type 2 (The Helper):** Their core fear is being unloved or unwanted by themselves. They seek to mitigate this fear by becoming indispensable to others.
- **Type 3 (The Achiever):** They fear being worthless or failing to be seen as flourishing and valuable. This fear drives their pursuit of achievement and recognition.
- **Type 4 (The Individualist):** Their core fear is of having no identity or personal significance. They respond to this fear by cultivating a unique identity.
- For these types, the central fear revolves around how others see and value them. Each type manages this fear differently: Type 2 by gaining love through helping, Type 3 through achievement, and Type 4 through individuality.

Types 5, 6, and 7 (The HEAD triad)

These types are primarily concerned with security and anxiety, focusing on the mental domain. The core fear is not to know.

- **Type 5 (The Investigator):** Their core fear is being useless, helpless, or incapable. They cope by accumulating knowledge and becoming self-sufficient.

- **Type 6 (The Loyalist):** They fear being without support or guidance, leading to constant vigilance for threats and searching for trustworthy alliances.
- **Type 7 (The Enthusiast):** Their core fear is being deprived, trapped in pain, or limited. They respond by seeking pleasure and avoiding pain.

In these types, the core fear concerns security and coping with uncertainty. Type 5 manages this through detachment and self-reliance, Type 6 through loyalty and security, and Type 7 by escaping from pain and seeking variety.

Understanding these core fears offers insight into the deeper emotional motivations of each type and highlights the commonalities between types within the same center. Each Enneagram type has coping mechanisms for dealing with stress and life challenges, deeply tied to its core characteristics.

Discovering that I am a Type 6 (The Loyalist) has profoundly reshaped how I navigate the challenges of consciousness and neuroplasticity. As a 6, my core fear of uncertainty and my deep desire for security often manifest as overthinking and seeking external guidance. This transformative self-awareness has revealed how these patterns limit my ability to embrace the unknown—a critical step in rewiring the brain. Understanding my type has taught me to recognize the anxiety-driven loops my mind falls into and to respond with compassion instead of resistance.

Facing these challenges head-on, I have leaned into practices that foster trust—trust in myself, others, and the process of neuroplasticity itself. Knowing that growth requires stepping into discomfort, I've begun reframing fear as an opportunity rather than a barrier. This perspective has allowed me to use my natural vigilance as a strength, channeling it into mindful awareness and intentional action. By aligning with my type's growth path, I've cultivated greater inner stability, enabling me to explore consciousness with courage and curiosity.

While the Enneagram is more descriptive than predictive, it offers a robust framework for understanding personality styles, motivations, and behaviors. However, it does not predict future behavior or outcomes in a strict sense. Like most personality models, the Enneagram does not have a direct, established link to genetic markers either (245). Personality is a complex interplay of genetics and environmental factors. While certain genetic predispositions might incline individuals towards certain personality traits, the Enneagram is not a tool for measuring genetic influence. It is more about understanding psychological and emotional patterns.

While popular and used in various contexts, the Enneagram model still lacks a solid empirical basis compared to the Big Five. However, when combined, both models are valuable for exploring the depth of one's character. They offer a rich framework for personal understanding and growth, focusing on motivational and emotional aspects of personality.

How Neuroplasticity Helps You Rewrite Your Story?

Neuroplasticity allows the brain's neurons (nerve cells) to adjust their activities in response to new situations, changes in their environment, or because of injury. There is good evidence supporting the fact that even personality traits can be rewired (246). The neurobiology of neuroplasticity involves several vital mechanisms that enable you to rewrite your personal story:

Synaptic Plasticity

This is the most well-known form of neuroplasticity. Synapses are the junctions where neurons communicate with each other. It refers to the ability of synapses to strengthen or weaken over time in response to increases or decreases in their activity. This is often encapsulated in the Hebb principle known as *"neurons that fire together, wire together,"* meaning that the more two neurons activate simultaneously, the stronger their connection becomes (247). This mechanism underlies learning and memory; therefore, consider synapses to be the new musical sheet of your biographical story. Imagine the edited script of your life without the resentment, fears, and all the memories that have haunted you and made you feel less important and less loved.

Neurogenesis

Once thought impossible in the adult brain, *neurogenesis* can generate new neurons in specific brain regions, such as the hippocampus, throughout life. These new neurons can integrate into existing neural circuits, contributing to brain plasticity. By generating new neurons, you multiply your opportunity to change the pathways that hold your personal story together permanently (248). This includes the formation of new neural pathways and the alteration or elimination of older ones. These changes can be driven by learning, experience, and recovery from brain injury. Structural changes can occur in neurons in all these cases. For example, a neuron might grow new dendrites (the branch-like extensions that receive signals from other neurons) to form new connections.

As discussed earlier, critical periods of neuroplasticity refer to specific times during an organism's life when the brain is particularly receptive to environmental stimuli, making it easier to acquire or modify skills and abilities (249). During these periods, the neural connections in the brain are more dynamic and malleable. This property is crucial to understanding the specific role stories play in that process under the influence of psychedelics.

First, psychedelics are believed to induce a state of heightened neuroplasticity temporarily, akin to reopening a critical period (250). This can allow for rapid and significant changes in neural networks, which might otherwise be less malleable in adulthood. The effect enables the reorganization of thought patterns and behaviors that are otherwise rigid and difficult to change.

Second, the ability of psychedelics to induce states of enhanced neuroplasticity is one of the fundamental reasons they are so valuable in treating psychological disorders, especially those rooted in rigid, and maladaptive thought patterns or behaviors that need to be rewired.

Finally, studies have shown that conditions like depression, PTSD, and addiction, which are often resistant to conventional treatments, can be more effectively addressed during these periods of heightened brain plasticity (251). Psychedelics may facilitate this process, allowing individuals to break free from entrenched neural pathways, such as those involved in addictive behaviors or negative thought cycles associated with mental health disorders. This can lead to more profound and lasting changes in perspective, behavior, and emotional processing, key focus areas in psychotherapy.

Undoubtedly, all forms of neuroplasticity we reviewed are fundamental in building psychological well-being and resilience. As we encounter challenges and learn to overcome them, our brain adapts, strengthening the neural pathways associated with coping and resilience. This process helps us construct a personal narrative centered around strength and perseverance while enhancing self-awareness and presence. This heightened awareness can lead to a deeper understanding of our life story and encourage a more conscious and intentional approach to writing our future chapters.

By understanding and utilizing the principles of neuroplasticity, you can reshape your behaviors, thoughts, emotions, and, ultimately, your personal story, leading to a more fulfilling and purposeful life.

While we have examined the value of leveraging neuroplasticity to rewrite your personal story, there are other ways to "nudge" your brain to accelerate and transform your thinking. *Manifestation*, a concept often associated with the law of attraction and positive psychology, can be a powerful option to enhance the rewriting of your personal story (252).

Manifestation involves focusing your thoughts, emotions, and energy on desired goals or outcomes,

Figure 87: The Power of Manifestation

believing that this focus can positively change your life. I have dismissed this approach for a long time since I could not find strong scientific evidence supporting the practice's benefits. However, by gradually moving to the camp of non-materialistic scientists, I have finally opened my mind to the possibility that this phenomenon warrants more attention and consideration.

Setting clear intentions or goals is crucial in rewriting one's personal story. It defines what aspects of the story one wishes to change or what new chapters to add. A core aspect of manifestation is cultivating a positive mindset and a strong belief in the possibility of achieving the desired changes. This positive outlook can be instrumental in overcoming limiting beliefs and negative thought patterns that might have shaped one's previous narrative.

Visualization is a common technique used in manifestation. It involves vividly imagining oneself in a scenario where the desired changes or goals are already achieved. This practice can stimulate the brain in ways similar to actually experiencing those events, leveraging neuroplasticity to reinforce new ways of thinking and acting. Many performance athletes use this technique to limit cognitive friction and quickly and precisely improve flow and speed (253). Manifestation also emphasizes aligning one's emotions with desired outcomes. Feeling joy, gratitude, or success as if the goals have already been achieved can create a more optimistic and proactive approach to life, influencing one's personal story.

While manifestation focuses on thoughts and beliefs, it often encourages concrete actions toward one's goals. This active engagement is crucial in rewriting one's story since it is about thinking and acting differently.

Consistently focusing on and manifesting new goals can lead to developing new habits and behaviors. Over time, these can become integral parts of one's redefined personal story.

So, manifestation can change how you perceive and interpret your experiences. By focusing on positive outcomes and lessons in every situation, you can reframe challenges as opportunities for growth, thereby rewriting your narrative from a more empowered perspective. The belief in the power of manifestation can create a self-fulfilling prophecy. When you genuinely believe you can achieve your goals, they may become more motivating and persistent, increasing the likelihood of success.

As remarked in my introduction, it is essential to note that the concept of manifestation, particularly in the context of the law of attraction, is not universally accepted in scientific circles and often lacks empirical support. William Walker Atkinson first coined the expression "*Law of Attraction*" in his 1906 book "*Thought Vibration or the Law of Attraction in the Thought World*" (254). Atkinson was a significant figure in the early New Thought movement, which emphasized the power of the mind in shaping one's reality and experiences. In his book, Atkinson explored the idea that thoughts are a form of energy and that positive thoughts can attract positive outcomes, while negative thoughts can attract negative ones. This concept laid the groundwork for what would later become a central tenet of the Law of Attraction as it is known today.

While Atkinson is credited with coining the term, the ideas behind the Law of Attraction have roots in much older philosophical and religious traditions, including those of the New Thought movement, which began in the 19th century, and even earlier spiritual and metaphysical concepts. It is associated with the belief that one's thoughts and intentions can influence one's reality, a topic that intersects today with various fields, including psychology, neuroscience, and quantum physics. Manifestation, as discussed in self-help and New Age contexts, is not widely endorsed in the scientific community. Yet, several scientists and researchers have interpreted work in related areas as supportive or relevant to these ideas. Some notable figures include Deepak Chopra. Mario Beauregard and many others (255, 256).

Chopra, an Indian-American author and alternative medicine advocate, is widely known for his work in mind-body medicine and spirituality. His approach to the power of manifestation is rooted in a synthesis of Ayurvedic principles, quantum physics, and consciousness studies. While well-known and influential, Chopra's work often stirs debate among scientists due to its blend of spirituality, science, and philosophy. Chopra's views on manifestation are primarily centered around the concept that through meditation and mindfulness, human consciousness can directly affect physical health and reality.

Chopra suggests that a non-local aspect of consciousness transcends space and time, which can influence our physical world. Although not the originator of this concept, Chopra has contributed to the popularity of the law of attraction. He advocates that positive or negative thoughts bring positive or negative experiences into a person's life. This principle is often linked to the practice of manifestation. He argues that changing one's thoughts and emotional patterns can manifest changes in physical health and life circumstances. Much of Chopra's philosophy is influenced by Eastern spiritual traditions, including concepts like dharma (purpose or duty), karma (action and consequence), and the interconnectedness of all things.

Chopra's work often crosses into the realm of spirituality and metaphysics, areas that do not align with empirical scientific methodologies. As such, many in the fields of self-help and spiritual growth embrace his theories on the power of manifestation, but they are viewed critically by many in the scientific community. Despite this, his contributions have sparked significant interest in exploring consciousness and its relationship to physical reality.

Meanwhile, Mario Beauregard is another prominent scientist who embraces non-materialistic views on the power of manifestation (15, 256). Beauregard is known for his attempts to bridge the gap between spirituality and science. He has extensively investigated phenomena often associated with spirituality (like mystical experiences) using neuroscientific methods, contributing to a broader understanding of these experiences. His books have had a profound impact on my reductionist perspectives. I greatly credit him for getting me out of the scientific closet of materialism. Beauregard is known for his research on the neuroscience of consciousness and neuroplasticity. His work explores the relationships between the mind and brain, mainly how consciousness and intention influence physical processes.

More importantly, Beauregard's work touches on multiple themes relevant to manifestation. For instance, Beauregard has extensively studied the neural correlates of consciousness, focusing on how conscious intention and emotional self-regulation can physically affect the brain. His research is significant in understanding the power of the mind over bodily processes. His work delves into the brain's ability to reorganize itself by forming new neural connections throughout life. Beauregard's research demonstrates how mental practices like meditation can lead to significant changes in brain structure and function, indirectly related to manifesting changes through focused thought and intention.

Additionally, Beauregard has explored how mental states can influence physical health. This includes studies on the placebo effect and how positive thinking and emotional well-being affect physiological processes. He criticizes strict materialistic perspectives in neuroscience that view mental processes as merely the byproducts of brain activity. Instead, Beauregard posits that consciousness can exert a top-down influence on brain function.

I recommend a few other essential thought leaders on manifestation. Dr. Bruce Lipton, a biologist, focuses on epigenetics and the idea that a person's beliefs and environment can influence genes and DNA (257). His theories have been influential in alternative medicine circles. Also, Dr. Emoto, known for his experiments on the effects of thoughts and emotions on the molecular structure of water, has been popular in alternative medicine and New Age circles. However, he has been widely criticized by the scientific community for lacking rigorous scientific methodology (258).

It is important to note that while these individuals have contributed to discussions about the power of thought and belief, their work varies widely regarding scientific rigor and acceptance within the mainstream scientific community. The concept of manifestation often presented in popular media is not typically recognized as scientifically validated. Most established scientific fields advocate for evidence-based approaches to understanding reality and human psychology. However, as a psychological tool, manifestation can foster a positive mindset, clarify goals, and motivate individuals toward personal change. In this way, it can be helpful in the broader process of personal growth and rewriting one's story.

By accepting that science and ancient healing traditions can help rewrite your story, we set our intentions toward learning, healing, and trust in the future. That mindset creates another mysterious yet undeniable phenomenon that catapults your healing process: *The placebo effect!*

The Placebo Effect

The placebo effect is a fascinating phenomenon in which a person experiences a fundamental change in their health or behavior due to their beliefs and expectations rather than an active component of treatment (259). It can play a significant role in rewriting personal stories and facilitating life changes.

The placebo effect is fundamentally driven by an individual's hope that a given intervention (even if it is inert) will be beneficial. This effect contributes value to rewriting personal stories, as believing in the power of change and positive outcomes can be a self-fulfilling prophecy, leading to actual improvements in one's life.

The placebo effect demonstrates the strong connection between the mind and the body. Believing that rewriting one's personal story will lead to positive changes can activate psychological and physiological processes that contribute to making those changes a reality. Just as placebos can reduce symptoms of discomfort by altering perception, the act of actively rewriting a personal narrative can alleviate feelings of negativity or stress. This shift in mindset can lead to improved emotional well-being. Believing in the effectiveness of a placebo

Figure 88: The Placebo Effect

can increase motivation and the likelihood of taking actions that align with one's new narrative. This can create a positive feedback loop where belief and action reinforce each other.

Similar to how a placebo can change one's subjective experience of symptoms, rewriting a personal story often involves reinterpreting past experiences and future possibilities. This cognitive reappraisal can lead to remarkable changes in behavior and outlook. Just as patients who experience positive effects from a placebo may feel more confident in their ability to manage their health, individuals who believe in their rewritten personal story may experience increased *self-efficacy*.

Self-efficacy refers to an individual's belief in their capacity to execute behaviors necessary to produce specific performance attainments. It reflects confidence in the ability to exert control over one's own motivation, behavior, and social environment. Individuals with high self-efficacy are more likely to set challenging goals, remain persistent in the face of difficulties, and recover more quickly from setbacks, believing in their ability to influence outcomes positively. This confidence can be pivotal in pursuing personal goals and embracing change.

The placebo effect can induce traceable changes in the brain, such as alterations in neurotransmitter release and brain activity. Similarly, changing one's narrative can lead to neuroplastic changes in the brain, reinforcing new ways of thinking and behaving. The placebo effect highlights the importance of subjective experience in determining our reality. In rewriting personal stories, an individual's subjective interpretation and perspective are critical to personal change.

While the placebo effect is often discussed in the context of medical treatment, its underlying principles—the power of belief, expectation, and the subjective interpretation of experience—are highly relevant to the process of personal change. Believing in the efficacy of rewriting one's personal story can catalyze profound life changes, much like a placebo can lead to real improvements in health outcomes.

Many studies have demonstrated the power of the placebo effect. In 1955, Harvard anesthesiologist Henry K. Beecher published a groundbreaking paper titled "The Powerful Placebo." He analyzed 15 clinical trials and found that about 35% of patients were relieved by a placebo. This study was instrumental in establishing the placebo effect as a recognized medical phenomenon (260).

Another landmark study published in the New England Journal of Medicine involved a group of patients with Parkinson's disease who underwent a placebo surgery (they did not receive the actual brain surgery they were expecting). Nonetheless, many patients showed significant improvement, demonstrating the placebo effect even in surgical interventions (261). More studies have also found that placebos were 80% as effective as the antidepressant medications *nefazodone* and psychotherapy in treating moderate to severe depression (262).

Finally, a notable study published in the New England Journal of Medicine had patients with osteoarthritis undergo either actual knee surgery or placebo surgery, where surgeons only made incisions but did not perform actual surgery. The results showed that the placebo surgery was as effective as the actual surgery in relieving pain and improving function (263).

Collectively, these studies underscore the powerful and sometimes surprising influence of the placebo effect, demonstrating that the belief in and expectation of treatment can produce fundamental, measurable changes in health outcomes. This phenomenon continues to be a subject of considerable interest and research, as it challenges our understanding of how healing occurs and the role of the mind in physical health.

- Rewiring your story means integrating new perspectives on your Self. Examining your biographical, genetic, and psychological story can help you heal old patterns that no longer serve you.
- Some stories lock you in projections that keep you in the vicious loop of rumination and sometimes even anger.
- A significant portion of your self-referential thinking occurs when the brain activates the default mode network (DMN), mostly during rest and introspection.
- The theory of mind (TOM) explains our ability to attribute mental states to ourselves and others. Engaging in this mentalizing process makes stories more captivating and memorable, as the human brain can recognize patterns and predict outcomes.
- Narrative transportation is a process, often unconscious, during which we become cognitively and emotionally immersed in a story, suspending disbelief and, therefore, accepting the story world as real for the duration of the engagement.
- Many forms of therapy focus on personal stories to achieve relief and healing by reframing and reinterpreting people's experiences.
- Personal stories play a central role in the success of Alcoholics Anonymous (AA), a global organization dedicated to helping individuals recover from alcoholism.
- In psychedelic-assisted therapy (PAT), personal stories play a significant role in both the therapeutic process and the patient's journey toward healing and self-discovery.
- Ayahuasca sharing circles, often rooted in Indigenous traditions, place a strong emphasis on personal stories.
- Personality traits significantly influence our personal stories and how we interpret and narrate them to others and ourselves.
- DNA plays a vital role in shaping our personality, operating through the complex interplay of genetics and environment.
- Epigenetic mechanisms alter DNA when environmental factors and experiences drive long-lasting changes, and they can be heritable.
- According to Dr. Stan Grof, pre-natal stages can deeply imprint on an individual's subconscious and influence their psychological development, behaviors, and patterns in adulthood.
- The Big Five personality traits model is a widely accepted framework for understanding human personality. It is based on openness, conscientiousness, extraversion, agreeableness, and neuroticism.

- The Enneagram model is a typology system that describes human personality based on nine distinct types.
- Neuroplasticity allows the neurons (nerve cells) in the brain to adjust their activities in response to new situations, changes in their environment, or as a result of injury.
- Manifestation involves focusing one's thoughts, emotions, and energy on desired goals or outcomes, believing that this focus can bring about positive changes in one's life.
- The placebo effect is a fascinating phenomenon in which a person experiences a real change in their health or behavior due to their beliefs and expectations rather than an active component of treatment.

Having rewritten your personal story to reflect a more empowered and authentic version of yourself, the next step is to bring that narrative to life. In *Open Your Life,* we will explore how to transform these new perspectives into daily rituals and practices that reprogram your mind, creating lasting change and a life filled with the flow, purpose, and joy you want.

CHAPTER FIVE

OPEN Your Life

"Habits are the brain's way of saving energy. To change a habit, you must consciously reprogram your brain's autopilot mode."

--David Eagleman, Neuroscientist and Author

Numerous articles and blog posts offer tips on managing stress, anxiety, and depression that may involve engaging with specific habits. However, many lack robust scientific evidence to support their recommendations. In contrast, while researching this book, I rigorously analyzed the science of rituals that can promote higher consciousness and improved neuroplasticity while delivering a soul-soothing experience.

If you truly want to clear your nervous system from the toxicity of stress, anxiety, depression, addiction, and trauma you may have accumulated for many years, engaging with *OPEN rituals* is the best way to maintain and strengthen the benefits of an *OPEN life*. However, it is important to note that my definition of *OPEN rituals* is unique and may not entirely align with the traditional psychological literature on rituals.

The reason for this distinction lies in the neurospiritual framework of *OPEN*, which calls for its own definition of rituals as intentional, consciousness-expanding practices designed to rewire the brain, enhance emotional resilience, and foster a deeper connection to the self and the world. Unlike conventional psychological or anthropological views of rituals that often emphasize sacred functions, *OPEN rituals* are specifically designed to promote neuroplasticity, flow, and personal transformation. This neurospiritual approach integrates elements of mindfulness, cognitive restructuring, and spiritual intention, distinguishing *OPEN rituals* as deliberate tools for elevating consciousness and fostering holistic well-being.

Meanwhile, abundant evidence indicates that routines are vital in regulating and optimizing healthy behaviors. So, when behavior is repeated mindfully (an *OPEN ritual*), the brain interprets it as necessary, reducing the need for cognitive effort and leaving no room for self-doubt. The basal ganglia, critical brain structures within the PRIMAL brain, are crucial in habit formation. These structures are interconnected with the cortex, influencing movement, perception, and judgment.

When Habits Can Become OPEN Rituals

Performing *OPEN rituals* provides structure and predictability, which can benefit cognitive processes in stressful or uncertain situations. More formal than habits, *OPEN rituals* may involve ceremonial or sacred routines. Regardless, plasticity habits can provide neurological benefits when formatted as rituals. They are often used to access positive emotional states or motivation. For example, many athletes engage in pre-performance rituals. For example, the golfer Tiger Woods always wears a red shirt on the final day of tournaments, believing it gives him a psychological edge. The basketball legend Michael

Jordan wore his University of North Carolina practice shorts under his Chicago Bulls uniform for every game, believing they brought him luck. The gymnast Simone Biles listens to the same music playlist before competitions and performs specific mental visualization exercises. These rituals help athletes maintain focus, build confidence, and reduce anxiety before high-pressure performances (264).

Rituals in general enhance performance by improving focus and confidence, likely due to their impact on the PRIMAL brain. Studies have shown that rituals relax PRIMAL and RATIONAL brains by preventing conscious motor control, which enhances reaction speed. They create a sense of predictability and reduce performance-related stress. Performing neuroplastic habits at scheduled times and with greater consciousness is advisable to convert them into rituals. This will increase calmness and predictability while relaxing the brain.

By imposing order and routine, rituals can help in managing anxiety and stress, which are known to impact neuroplasticity negatively. A study by Canadian Psychologist Hobson and his colleagues suggests that structured rituals can reduce cognitive load and improve executive functioning ((265)). Repetition is key to ritual formation as it strengthens neural pathways, making certain behaviors more automatic and less cognitively taxing. Dr. Pascal Boyer and Dr. Pierre Liénard have collaborated on research examining ritualized behavior's cognitive and evolutionary aspects, contributing significantly to understanding how such practices influence human psychology and culture.

Meanwhile, rituals often serve as a means of emotional regulation. For example, mourning rituals can help individuals process grief, while daily gratitude practices can enhance overall well-being. A Harvard study by Norton and Gino confirmed that rituals can help mitigate grief and improve emotional recovery after loss (322). Many rituals are social and can strengthen community bonds and empathy, engaging neural networks associated with social cognition. A study by Anthropologist Xygalatas et al. showed that participants who engaged in intense ritual activities had higher pain thresholds and reported feeling more socially connected (323).

So, rituals can induce relaxation and reduce stress, which is beneficial for overall brain health and neuroplasticity. Religious or spiritual rituals often incorporate mindfulness and meditative elements. This can improve attention, focus, and mental clarity. A study by Kozhevnikov et al. found that Tibetan Buddhist meditation practices (which can be considered ritualistic) enhance visuospatial abilities and neuroplasticity (324). Finally, performing rituals can aid in memory consolidation and learning. For example, the ritual of reviewing and summarizing information can enhance academic performance. A study by

Legare and Souza discussed how ritualistic actions could improve attention and memory, particularly in children (325).

Meanwhile, the role of *self-love* is critical to the process of ritualizing a formal and regular practice. Self-love encourages a positive relationship with oneself, making it easier to commit to and maintain new habits that contribute to personal growth and well-being. Self-love is crucial because it involves treating oneself with care and kindness and without self-judgment. Accepting imperfections and past failures is central to maintaining a self-love perspective. This mindset facilitates personal growth and self-actualization processes, underpinned by neurochemical impacts that motivate transformation. Indeed, self-love enhances dopamine production, fueling the motivation and commitment necessary for changing one's life and fostering the acceptance and practice of self-love habits, which are essential for developing and sustaining new, healthy behaviors.

Grasping the purpose behind a ritual can deepen one's dedication to it. Resistance and discomfort are natural when establishing new rituals, but embracing these challenges is crucial for growth. When I published The Serenity Code in 2020, I initially believed that self-love habits alone could reshape the brain for greater peace and happiness. Now, I see that self-love is an enriching outcome of ritual practice, reinforcing the neural pathways that support these behaviors. Engaging in rituals, however, goes beyond cultivating self-love; it also targets the neurobiological roots of harmful behavioral patterns. These destructive patterns may remain embedded in the nervous system, even when one has elevated their consciousness and rewritten their personal narrative.

Below are 25 *OPEN* rituals that can help you maintain the highest level of consciousness and solidify the neuropathways involved in regulating the peaks and valleys of our nervous system. I will introduce them first, and then go into more details for each one. As you discover them, feel into each one and how it may intuitively resonate with you. The idea is not to pursue all 25 but rather create a plan of two or three that aligns with your personality, challenges, and goals.

- **PRIMAL rituals (8)** encompass calming the survival-centric brain while enhancing our connection with concrete, physical objects or concepts such as our body (3), breath, sleep, food, nature, and pets. These practices are exceptional for grounding you in the present, soothing your nervous system, and stimulating neurochemicals that promote tranquility. They reassure the PRIMAL brain that constant vigilance is unnecessary, signaling consistent, safe outcomes.

- **RATIONAL rituals (11)** focus on generating positive emotions while turning off your mind's chatter. These rituals reconnect you with the pleasure of laughing, listening to evocative music, playing an instrument, laughing, watching movies, gardening, cooking, playing video games, learning, traveling, and creating art. RATIONAL rituals stimulate neurotransmitters that reward the entire brain. These activities involve brain regions that process humor, melody, and narrative transportation, creating an orchestra of neuronal responses under your direction.

- **SPIRITUAL rituals (6)** produce transcendent experiences that suspend the role of the ego, effectively reducing the energy trapped in the Default Mode Network (DMN), and releasing more consciousness and flow in your nervous system. Engaging with your spiritual self can release compounds that contribute to overall well-being, quieting both the PRIMAL and RATIONAL brain systems while stimulating the emergence of a state of profound spiritual significance. These rituals, which I have identified here, include meditation, prayer, dreams, mantras, stargazing, and the responsible use of entheogens.

To create a step-by-step program for cultivating an *OPEN* life, begin by integrating PRIMAL, RATIONAL, and SPIRITUAL rituals into your daily, weekly, and monthly routines. Vary the mix and frequency of these rituals to align with your personal goals. Take my OPEN *Assessment* (OA) to gauge the severity of your SADAT stressors, assess the current dominance of your PRIMAL, RATIONAL, and SPIRITUAL brain systems in your life; identify the character traits that are at play in critical coping mechanisms, and assess your current engagement with positive emotions, self-efficacy, meta-consciousness, plasticity. The complimentary assessment takes about 15-20 minutes to complete at **https://su.vc/eatphssm.**

Once you know your scores, select two or three *OPEN rituals* you can practice each week, gradually building momentum and reaping the benefits over time. Remember, incremental changes are more sustainable than drastic ones in fostering lasting transformation. You will be amazed how *"opening up to your rituals"* positively impacts mood, stress levels, anxiety, and peace within a few days. Each ritual becomes a powerful, conscious practice that fosters deeper awareness, grounding, and potential for lasting transformation.

The PRIMAL brain automates tasks to conserve energy and reduce strain on its key system, the autonomic nervous system. Rituals that target the PRIMAL brain are designed to release the toxic loops activated by elevating levels of stress and anxiety.

Open Up to Your Body (3)

Figure 89: Open Up to Your Body (AI)

I discussed physical exercise earlier as an effective way to raise your consciousness. Exercise plays a significant role in maintaining neuroplasticity as well. Aerobic activities, like jogging, swimming, or cycling, not only promote the release of endorphins to alleviate pain and reduce stress (266). They can also enhance neurogenesis.

Indeed, exercise stimulates the production of new neurons, particularly in the hippocampus, a region critical for memory and learning. This process is a crucial aspect of neuroplasticity (267). Additionally, physical activity promotes the strengthening of synapses (the connections between neurons), which is essential for learning and memory.

Exercise also increases the production of proteins that support neuron survival and growth and facilitate neuroplasticity. Regular exercise can also reduce brain inflammation linked to neurodegenerative diseases and cognitive decline, thus supporting neuroplasticity and overall brain health. Therefore, regular physical activity reduces age-related cognitive decline, helping maintain neuroplasticity and consciousness in older adults. In addition, exercise has neuroprotective effects that can reduce the risk of developing Alzheimer's and Parkinson's, both of which reduce consciousness and impair cognitive functions.

Of course, physical activity is effective in relieving symptoms of depression and can positively affect consciousness and mental function. So, exercising profoundly affects the brain's neuroplasticity. It improves brain health and function through various mechanisms, including increased blood flow, neurotransmitter regulation, neurogenesis, synaptic strengthening, and the production of neurotrophic factors. Regular physical activity is essential for maintaining and enhancing brain health and cognitive function across the lifespan.

Let us now review the benefits of three rituals that keep your body and mind in top shape: yoga, dancing, and running.

Open Up to Yoga

Practicing yoga is one of the easiest ways to exercise while improving your mental health and neuroplasticity (268). Meanwhile, yoga's journey from ancient ritual to modern exercise reflects its adaptability and enduring relevance. It has evolved from a set of esoteric practices in Indian philosophy to a widespread, diverse range of practices around the globe.

Figure 90: Open Up to Yoga (AI)

Practicing yoga has several neuroscience-based benefits that contribute to enhancing neuroplasticity. Yoga has been shown to improve connectivity in various brain networks, including those involved in attention, executive function, and sensory processing. Enhanced connectivity is a sign of healthy neuroplasticity. Studies suggest that yoga practitioners have increased gray matter volume in brain areas involved in memory processing, emotional regulation, and decision-making, indicative of improved neuroplasticity (269). Yoga can also stimulate the production of proteins which supports neuron survival, differentiation, and synaptic plasticity.

Yoga may help slow down cognitive decline associated with aging, thereby preserving neuroplasticity and consciousness in older adults (270). There is also good evidence suggesting that yoga might have neuroprotective effects that can reduce the risk of neurodegenerative diseases. Yoga's combination of physical postures, controlled breathing, and meditation or relaxation techniques makes it a comprehensive practice for brain health. Its benefits for stress reduction, mindfulness, emotional regulation, and cognitive function all contribute to maintaining a heightened consciousness and fostering neuroplasticity.

Figure 91: Open Up to Dancing (AI)

Dancing offers a unique combination of physical, cognitive, and emotional engagement, making it valuable for maintaining consciousness and enhancing neuroplasticity. The neuroscience behind the benefits of dancing encompasses many aspects (271). Dancing sometimes involves learning and memorizing routines, which stimulates the hippocampus, the PRIMAL brain region associated with long-term memory and learning. It requires planning, coordination, multitasking, and engaging the RATIONAL prefrontal cortex.

Dancing improves connectivity in various brain regions. It integrates sensory and motor signals, enhancing the neural networks involved in these processes. Dancing engages several PRIMAL brain areas. It stimulates the basal ganglia, a group of structures linked to control voluntary motor movements, procedural learning, routine behaviors, and emotion. Dancing improves balance and spatial awareness, skills governed by the cerebellum, which are very important as we age (272). Dancing triggers the release of endorphins, the body's natural mood lifters, which can reduce stress and anxiety. Since dance allows for emotional expression, it can be a form of therapy, helping process emotions.

Finally, observing and mirroring dance movements can activate mirror neurons, which is important for empathy and understanding others. Dance combines auditory (music), visual (movement), and proprioceptive (sense of body position) inputs, providing a rich multisensory experience that can enhance sensory integration. The focus required in dancing on movement and music can create a state of flow, similar to mindfulness, aiding concentration and a heightened state of consciousness.

In summary, dancing is a multifaceted activity that benefits the brain by maintaining consciousness and enhancing neuroplasticity.

I have been a runner my entire life, completed five marathons, and covered close to 100,000 miles (thanks to the data of my beloved Garmin watch!). So, I am partial to this ritual. Running, a form of aerobic exercise, has numerous benefits for brain health, particularly in maintaining consciousness and enhancing neuroplasticity (273).

Figure 92: Open Up to Running (AI)

Running increases blood flow to the brain, delivering more oxygen and nutrients essential for optimal brain function and the maintenance of consciousness. Improved blood flow also supports neurovascular health, which is crucial for cognitive function and neural efficiency. Running has been shown to stimulate the growth of new neurons in the hippocampus, thereby enhancing neuroplasticity. Running also increases the production proteins supporting the survival and growth of neurons.

Meanwhile, regular running can lower cortisol levels, the body's stress hormone, which is beneficial for cognitive function and consciousness. Running triggers the release of endorphins and natural mood elevators, which can help reduce stress and anxiety, contributing to emotional stability. Running can improve executive functions like planning, organizing, multitasking, and decision-making.

Aerobic exercise improves attention and concentration, key components of consciousness. Regular runners often experience better quality sleep, which is crucial for brain health, cognitive function, and consciousness. Running can slow down age-related cognitive decline and reduce the risk of neurodegenerative diseases like Alzheimer's, thereby preserving neuroplasticity and consciousness, especially in older adults (274).

In summary, running contributes significantly to brain health by enhancing cerebral blood flow, promoting neurogenesis and neuroplasticity, reducing stress, improving cognitive functions, aiding sleep, slowing cognitive decline, and boosting mood. These benefits collectively support the maintenance of consciousness and the enhancement of neuroplasticity.

Meanwhile, any form of physical exercise recruits our breath extensively. Controlled breathing techniques help manage the body's stress response and are particularly useful in dealing with anxiety and panic attacks. I have already discussed the value of *Holotropic Breathwork* as a consciousness-raising practice. Now, I will explore more specifically how breath sustains high neuroplasticity levels.

Open Up to Your Breath

Figure 93; Open Up to Your Breath (AI)

Conscious breathing and its positive health impacts are a significant focus among health and wellness researchers. This practice is fundamental to self-love, but also simply because our survival depends on breathing; we cannot live without air for more than a few minutes.

The most effective way to engage with our breath is through mindfulness, which involves dedicated attention and energy towards our breathing. French philosopher and mindfulness expert Christophe André emphasizes that breath is "*the most powerful means to connect to the present moment (275).*" André notes that our breathing patterns are often disrupted during depression or anxiety. He highlights the importance of focusing on our breath, pointing out its unique quality of being "*simultaneously inside and outside of us,*" thereby blurring the lines between the self and the external world. Clearly, mindful breathing is a pathway to connecting with something bigger than oneself.

Practicing breathwork, or controlled breathing exercises, has several neuroscientific benefits that enhance neuroplasticity. Intentional breathwork can help balance the autonomic nervous system's sympathetic (fight-or-flight) and parasympathetic (rest-and-digest) branches. Controlled breathing, prolonged and deep breaths, activates the parasympathetic system, promoting relaxation and reducing stress. As a result, breathwork can lower cortisol levels, the body's primary stress hormone, reducing stress and its adverse effects on the brain and body. Controlled breathing exercises can also decrease activation in the amygdala, the brain region involved in emotional processing, particularly fear and anxiety responses.

In a significant study, participants learned "*attention to breath*" (ATB) techniques, focusing on body posture and breath sensations (276). After two weeks of practice, they viewed aversive images while undergoing fMRI scans. The study showed that ATB techniques reduced the emotional impact of these images. This effect was observed in both the anterior cingulate cortex (ACC), which is involved in emotional processing, and the parietal cortex, which integrates sensory information. The key finding is that ATB can diminish PRIMAL brain activity and enhance emotional control by engaging the RATIONAL brain, specifically through the prefrontal-parietal cortical network.

This process allows the RATIONAL brain to moderate the PRIMAL brain's emotional responses. EEG studies further indicate that breathing techniques can increase alpha waves and decrease theta power (277). Alpha waves, typical of a state of calm and relaxation, are particularly effective in reducing symptoms of generalized anxiety disorder. This has led to a growing interest in breath-focused practices like meditation and yoga, which incorporate breathing as a core component. Moreover, breath practices can stimulate the vagal nerve, enhancing vagal tone. Vagal tone refers to the ability of the vagus nerve to regulate the parasympathetic nervous system, acting as a "brake" on stress responses to promote relaxation, resilience, and emotional balance.

Specific breathing techniques can enhance alertness and improve attention. For instance, rapid breathing might stimulate increased arousal and readiness, essential for heightened consciousness. Breathwork often fosters mindfulness, leading to improved mental clarity and heightened consciousness. Some studies suggest that controlled breathing exercises can increase critical proteins supporting brain plasticity, protecting the survival of existing neurons, and promoting the growth of new neurons and synapses (278).

Regular breathwork may improve connectivity in brain regions associated with emotional control and the autonomic nervous system. These areas are crucial for memory, learning, and executive function, potentially enhancing memory and decision-making abilities. Finally, effective breathing techniques can enhance brain oxygenation, vital for optimal brain function. Finally, controlled breathing can help regulate blood pressure, ensuring stable blood flow to the brain.

In conclusion, breathwork offers a range of neurological benefits that contribute to maintaining consciousness and enhancing neuroplasticity. Through its impact on the autonomic nervous system, stress reduction, cognitive function, brain structure and function, oxygenation, and mental health, breathwork is a powerful tool for promoting brain health and resilience.

311

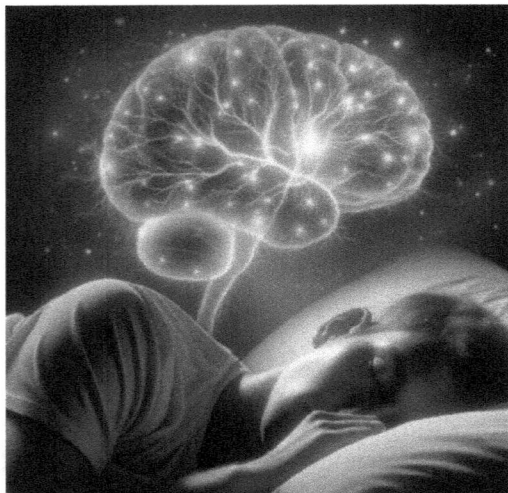

Figure 94: Open Up to Your Sleep (AI)

The neuroscience of the relationship between sleep, consciousness, and neuroplasticity is extensive (279, 280). Sleep is critical in various brain functions, affecting everything from cognitive processes to overall brain health (281). It is essential for restoring and rejuvenating the brain by clearing metabolic waste products that accumulate during waking hours, including *beta-amyloid*, a protein associated with Alzheimer's disease.

During sleep, during which we can observe rapid eye movements (REM sleep), the brain processes and consolidates memories from the day. This is vital for learning and retaining information. Therefore, adequate sleep is crucial for various aspects of cognitive function, including attention, problem-solving, and decision-making. Sleep deprivation can lead to impaired consciousness and cognitive abilities.

Sleep is believed to be involved in synaptic homeostasis, which balances synaptic strengthening (significant for learning and memory) and weakening (necessary to avoid saturation and maintain neural efficiency). Some evidence even suggests that sleep may support the formation of new neurons in the hippocampus. Sleep, particularly REM sleep, plays a role in emotional processing and regulation. It helps the brain process emotional experiences, contributing to emotional stability and mental health.

Good sleep rituals can reduce stress and anxiety, positively impacting overall mental health and consciousness. Regular, healthy sleep patterns are also associated with a lower risk of developing neurodegenerative diseases like Alzheimer's and Parkinson's (282).

Meanwhile, sleep facilitates brain plasticity and repair, aiding the brain in recovering from daily neural wear and tears. Sleep regulates various hormones, including those involved in growth, appetite, and metabolism, essential for brain health. The *glymphatic system* is a waste clearance system in the brain, functionally similar to the lymphatic system in the rest of the body. This system was only recently discovered and is named for its dependence on glial cells and its similarity to the lymphatic system (283). It works by circulating cerebrospinal fluid through the brain's tissue, flushing waste into the bloodstream, where the liver can process it. This system helps remove toxic byproducts that accumulate in the brain, potentially preventing neurodegenerative diseases.

During my retreat in the Amazon, one of the most profound lessons I learned was the importance of honoring the body's natural rhythms, particularly the restorative power of sleep. Immersed in the ancient healing traditions of the Shipibo people, I experienced the deep, rejuvenating effect of aligning with day and night cycles. Sleep became a sacred ritual when the mind could release its hold on the day's events, allowing the body and spirit to heal.

However, the retreat also taught me something surprising: temporary sleep deprivation could open gateways to transcendent states of consciousness under specific circumstances. In one guided session, a night of intentional wakefulness led to heightened clarity, creative insights, and an extraordinary sense of interconnectedness with the natural world.

This paradoxical effect is rooted in neuroscience. Sleep deprivation, when controlled and short-term, can stimulate the release of neurotransmitters like dopamine and serotonin, intensifying focus and emotional receptivity. It disrupts habitual neural patterns, enabling access to deeper or altered states of awareness. These experiences deepened my respect for balance. While restorative sleep is vital for overall well-being, temporary disruptions can unlock transformative perspectives when approached mindfully, showing that our relationship with rest and wakefulness is as dynamic as the mind itself.

In summary, sleep is vital for maintaining consciousness and enhancing neuroplasticity. It contributes to brain health and function through memory consolidation, synaptic homeostasis, emotional regulation, prevention of neurodegenerative diseases, and maintenance of metabolic health. Regular, quality sleep is a cornerstone of cognitive health and overall well-being.

Open Up to Your Food

The neuroscience behind the value of consuming specific foods for maintaining consciousness and enhancing neuroplasticity is grounded in how nutrients affect brain function and structure. Certain foods provide essential nutrients that support brain health, cognitive function, and neuroplasticity. Meanwhile, practicing mindfulness while eating (referred to as *mindful eating* by scholars) enhances digestion by promoting a relaxed state that optimizes nutrient absorption and reduces stress-related digestive issues. When we slow down and fully engage with the food we put in our bodies—paying attention to flavors, textures, and sensations—we activate the parasympathetic nervous system, which supports efficient digestion and overall gut health.

Omega-3 fatty acids are crucial for brain health, especially for glial cells, essential for maintaining a properly functioning nervous system. They are found in fatty fish (like salmon, mackerel, and sardines), flaxseeds, chia seeds, and walnuts. They are integral components of neuronal membranes and play a key role in neuronal function and neuroplasticity. DHA, a type of omega-3, is essential for maintaining neuronal structure and function.

Figure 95: Open Up to Your Food (AI)

Antioxidants combat oxidative stress, which can damage brain cells. They can be found in berries, nuts, dark chocolate, and green leafy vegetables. They help maintain cognitive function and are associated with a reduced risk of neurodegenerative diseases. Flavonoids, a type of antioxidant found particularly in berries, have been shown to enhance memory and learning. *B vitamins*, including B6, B12, and *folic acid*, are vital for brain health. They can be found in whole grains, leafy greens, dairy products, and meat. They help reduce homocysteine levels, high levels of which are linked to cognitive decline and Alzheimer's disease.

B vitamins are also essential for neurotransmitter synthesis and energy metabolism in the brain. *Curcumin* has anti-inflammatory and antioxidant properties. It can be found in turmeric. It can cross the blood-brain barrier, is beneficial in neurodegenerative diseases, and enhances neuroplasticity. Vitamins C and E also act as antioxidants and protect the brain from oxidative stress. They can be found in fish oil, citrus fruits, almonds, sunflower seeds, and flaxseed oil. Vitamin E is vital for cognitive function in older people. A diet rich in omega-3 fatty acids, antioxidants, and other nutrients supports brain health. Meanwhile, polyunsaturated fatty acids (PUFAs) are essential for maintaining the fluidity of cell membranes, which is crucial for synaptic plasticity and neuronal communication. They can be found in fish oil and flaxseed oil.

As the emerging field of *Nutritional Cognitive Neuroscience* uncovers specific foods and nutrients that affect our brains, new studies also demonstrate the relationship between our eating habits and brain longevity. A comprehensive

314

battery of well-established cognitive and brain imaging measures was recently administered to assess brain health, along with 13 blood-based biomarkers of diet and nutrition (284). The findings of this study revealed distinct patterns of aging, categorized into phenotypes of brain health. A phenotype refers to an organism's observable characteristics or traits, which result from the interaction between its genetic makeup (genotype) and environmental influences. A phenotype typically includes physical attributes, behavior, and physiological responses. One phenotype of the study demonstrated an accelerated rate of aging, while the other exhibited slower-than-expected aging. A statistical analysis of these two phenotypes revealed a nutrient profile with higher concentrations of specific fatty acids, antioxidants, and vitamins correlating with better cognitive scores and delayed brain aging

Finally, a study published in *BMC Psychology* examined the socio-cognitive factors influencing the adoption of mindful eating (ME) among adults with varying levels of practice experience (285). The researchers found that individuals with low mindful eating engagement exhibited higher tendencies toward mindless eating and lower internal awareness than those with medium or high ME engagement. These findings suggest that enhancing internal awareness and reducing habitual mindless eating are crucial for adopting mindful eating practices. This study also demonstrates that mindful eating can improve digestion and overall well-being (286).

Open Up to Nature

Figure 96: Open Up to Nature (AI)

Spending time in nature has been shown to reduce stress and improve mood. I discussed forest bathing earlier as a well-documented practice to enhance consciousness and lower the activity of the sympathetic nervous system. Natural settings, in general, can all have a calming effect, offering a respite from stressful stimuli and improving mood. Neuroscientific findings support the value of being in nature for maintaining consciousness and neuroplasticity. Exposure to nature positively impacts the brain, enhancing cognitive function, emotional well-being, and neural connectivity. Spending time in nature can lower cortisol levels, the body's primary stress hormone, reduce stress, and improve cognitive functioning and mental clarity.

Mindful walking in nature with focused breathing enhances physical and psychological well-being. It allows for more profound relaxation, reduces stress levels, and improves mood by promoting the release of endorphins. This practice encourages present-moment awareness, connecting you with the environment, which can lead to a more profound sense of peace and grounding. Additionally, it supports improved concentration and mindfulness, making cultivating a habit of presence and appreciation for the natural world easier.

Nature exposure can lead to decreased activity in the amygdala, moderating the brain's emotional response center, particularly in processing fear and anxiety. *Attention Restoration Theory* suggests that natural environments have a therapeutic effect on cognitive resources, especially attention (287). *Attention Restoration Theory* claims that exposure to natural environments can help restore our attentional capacities taxed by modern life. According to *Attention Restoration Theory*, spending time in nature allows individuals to experience being away from everyday stresses, engage in environments that match their intrinsic motivations, and find interest in engaging but not overwhelming stimuli. These experiences help shift attention in an "involuntary" or "indirect" manner, giving the more strained "voluntary" or "directed" attention systems a chance to recover and restore. Natural settings, with their unique aesthetic and ability to facilitate reflection, are particularly beneficial for this restorative process.

Attention Restoration Theory is grounded in the distinction between "top-down" directed attention and "bottom-up" involuntary attention. The first is linked to higher-order mental functions and becomes exhausted from suppressing distractions. Bottom-up attention is less cognitively demanding and mainly triggered below our awareness. Studies have shown that nature can positively impact directed attention tasks that require significant cognitive resources, like the *Backwards Digit Span test*, but might not affect simpler attentional processes as much.

The test is a cognitive task used to assess working memory, particularly the central executive component responsible for manipulating and organizing mental information. In this test, participants hear a sequence of numbers and are asked to repeat them in reverse order. The task's difficulty increases as the length of the number sequence increases. This test measures the ability to hold and manipulate information temporarily, an essential function of *working memory*. The Backwards Digit Span test is widely used in psychological assessments and research to evaluate attention, concentration, and memory capabilities.

More research supports the idea that exposure to natural environments can improve cognitive performance by restoring attentional resources. Many peer-reviewed papers support the benefits of nature on our nervous system (288). For instance, studies show that nature exposure can enhance executive functions, such as problem-solving, planning, and impulse control, by providing a calming and less mentally taxing environment. Nature can boost levels of serotonin and dopamine, neurotransmitters linked to feelings of happiness and well-being, which is why regular exposure to nature has been associated with reduced symptoms of depression and improved mental health. Time in nature can also increase levels of proteins vital for neuroplasticity, the growth of new neurons, and the strengthening of synapses.

Natural settings often promote a state of mindfulness, which is beneficial for mental clarity and can positively impact the hippocampus, a brain area crucial for memory and learning. Nature gently engages the senses, which can benefit sensory processing and cognitive function. Nature usually provides settings for social interactions, which are essential for emotional well-being and can contribute to cognitive health. Being in nature often involves physical activity, contributing to brain health and cognitive function.

A study by Stanford researchers found that walking in natural environments, as opposed to urban settings, can significantly decrease activity in a brain region associated with depression, suggesting that nature walks could lower the risk of depression (289). This supports the idea that exposure to nature can boost mental health by potentially influencing neurotransmitter levels related to happiness and well-being.

In summary, being in nature can significantly contribute to maintaining consciousness and enhancing neuroplasticity. It provides a restorative environment that reduces stress, improves mood, boosts cognitive function, and stimulates sensory processing. Combining these factors makes nature a powerful tool for promoting overall brain health and well-being.

Open Up to Your Pets

While the direct relationship between pet ownership and neuroplasticity has not been extensively studied, the benefits of interacting with pets can indirectly support brain health and cognitive function. Interacting with pets can lower cortisol levels, the body's stress hormone (290). (290) This benefits cognitive functioning and neuroplasticity, as chronic stress can impair the brain's ability to form new connections.

Pets often can aid in regulating emotions and reducing symptoms of mental health issues like depression and anxiety, indirectly supporting cognitive health. For instance, Barker et al. found that pet ownership can provide emotional support comparable to human support in managing mental health conditions (291). Meanwhile, caring for a pet requires social interaction and fosters empathy, which engages and strengthens brain regions associated with social cognition and emotional intelligence. A study by Dr.

Figure 97: Open Up to Your Pets (AI)

McCardle et al. showed that interacting with animals can enhance interpersonal skills, particularly in children (292).

Additionally, caring for a pet, including maintaining a routine for feeding, exercise, and care, can provide cognitive structure, aiding in executive functioning and planning skills. A study by Friedmann and his colleagues indicated that pet owners exhibit better cognitive function, partly due to the structured routine and responsibility of pet care (293). Also, pet ownership, particularly of dogs, often involves physical activity like walking, which, as I mentioned earlier, can improve overall health, including brain health. A study by Curl et al. found that dog owners who walked their dogs regularly had lower obesity rates and better physical health (294).

Finally, spending time with pets can promote a state of mindfulness and presence, which is beneficial for mental clarity and neuroplasticity. Interacting with pets involves various sensory experiences (touch, sound, sight), which can stimulate the sensory processing areas of the brain.

After returning from my transformative journey in Peru, I made a bold decision that marked a profound shift in my life: I adopted a puppy! For decades, I had carried a deep-seated fear of dogs—a fear born from a trauma I shared in Chapter ONE. An event that lasted less than five minutes 27 years ago, filled with barking and chaos, had etched itself into my memory, creating a wave of anxiety every time I encountered a dog, regardless of the size and breed. It was an irrational fear I had long accepted as part of my identity.

But something changed in Peru. The healing ceremonies, the introspection, and the realization of my capacity for growth planted a seed of courage within me. I returned home feeling more open and willing to confront the shadows that shaped my life. And so, in an act of defiance against my fear, I welcomed a mutt into our home.

The day we brought Kona home was as terrifying as it was exhilarating. Even at 11 pounds, her tiny paws and curious eyes represented a vulnerability I had not expected. As I adjusted to her ongoing unruly playfulness, those first days were filled with challenges. I flinched at her sudden movements and winced when her bark echoed through the house. But with each passing moment, I found myself softening. Her innocence and unwavering trust began to dismantle the walls I had built around my fear.

Caring for her became a daily healing ritual. The simple act of feeding her, walking her, and watching her sleep helped me rewire the trauma that had held me captive for so long. She taught me to be present and embrace life's unpredictability with an open heart. Her love and her presence were a reminder that fear could be transformed into connection.

Welcoming Kona into my life was among the most courageous and healing decisions ever. It was not just about overcoming my fear of dogs—it was about proving to myself that I could rewrite the narratives of my past. Today, she is more than just a pet; she symbolizes my resilience and willingness to embrace change. Through her, I have discovered a new capacity for love, trust, and joy that I never thought possible.

Figure 98: Kona

Meanwhile, owning and caring for our pets (one dog and five cats) is an essential routine in my household. My wife and I probably spend nearly two hours a day playing with, feeding, and grooming our four-legged friends. Being present with our animals is powerful because they are here to help us heal and stay open. While direct research on pets and neuroplasticity is limited, the broader cognitive and emotional benefits of pet ownership, such as stress reduction, emotional support, social interaction, routine, physical health, mindfulness, and sensory stimulation, can all contribute to a healthy brain.

By targeting the RATIONAL brain more specifically, you can lower the default mode network's (DMN) activity while providing us more flow and ease in our day-to-day lives. They can also trigger the bottom-up effect, without which the RATIONAL brain does not participate actively in processing any event. RATIONAL rituals all trigger neurotransmitters that stimulate and delight the most evolved part of your brain.

The electrochemical responses of the neurons recruited in these activities have remarkable healing properties and a direct effect on raising your ability to reprogram critical neuropathways. The 11 rituals I will discuss next can improve your consciousness in many ways. However, the list is certainly not exhaustive. Let your imagination and intuition guide you toward more opportunities to nurture *your brain.*

Open Up to Evocative Music

Figure 99: Open Up to Evocative Music (AI)

Evocative music is music capable of invoking strong emotions, memories, or imagery in the listener. This type of music can transport individuals to different times and places, evoke nostalgia, joy, sadness, or excitement, and create vivid mental pictures or emotional experiences. The power of evocative music lies in its ability to connect with listeners on a deep emotional level, often transcending words and conscious thought.

Evocative music is a profound emotional conduit, enriching our lives beyond mere auditory pleasure. Deliberate, focused engagement with music as a ritualistic practice can rewire our brains. Evocative music triggers many emotions, from joy to sorrow, excitement to tranquility. It enhances focused attention—a property many surgeons utilize to aid concentration (266). Athletes, too, use music to boost concentration, motivation, and performance (295).

Historically viewed as a form of medicine, the therapeutic use of music in clinical settings has recently gained traction, demonstrating its efficacy in alleviating pain, fostering relaxation, and boosting self-confidence (296). Often, we unconsciously use music to regulate our moods. The exploration of music's physical and psychological benefits is still developing, but it is clear that music profoundly influences our neurochemistry (297). Due to its diversity and complexity, it is challenging to investigate music's impact on our affective, cognitive, and behavioral responses.

I discussed earlier the specific value of Icaros during Ayahuasca ceremonies and the physiological and spiritual role of sound waves in amplifying a psychoactive experience. Music or songs used in psychedelic sessions can quickly arouse our nervous system, with notable benefits, including activating the reward system. Indeed, music can be a rewarding stimulus akin to other rewards such as love, food, or sex. Dopamine, a key neurotransmitter in reward processing, plays a significant role in both the anticipatory and consummatory phases of the rewarding effect of music. Music's rewarding nature activates crucial areas of the mesocorticolimbic system (PRIMAL).

Music also stimulates norepinephrine and *endogenous opioids* (endorphins). The term "endogenous" indicates that these opioids are produced within the body, distinguishing them from exogenous opioids, which are introduced from external sources, such as morphine and other opioid medications. Endogenous opioids are crucial for enhancing both desire and pleasure associated with stimuli.

Meanwhile, music's ability to mitigate stress is well-documented, especially in clinical settings. It can reduce pre-surgery anxiety in patients and lessen the pain post-surgery. Music's stress-relieving properties extend to everyday scenarios, like challenging tasks at work or school. Stress and aging adversely affect immune responses, potentially leading to systemic inflammation. Since music can generate positive emotions, it logically follows that it can bolster the immune system (298). Animal studies have shown that music can reduce the progression of lung metastases caused by carcinosarcoma cells. Similarly, relaxing music has been found to decrease stress in both critically ill patients and healthy individuals.

But there is more. Listening to music activates the auditory cortex and stimulates complex neural networks that process sound, rhythm, melody, and harmony. Music engages auditory processing and memory, attention, emotion, and motor functions, leading to enhanced cross-modal plasticity in the brain. Dopamine release triggered by music supports positive mood states and motivation conducive to learning and memory. Music can also enable the release of serotonin and endorphins, enhancing mood and reducing stress and anxiety. As a result of these rich chemical responses, emotionally charged music can significantly improve memory encoding and retrieval processes.

Meanwhile, music can activate mirror neurons involved in empathy and understanding others' emotions (299). *Rhythmic Auditory Stimulation* is a technique that uses the rhythmic components of music to improve movement in rehabilitation settings (300). It can activate the motor cortex, aiding motor skill development, coordination, and rehabilitation, especially in stroke patients.

Finally, music can stimulate the production of proteins improving neurogenesis and synaptic plasticity. Regular engagement with music can enhance synaptic connectivity, particularly in the auditory cortex and related areas. Listening to music enhances pattern recognition skills, essential for cognitive functions and learning. It activates the limbic system, the brain's emotional center, aiding in emotional processing and empathy. And, of course, music often has a social component, enhancing social cognition and interaction, which is essential for overall brain health.

When I came to Peru, I had a profound realization: I was completely addicted to the news. Every morning, I would reach for my phone or laptop, scrolling through headlines, updates, and opinions as if they were essential fuel for the day. But instead of energizing me, it left me feeling anxious, reactive, and drained before I started. That experience was a wake-up call. In Peru's slower, more intentional pace of life, I began to see how this habit robbed me of clarity and creativity. I decided to make a change.

Now, my mornings look very different. Instead of news, I start my day writing while listening to evocative music. This combination helps me center myself and tap into a deeper part of my mind. Writing clears the clutter of my thoughts, while the music stirs my emotions and sets a tone of inspiration. It is not just a habit but a ritual that has transformed how I approach daily. By stepping away from the constant buzz of information, I have gained something far more valuable: a calm, focused mind and a creative start to the day.

In summary, listening to evocative music can significantly enhance neuroplasticity through various neurobiological mechanisms, including activating multiple brain regions, neurotransmitter release, stress reduction, enhancement of learning and memory, emotional processing, and improvement in motor skills. The integration of these processes contributes to the overall positive impact of music on brain plasticity.

Open Up to an Instrument

Playing a musical instrument can indeed increase neuroplasticity, and the neurobiology behind this mechanism is multifaceted and profound. Music-making involves complex processes in the brain, impacting various aspects of neuroplasticity (301)Playing an instrument requires fine motor skills, which activate and strengthen neural pathways in the motor cortex. The coordination of auditory feedback with motor actions during instrument playing enhances the connections between auditory and motor regions in the brain. Playing music involves skills like memory, attention, and problem-solving, engaging the prefrontal cortex and improving executive functions.

Figure 100: Open Up to an Instrument (AI

Musicians are often required to read music while playing their instrument alone or with other musicians. All these tasks improve multitasking abilities and working memory. Learning and playing music refines auditory processing abilities, making musicians more adept at distinguishing sounds, a skill that also translates to better language abilities (302). Musicians often show enhanced neuroplastic changes in brain regions associated with language processing.

Playing an instrument can engage the brain's emotional centers, like the limbic system, enhancing emotional processing and empathy. Engaging in complex activities like playing an instrument can stimulate the growth of new neurons, particularly in the hippocampus, which is vital for memory and learning. It requires the integration of sensory information (auditory, visual, tactile), leading to increased cross-modal plasticity in the brain.

Playing my handpan in Peru was a profound and joyful experience, unlike anything I had felt before. The stunning beauty and tranquility of the surrounding nature seemed to amplify the instrument's resonance, turning each note into something deeply sacred. The act of playing felt like a dialogue with my surroundings. The soft, meditative tones of the handpan blended seamlessly with the rustling leaves, the distant hum of wildlife, and the whispers of the jungle. It was as if the instrument and the environment were in perfect harmony, creating a grounding and transcendent soundscape. What made it even more special was the simplicity of those moments. Sitting under the open sky, often surrounded by mountains or near a flowing stream, I felt entirely present. Each note carried a sense of gratitude—for the peace of the moment, the gift of music, and the connection I felt to something greater. Playing the handpan in Peru was about creating music and being part of something timeless and universal. It became a meditation, a prayer, and a celebration all at once—a gift that will always stay with me.

Meanwhile, studies have shown that musicians often have increased cortical thickness in areas related to playing music, indicating structural changes in the brain (303). For instance, musicians typically have a larger *corpus callosum*, the bridge between the brain's two hemispheres, suggesting improved interhemispheric communication. The level of musical training can even affect the speed of postoperative language recovery (304, 305). Finally, music releases dopamine, which is associated with pleasure and reward. It enhances mood and reinforces learning and memory processes. Playing music can reduce stress and anxiety, creating a more conducive environment for neuroplasticity.

In conclusion, playing a musical instrument involves and enhances a broad range of brain functions and areas, leading to increased neuroplasticity. These benefits range from improved motor and auditory skills to enhanced cognitive functions, emotional processing, and structural brain changes, all contributing to the brain's ability to reorganize and form new neural connections.

Open Up to Laughter

Figure 101: Open Up to Laughter (AI)

Laughter, an instinctual human response, offers many affective, emotional, and cognitive benefits. This spontaneous vocal expression, often associated with playfulness and humor, can strongly influence our emotions and social interactions. Historically, thinkers like Plato, Aristotle, Descartes, Darwin, and Freud have pondered over laughter's role in human nature. However, the neuroscience of laughter is relatively recent.

Anthropologists view laughter as a primordial auditory signal, possibly linked to early forms of communication like calls, cries, and songs. Intriguingly, genuine laughter is an instinctual reaction, usually triggered by an act, thought, or feeling. This highlights its roots in the PRIMAL brain, though the RATIONAL brain's activation often amplifies it. As Dr. Robert Provine points out, people frequently incorrectly rationalize why they laugh, assuming it to be a conscious, deliberate choice (306). His research also demonstrated that laughter is not just about humor but plays a fundamental role in group dynamics and nonverbal communication.

Laughter's distinct pattern arises from the PRIMAL brain's neuromuscular control over our vocal system. That explains why forced laughter often feels unnatural. Contrary to common belief, laughter is not exclusive to humans, further underscoring its evolutionary significance and the PRIMAL brain's dominant role in triggering it. However, laughter is also regulated by higher-level linguistic processes controlled by the RATIONAL brain, as it does not disrupt sentence structure and typically accentuates the emotional tone of speech. Laughter's influence on neurophysiology is extensive, impacting affective, emotional, and cognitive brain networks.

Laughter's contagiousness is key to its appeal as a self-love practice. We can tap into laughter's healing power simply by observing humorous situations. This response is partly due to *mirror neurons*, which mimic the neural activity of behaviors or thoughts we observe in others. Watching funny or surprising events, especially involving others in mishaps, can evoke a sense of relief and joy (307). Humor often arises from unexpected deviations in stories or events, catching us off guard and triggering a surprised response. I suggest that the basis of what our brain finds funny is rooted in humor's disruptive nature, often presenting information or a twist we do not expect.

Meanwhile, laughter primarily evokes joy, a PRIMAL brain-based emotion. Shared laughter enhances social bonding and engagement, while higher brain structures from the RATIONAL brain are also engaged. PET-MR scans indicate that laughter boosts endorphin release in the brain. Endorphins, produced by the pituitary gland and hypothalamus, elevate pain tolerance and overall well-being. Laughter in social settings signals safety and happiness to the PRIMAL brain. Similarly, laughing alone while watching a show or simply reading jokes releases endorphins, strengthening self-relationship as an act of self-love.

Research from the exciting field of *embodied cognition* suggests that physical expressions like smiling can internally reinforce our emotional experiences. Embodied cognition emphasizes the role of the body (or physical responses like smiling or laughing) in shaping the mind (308). It challenges traditional cognitive theories that treat the mind as a processor of abstract information disconnected from the physical body. Instead, embodied cognition suggests that cognitive processes are deeply rooted in the body's interactions with the world. This feedback loop between the body and emotions explains why even forced laughter or smiling can convince the brain of happiness (309).

Laughing can indeed contribute to increasing neuroplasticity, and the neurobiological mechanisms behind this involve various processes in the brain.

Laughing reduces cortisol levels, which is beneficial for cognitive functioning and neuroplasticity, as chronic stress can inhibit the formation of new neural connections. By reducing stress and increasing feelings of pleasure, laughter can create a more conducive environment for learning and memory. Indeed, the positive emotional state of laughter can lead to enhanced attention and concentration, facilitating cognitive processing. Laughter can also activate mirror neurons, essential for empathy and understanding social cues and potentially enhance social cognition (310).

Meanwhile, laughter can lead to synchronized gamma wave activity in the brain. Gamma waves are associated with higher cognitive functions like perception, problem-solving, and consciousness (311). This may explain why laughter can broaden perspective and enhance creativity, which are mental aspects of effective problem-solving and adaptability. Meanwhile,

Finally, laughter improves blood flow and increases body oxygenation, including the brain, which can contribute to better brain health and function. After laughing, muscles remain relaxed for up to 45 minutes. Our laughing muscles include facial muscles like the *zygomaticus major*, which lifts the corners of the mouth, and the *orbicularis oculi*, which creates crinkles around the eyes for genuine laughter. The diaphragm and intercostal muscles drive the rhythmic breathing pattern of laughter, while the abdominal muscles contract during deep, uncontrollable laughs. This creates physical relaxation that can help alleviate the physical symptoms of stress.

Every evening, as the day winds down, I retreat to my favorite spot, ready for my nightly ritual: watching a funny show from France. It is not just any show—it is a bold, irreverent satire that leaves no topic untouched, from the absurdities of daily life to the weightiest, most serious global issues. Through the lens of brilliant satirist Philippe Caverivière (often called the French Trevor Noah!), even the gravest subjects are transformed into something hilarious yet thought-provoking.

As the first joke lands, I feel the corners of my mouth lift almost involuntarily. By the second or third punchline, I am in full laughter—deep, belly-shaking, uninhibited. My wife often sits nearby, quietly amused, watching me with a smile. She laughs at times but mostly observes, seeing how the stress and tension of my day visibly dissolve with each burst of laughter.

This ritual is powerful because it is more than just a moment of joy; it is a workout for my brain and a balm for my soul. The humor challenges me to think differently, to reframe even the most difficult topics through a lens of absurdity and wit. It is a kind of mental gymnastics, engaging neuroplasticity as my brain makes new connections and finds fresh perspectives. At the same time, laughing releases endorphins, flooding me with relief and well-being.

By the end of the episode, I am lighter, clearer, and more connected to a sense of joy. This nightly ritual has become more than just entertainment—it is a way to reconnect with the playful, open parts of myself that the day's demands often bury. It reminds us of the power of humor to make us laugh and heal, transform, and open us up to life.

In summary, laughter can positively impact neuroplasticity by releasing neurotransmitters that enhance mood and reduce stress, improve cognitive functions, foster social connectivity, synchronize brainwave activity, provide physical relaxation, and improve creativity. These processes collectively contribute to a brain environment conducive to neuroplasticity.

Open Up to Uplifting Movies

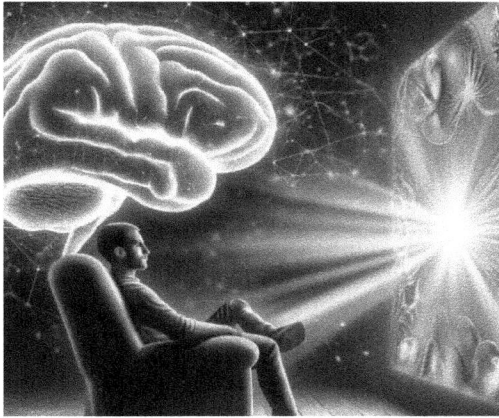

Figure 102: Open Up to Uplifting Movies (AI)

I have spent over two decades exploring how stories impact our brains, leading me to view connecting with movies as an essential plasticity habit. Creating and sharing stories is an ancient human tradition I extensively discussed in Chapter FOUR. It likely began as early as humans gathered around fires, telling tales of survival, debating survival strategies, or explaining weather patterns.

But what makes movies so captivating, and how do they benefit our brain's plasticity? Movies based on uplifting stories improve our likelihood of remembering information. Their narrative structure aids in recalling valuable lessons. Narrative transportation immerses us in a story as if we were part of it. Moreover, stories can subtly influence our beliefs, often bypassing rational counterarguments. As I explain in *The Persuasion Code*, framing a sales pitch as a story can be a highly effective tactic.

Compelling stories have a unique narrative architecture, simplifying the retrieval of the entire plot from a few key elements. This efficiency makes stories particularly brain-friendly. They captivate our attention at the outset and often culminate in an emotional climax I call *an emotional lift*. Emotional lifts are moments of intense emotional transition during which we encode some of the most significant events of our lives. Over the last two decades, my research has confirmed that the most impactful stories always include potent emotional lifts, often transitioning from negative to positive emotions.

According to neuroscientist Antonio Damasio, storytelling is a natural, implicit brain function integral to human culture (61). He posits that storytelling organizes human experience into coherent and meaningful narratives as an extension of the brain's inherent functions. This perspective highlights the importance of narrative in understanding the self and the world, suggesting that storytelling is not just a cultural artifact but a fundamental aspect of human cognition and emotional life.

We experience life vicariously through movies, amplified by our mirror neurons. Curiously, we quickly empathize with characters and feel their emotions like our own since the PRIMAL brain does not distinguish between fiction and reality. Movies often evoke many emotions, activating the limbic system. This emotional engagement can enhance empathy and emotional intelligence.

So, watching movies can positively influence neuroplasticity through various neurobiological mechanisms (312, 313). Movies provide a rich sensory experience that engages the brain's auditory, visual, and emotional processing systems, enhancing cognitive flexibility and multisensory integration. Following complex narratives and visual sequences in movies requires sustained attention and concentration, which can strengthen these cognitive abilities.

Meanwhile, watching enjoyable or relaxing movies can reduce cortisol levels, the body's stress hormone. Movies that evoke positive emotions can release neurotransmitters like dopamine and serotonin, enhancing mood and cognitive function. This is why people who suffer from stress and anxiety can benefit from movies as long as they apply discernment in their viewing habits. Violent or depressing movies will not trigger the same production of neurochemicals as calm, positive, and inspiring stories.

Also, engaging storylines and characters can encode memory banks automatically, particularly when they resonate emotionally or personally with the viewer. So does recalling and reflecting on movie scenes and narratives (314). This means that we can store disturbing images or movie sequences we do not intend to memorize, often revisiting the content through our nightmares.

Finally, movies stimulate imagination, encouraging creative thinking and exploring new ideas and perspectives. Educational or thought-provoking movies can provide new information and insights, contributing to learning and acquiring knowledge that can stimulate the production of Brain-Derived Neurotrophic Factor (BDNF). This process supports neuron survival, differentiation, and synaptic plasticity.

In summary, watching uplifting movies can enhance neuroplasticity by engaging brain functions such as cognitive and emotional processing, memory formation, stress reduction, social cognition, creativity, and learning. These processes collectively stimulate the brain, contributing to its ability to adapt and reorganize neural pathways.

Open Up to Gardening

Gardening can positively influence neuroplasticity through several neurobiological mechanisms. It is a multifaceted activity that uniquely engages the brain, supporting the development and strengthening of neural connections (315). Gardening involves touch, smell, sight, and sometimes taste. This multisensory interaction stimulates various brain regions while enhancing sensory processing and integration.

Figure 103: Open Up to Gardening (AI)

Gardening activities can improve neural connectivity, especially in areas responsible for sensory integration. Planning a garden layout, remembering to tend the plants, and scheduling tasks engage executive functions like planning, problem-solving, and working memory. Indeed, focusing on gardening tasks can improve attentional capacities and concentration, engaging and strengthening the prefrontal cortex. Handling tiny seeds, planting, weeding, and pruning requires precise motor skills and hand-eye coordination.

Gardening often involves physical activities like digging, lifting, and raking, which benefit overall health and enhance brain health through increased blood flow. The calming nature of gardening can reduce stress and promote a state of mindfulness, contributing to emotional regulation and a healthy brain environment conducive to neuroplasticity. Gardening can activate the brain's reward pathways, releasing neurotransmitters like dopamine, a critical molecule associated with pleasure and motivation. Learning about different plants, gardening techniques, and ecosystems provides novel cognitive challenges essential for brain development and plasticity. Community gardening or sharing gardening experiences can enhance social interaction, which is necessary for emotional health and mental function (316). Engaging with nature through gardening can improve mood, cognitive function, and overall well-being. Growing your own food can lead to healthier eating habits, benefiting brain health.

For my wife and I, tending to the nature around our property has become a cherished daily ritual that is as vital for our relationship as it is for the land itself. Each morning or evening, we step outside together, hands in the soil, pruning shears in hand, or simply walking among the trees and flowers. This shared practice is more than a routine—it is a profound way to connect with nature and each other.

The act of gardening fosters a partnership in every sense of the word. Together, we experience the joys of watching the flowers we planted bloom or savoring the fresh fruits of our labor. We celebrate the triumphs, like a flourishing tree that shades our home or a newly sprouted vegetable. These moments bring laughter, gratitude, and a sense of accomplishment that deepens our bond.

At the same time, the challenges we face in caring for our landscape—battling plant disease, deciding to cut down a tree, or witnessing the subtle but undeniable effects of climate change—demand teamwork and communication. Navigating these difficulties together strengthens our relationship, reminding us that growth, like love, requires both effort and patience.

This ritual is grounding for both of us. It is a time to disconnect from the busyness of life and reconnect with what truly matters: the natural world and each other. As we nurture the land, we encourage our partnership, building a rhythm of care, cooperation, and shared purpose that sustains us in more ways than one. For us, this daily communion with nature is not just good for the environment—it's vital for the health and harmony of our relationship.

In summary, gardening is a rich and diverse activity that engages multiple brain functions and regions. It involves sensory stimulation, cognitive and motor skills, stress reduction, emotional regulation, learning, social interaction, and a connection with nature. These aspects contribute to the enhancement of neuroplasticity, supporting the brain's ability to adapt and develop new neural pathways.

Open Up to Cooking

Cooking can positively influence neuroplasticity because it is a daily necessity and a complex cognitive and sensory process that engages multiple brain areas and functions, thereby contributing to the brain's ability to form new neural connections (317). Cooking involves using all our senses —sight, smell, taste, touch, and even hearing, like listening to the sizzle of food in a pan!

This multisensory engagement stimulates different brain regions and enhances sensory integration. Coordinating these sensory inputs requires complex neural connectivity, enhancing the brain's ability to process and integrate diverse sensory information.

Cooking requires planning, decision-making, multitasking, and time management, which engage the prefrontal cortex and improve executive functions. Following recipes and keeping track of various ingredients and steps in the cooking process involves working memory, which is essential for learning and cognitive flexibility. Chopping, stirring, and other cooking actions require precise hand movements,

Figure 104: Open Up to Cooking (AI)

which enhance fine motor skills and coordination, engaging the motor cortex and associated neural pathways. Experimenting with recipes and ingredients fosters creativity, stimulating neural networks in creative thinking and problem-solving.

When approached as a mindful activity, cooking offers relaxation and stress reduction. This aspect of cooking is particularly beneficial for overall brain health and neuroplasticity. (318). Cooking, especially when it leads to a successful and enjoyable meal, can activate the brain's reward system, releasing neurotransmitters like dopamine. Learning new cooking techniques or recipes provides novel cognitive challenges, essential for continued brain development and plasticity.

Cooking and sharing meals can also facilitate social interaction, enhancing emotional health and mental function. Finally, cooking different cuisines can serve as a means of cultural exploration, broadening knowledge and perspectives. Also, cooking at home often leads to healthier eating habits, directly benefiting brain health and function.

My passion for cooking began during the pandemic in 2020 when the world slowed down, and I spent more time in the kitchen. What started as a necessity quickly transformed into a deeply fulfilling ritual—a creative outlet that engages my senses, honors the food we consume, and connects me to something larger than myself.

Each evening, cooking is more than just preparing a meal; it's an opportunity to create. The kitchen becomes my canvas, the ingredients my palette. I love experimenting with flavors, textures, and techniques, blending the familiar with the adventurous. It is a process that invites focus and mindfulness, turning ordinary moments into something extraordinary.

Cooking also awakens my senses—the sizzle of garlic in a pan, the vibrant colors of fresh vegetables, the aroma of herbs and spices, and the satisfying feel of dough beneath my hands. These sensory experiences ground me in the present, reminding me to savor the food and the act of making it.

Most importantly, cooking has become a way to honor the food we put in our bodies. I am conscious of where it comes from, the effort it took to grow or produce, and how it nourishes us. Preparing a meal is a small act of gratitude, respecting the earth and the people who made the ingredients possible.

This nightly ritual, born out of a challenging time, has become one of the great joys of my life. It is not just about eating—it is about connecting to creativity, nature, and the love that goes into feeding ourselves and those around us.

In summary, cooking engages and enhances various brain functions, including sensory integration, cognitive abilities, motor skills, creativity, emotional well-being, and social interaction, all contributing to neuroplasticity. These processes stimulate and challenge the brain, supporting its ability to adapt and develop new neural connections.

Open Up to Board Games

Playing strategic board games like chess or Go can significantly increase neuroplasticity (319). These games require a combination of cognitive skills, including strategic planning, problem-solving, and memory, all of which contribute to developing and strengthening neural pathways.

Research on the benefits of playing board games has shown positive mental and physical health impacts. A systematic literature

Figure 105: Open Up to Board Games (AI)

review and meta-analysis highlighted that non-digital board games, can significantly improve health outcomes across diverse populations (320). The review encompassed 21 studies, revealing that board games can enhance plasticity and potentially modify health-related habits and outcomes. Meanwhile, non-digital games like chess and Go require players to plan, strategize, anticipate opponents' moves, and make decisions, engaging the prefrontal cortex responsible for executive functions. These games demand high-level problem-solving skills and logical thinking, stimulating neural networks associated with these cognitive processes.

Remembering previous games, anticipating opponents' strategies, and tracking complex sequences of moves involve both working memory and long-term memory systems. Regularly playing these games enhances pattern recognition skills, which are crucial for developing effective game strategies and making predictions.

Additionally, focusing on a game for extended periods improves concentration and attentional control, engaging and strengthening the brain's attention networks. Intellectually stimulating activities like chess or Go can provide a healthy outlet for stress and contribute to emotional regulation. Winning or successfully executing a strategy in the game can activate the brain's reward system, releasing neurotransmitters like dopamine. The cognitive challenges posed by these games can stimulate the growth of new neurons and synaptic plasticity, particularly in regions involved in complex thought and memory (321).

Meanwhile, playing these games with others can enhance social cognition and empathy as players learn to anticipate and understand opponents' thoughts and strategies. Competitive play can further stimulate cognitive and emotional engagement, improving the overall benefits for the brain. Most games require thinking several moves ahead, enhancing strategic planning, foresight, and critical cognitive skills. Also, the focus needed in chess and Go can create a flow state, similar to mindfulness, aiding mental clarity and presence.

As I shared in Chapter ONE, I watched the documentary *AlphaGo* and found it incredibly insightful regarding how board games and AI may enhance our plasticity. AlphaGo was released in 2017. The film chronicles the journey of the artificial intelligence program developed by Google DeepMind and its historic match against Lee Sedol, one of the world's top Go players. The documentary showcases the capabilities of AI in mastering a game that requires high levels of intuition, strategic thinking, and creativity, traits traditionally considered exclusive to human intelligence. The documentary demonstrates how AI could extend, complement, and sometimes surpass human cognitive abilities nearly eight years ago. We know now that AI capabilities, especially generative AI, have met and exceeded those predictions in many ways.

At the time, though, AlphaGo's success was already based on machine learning and neural networks, which mirror certain aspects of human brain function. The AI program learned from vast amounts of Go game records and improved through reinforcement learning, a concept analogous to how humans learn and adapt through experience.

The documentary indirectly highlights the role of human neuroplasticity. AI challenges, like the program AlphaGo, push human players to adjust their strategies and thought processes, demonstrating the human brain's capacity to learn, adapt, and create strategies in response to novel challenges.

Furthermore, the matches between AlphaGo and Lee Sedol highlighted the interplay between intuition and analytical thinking. While AI relies on algorithms and probability, human players often depend on guesses, prior experience, and deep understanding rooted in neuroplasticity. The film suggests that AI's development could enhance learning tools and methodologies, potentially aiding human cognitive development. The documentary also delves into AI's emotional and psychological impacts on humans, particularly on experts in fields where AI shows superior performance. It touches on how humans could cope with or adapt to these emerging technologies with significant neuroplastic changes.

In summary, playing strategic board games like chess or Go involves a complex interplay of cognitive processes, including executive functions, problem-solving, memory, concentration, and strategic thinking. These activities stimulate and challenge the brain, leading to increased neuroplasticity by developing new neural connections, enhancing cognitive skills, and improving brain health overall.

Open Up to Non-violent Video Games

Research indicates that non-violent video games can enhance neuroplasticity, enabling the brain to reorganize itself by forming new neural connections. Studies suggest that certain types of video games, especially those requiring strategy, problem-solving, and spatial navigation, can improve cognitive functions like attention, memory, and problem-solving skills.

For example, a study by Dr. Gregory Clemenson and Dr. Craig Stark (322) demonstrated that playing 3D video games could enhance hippocampal-associated memory. It is already known that exposing animals to a more stimulating environment can improve hippocampal function and performance while mediating memory tasks.

Another research by Dr. Joaquim Anguera and his colleagues found that a custom-designed video game improved multitasking abilities in older adults (323). This study revealed that cognitive control, crucial for goal-oriented interaction with our environment, declines with age, impacting multitasking abilities. However, using a custom 3D video game called NeuroRacer designed to study and improve cognitive function, researchers found that older adults could improve their multitasking skills to surpass those

Figure 106: Open Up to Non-violent Video Games (AI)

of untrained young adults, with benefits lasting for six months. This improvement also enhanced untrained cognitive abilities like sustained attention and working memory, indicating the potential of specific video games to strengthen the aging brain's cognitive control system and demonstrate the brain's capacity for neuroplasticity.

Open Up to Learning

Synaptic plasticity is the physical basis for learning and memory. Therefore, learning is only possible through neuroplasticity. The learning process involves several neurobiological mechanisms contributing to this phenomenon (324). Learning new information or skills leads to new synaptic connections between neurons.

Meanwhile, repeatedly practicing or revisiting learned material strengthens synaptic connections, a concept known as

Figure 107: Open Up to Learning (AI)

long-term potentiation, a neurobiological process by which repeated stimulation of neurons strengthens synaptic connections, enhancing learning and memory storage in the brain. A famous study by Dr. Eleanor Maguire and her colleagues discovered that London taxi drivers, who must memorize the city's complex layout, have larger hippocampi than the general population (325). This study also suggests that extensive spatial navigation experience (repetitions!) can lead to a localized increase in the hippocampus volume, a brain area critical for memory and navigation.

Remarkably, certain types of learning, especially those involving complex cognitive tasks or spatial navigation, can stimulate the growth of new neurons, particularly in the hippocampus. This process, called neurogenesis, can enhance learning by rewiring neural networks while enabling more efficient processing and retrieval of information. The brain constantly allocates resources (like blood flow and glucose) to activate areas involved in the learning process. Also, neurotransmitters play critical roles in learning. For instance, dopamine is associated with reward and motivation, enhancing learning, while acetylcholine is involved in attention and memory consolidation.

Meanwhile, many studies have demonstrated how learning is used in therapeutic interventions to repair and restore critical motor or mental functions (326). Learning incredibly complex and challenging material can also engage and strengthen executive functions of the RATIONAL brain, like problem-solving, critical thinking, and decision-making. Hence, learning promotes cognitive flexibility, allowing the brain to adapt to new situations and solve problems more efficiently. Continuous learning and mental stimulation can also help protect the brain against cognitive decline associated with aging and neurodegenerative diseases. Finally, learning can reduce stress levels, benefiting overall brain health and cognitive function.

In summary, learning enhances neuroplasticity through synaptic plasticity, neurogenesis, remodeling of neural circuits, neurotransmitter modulation, improvement in cognitive functions, reduction in cognitive decline, emotional regulation, and increased brain connectivity. These processes contribute to the brain's adaptability, resilience, and capacity for continued growth and development.

Open Up to Traveling

The relationship between neuroplasticity and travel involves how new experiences, such as those encountered while traveling, can stimulate the brain and lead to neural changes. Because humans have traveled to every continent, traveling must have some biological basis.

Research has shown that the spatial navigation required when exploring new areas can lead to structural changes in the brain (327). The study on London taxi drivers I mentioned earlier found that navigating the city's streets was associated with increased gray matter volume in the hippocampus. Similarly, traveling stimulates the brain by exposing individuals to new experiences, environments, cultures, and challenges, strengthening brain circuits associated with memory and learning.

Figure 108: Open Up to Traveling (AI)

Meanwhile, novelty is a crucial driver of neuroplasticity, as it encourages the brain to adapt to new information and situations. Novel experiences can promote the growth of new neurons and the formation of new neural connections, enhancing cognitive flexibility. Navigating unfamiliar places, learning new languages, and adapting to different cultures require planning, problem-solving, and decision-making, all of which engage the prefrontal cortex and improve executive functions.

Adjusting to new environments and customs enhances cognitive flexibility and open-mindedness. Also, traveling provides diverse and rich sensory experiences, from new sights and sounds to tastes and smells. This multisensory input stimulates various brain areas and enhances sensory processing and integration. Travel often involves intense emotional experiences, both positive and challenging, which engage the limbic system and contribute to emotional resilience.

Traveling can provide a break from daily routines and stress, promoting relaxation and emotional well-being. Being open to meeting new people and engaging in cultural exchange can improve social skills and empathy, stimulating areas of the brain involved in social cognition. Exposure to new languages during travel challenges the language centers of the brain, contributing to linguistic skills and cognitive flexibility.

Travel often involves higher levels of physical activity, such as walking, hiking, or exploring, which benefits overall health and brain function. Being in new and engaging environments can promote a state of mindfulness, where one is fully present and aware, which is beneficial for mental clarity and consciousness. New experiences, cultures, and environments can inspire creativity, stimulating neural networks associated with creative thought and innovation.

Traveling to 43 countries across six continents has been an incredibly transformative experience. It has genuinely "opened me up" to different cultures, traditions, and ways of being that I could never have fully understood by staying in one place. Exploring Europe, Asia, Australia, Southeast Asia, South America, and beyond has taught me to see the world through a broader lens and embrace the beautiful diversity of humanity.

Each journey has challenged my assumptions and broadened my perspective. Whether connecting with locals in a small Greek village, meditating in a serene temple in Thailand, or attending Sufi dancing ceremonies in Turkey, I learned the importance of stepping outside my comfort zone. These experiences rewired not just how I think but how I feel and relate to others, fostering a deeper sense of empathy and adaptability.

Travel has also helped me develop a sense of flow and openness. The unpredictability of a new environment forces me to let go of control and be in the moment—navigating unfamiliar streets in Sao Paulo, embracing a new language, or participating in foreign rituals. This surrender to the unknown has brought me some of my life's greatest joys and lessons.

Above all, traveling has been a mirror for self-discovery. By immersing myself in so many different cultures, I've had the opportunity to reflect on my narratives and habits. It has helped me rewrite parts of my story, let go of patterns that no longer serve me, and cultivate a deeper sense of gratitude for the rich tapestry of life. This journey has been more than just seeing the world—it's been about opening my heart and mind to everything it offers.

In summary, traveling can enhance neuroplasticity through exposure to novel experiences, cognitive challenges, rich sensory stimulation, emotional engagement, social interactions, physical activity, mindfulness, and creative inspiration. These varied and dynamic experiences encourage the brain to adapt, reorganize, and strengthen neural connections, improving cognitive function, flexibility, and overall brain health. Although direct studies linking travel with increased brain plasticity in humans are scarce, the components of travel experiences align with activities that stimulate the brain and potentially enhance its plasticity. These findings underscore the value of engaging in diverse and novel experiences to boost cognitive functions and contribute to brain health over the lifespan.

Creating art can significantly enhance neuroplasticity through a variety of neurobiological mechanisms. For example, Schlegel and his colleagues explored the neural effects of visual art production on structural brain connectivity. Using diffusion tensor imaging (DTI), they found that art students engaged in intensive visual art training showed significant changes in white matter structure, suggesting that creating art influences brain connectivity and

Figure 109: Open Up to Art (AI)

plasticity (328). Therefore, artistic activities stimulate multiple brain areas, encouraging the formation and strengthening of neural connections. Art creation involves visual, tactile, and sometimes auditory senses. This multisensory engagement stimulates the corresponding sensory areas of the brain, enhancing sensory processing and integration. Coordinating these sensory inputs requires complex neural connectivity, which can improve with continued artistic practice.

Artistic creation involves decision-making, problem-solving, and planning, all of which engage and strengthen the brain's executive functions. Creating art requires visualizing the end product and understanding spatial relationships, which engages and enhances the brain's spatial reasoning abilities. Drawing, painting, sculpting, and other art forms require precise hand-eye coordination and fine motor skills, engaging the motor cortex and associated neural pathways. Art encourages creative thinking and innovation, stimulating neural networks related to creativity. Artistic activities promote divergent thinking, which generates multiple solutions to a problem, enhancing cognitive flexibility. Art allows for emotional expression and processing, engaging the limbic system in emotion regulation.

Dr. Anne Bolwerk and her colleagues conducted a study to investigate the impact of producing visual art on psychological resilience in adults. Their findings suggested that creating visual art can lead to functional brain changes, particularly in areas associated with emotional processing and stress regulation (329). This study supports the idea that artistic activities can enhance neuroplasticity and emotional well-being.

Meanwhile, engaging in art can be therapeutic and relaxing, reducing stress and creating a positive environment for neuroplasticity. Creating art can release dopamine, a neurotransmitter associated with pleasure and reward, enhancing mood and motivation (330). Art can serve as a medium for self-reflection and personal insight. Engaging in complex, creative activities like art can stimulate the production of proteins which supports neuron survival and synaptic plasticity.

In summary, creating art enhances neuroplasticity through multisensory stimulation, cognitive enhancement, motor skill development, creativity and problem-solving, emotional regulation, neurotransmitter modulation, reflective thinking, and synaptic plasticity. These processes collectively stimulate the brain, contributing to its adaptability and developing new neural pathways.

Spiritual practices are defined here as activities that deepen our connections to the self, a higher power, nature, or the entire cosmos. These practices are often, but not exclusively, rooted in religious or philosophical traditions and are intended to enrich the practitioner's inner life, offer a path to transcendence, or enhance personal growth. Many scholars believe that spiritual practices may have evolved to counterbalance the PRIMAL brain's overactivity, which is constantly focused on survival.

For instance, Dr. Andrew Newberg and J. Eversen explored how meditation and other spiritual practices might influence the brain's functioning, suggesting that these practices have developed to impact brain areas associated with attention, emotion, and stress response (174). They argue that spiritual practices can lead to relaxation, reduced stress, and enhanced mood by affecting the autonomic nervous system, altering neurotransmitter levels, and improving neural connectivity.

The hypothesis presented by Newberg and Iversen supports the idea that spiritual practices, including meditation, prayer, and ritual, are core adaptive mechanisms for calming the mind, reducing stress, and enhancing social cohesion and personal well-being. This perspective also aligns with broader evolutionary psychology and neuroscience theories that suggest human behaviors and practices, including those related to spirituality, have evolved to promote survival and adaptation.

Meanwhile, engaging in spiritual practices can lead to lasting changes in the brain, particularly in networks governing emotions and self-awareness. These practices often foster peace, social awareness, and compassion. Even short-term engagement in these activities can initiate neural changes, and practices like meditation have been linked to slowing the aging process and enhancing memory through brain plasticity.

I have already discussed how spiritual practices can strengthen neurological circuits of consciousness, empathy, and social awareness. For instance, the anterior cingulate cortex, part of the PRIMAL brain, becomes more active during spiritual practices, enhancing empathy and emotional regulation. This also benefits aging, as the anterior cingulate cortex, plays a role in learning, memory, and reducing anxiety. Research into the sense of oneness during spiritual practices has shown that activity in the parietal lobe decreases. This area integrates sensory information for spatial awareness; reduced activity may diminish the connection with one's sense of self, possibly leading to a feeling of unity with the focus of contemplation.

More importantly, spiritual practices alter neurochemistry and improve overall neuroplasticity. Therefore, I will highlight several forms of spiritual practice documented to foster new connections or neurogenesis.

Open Up to Meditation

Figure 110: Open Up to Meditation (AI)

As discussed earlier, meditation has profound benefits in raising consciousness, but also neuroplasticity (174, 331). Regular meditation improves attention and concentration, strengthening neural circuits involved in these cognitive processes. It enhances executive functions, such as planning, problem-solving, and decision-making, by engaging the prefrontal cortex.

Meditation helps reduce cortisol levels, the stress hormone, creating a more conducive environment for neuroplasticity, as chronic stress can hinder brain plasticity. Meditation, particularly mindfulness meditation, encourages heightened awareness and focus on the present moment, which can lead to increased mental clarity and neural connectivity.

Studies have shown that meditation can increase the density of gray matter in various parts of the brain, including areas involved in memory, emotional regulation, and self-awareness. Regular meditation has also been associated with increased cortical thickness, particularly in attention and sensory processing regions.

Meditation can improve emotional regulation by affecting the limbic system, particularly the amygdala, a critical brain area for emotion processing and response. Meditation has been shown to slow down the process of brain atrophy that naturally occurs with aging, thereby preserving cognitive function. Certain meditation practices promote synaptic plasticity, facilitating learning new information and adapting to new experiences.

Repeated meditation reinforces specific neural pathways, making them more efficient, particularly in areas related to focus and calmness. Meditation has been found to reduce activity in the default mode network (DMN), the brain network associated with mind-wandering and self-referential thoughts. This can lead to increased focus and reduced symptoms of anxiety and depression. Regular meditation can also enhance connectivity between brain regions, improving overall cognitive function.

In summary, meditation significantly impacts neuroplasticity through structural and functional changes in the brain. These changes contribute to improved cognitive functions, emotional regulation, stress reduction, and overall brain health, demonstrating the powerful influence of meditation on the brain's adaptability and resilience.

Open Up to Prayers

Figure 111: Open Up to Prayers (AI)

Whether as a meditative, contemplative activity, praying engages multiple cognitive and emotional processes in the brain (332). Therefore, praying can contribute to enhanced neuroplasticity. Prayer often requires focused attention and concentration, which can strengthen these cognitive abilities by improving the neural circuits involved. Prayers that require planning, reflection, and contemplation can engage and improve executive functions managed by the prefrontal cortex.

Dr. Andrew Newberg and his colleagues examined the effects of different types of meditation and prayer on neurological function (333). They found that these practices could significantly change brain activity, particularly in areas associated with attention, emotion regulation, and self-awareness. While this study focuses on brain activity rather than neuroplastic changes, it suggests that regular engagement in prayer could influence the brain in ways that promote neuroplasticity over time.

Prayer can also have a calming effect, engaging and regulating the limbic system involved in emotional processing. Dr. Michael Inzlicht and his colleagues explored the neural mechanisms behind the cognitive processing of religious belief, including prayer regulation (334). Their work highlights the role of the anterior cingulate cortex (ACC), a brain region associated with emotional regulation, error detection, and cognitive control. While not directly studying neuroplasticity, their findings hint at the potential for prayer to engage and

343

strengthen neural circuits involved in cognition and emotions. Thus, as a stress-reducing practice, prayer can lower cortisol levels, creating a healthier environment for neuroplasticity.

Prayer can be a form of mindfulness, promoting heightened awareness and focus on the present moment, leading to increased mental clarity and neural connectivity. For many, prayer involves empathizing and connecting with others through intercessory prayer or communal practices, enhancing the neural networks involved in social cognition and empathy. Dr. Ufle Schjødt et al. investigated the neural underpinnings of prayer using functional magnetic resonance imaging (fMRI). They observed that praying activated brain regions involved in social interaction, empathy, and theory of mind, which are critical for understanding and interacting with others (335). Their study suggests that prayer engages complex cognitive and emotional networks in the brain, which could, through regular practice, lead to changes in brain structure and function indicative of neuroplasticity.

Meanwhile, prayer, mainly when it provides a sense of peace or joy, can release neurotransmitters associated with reward and well-being. Regular prayer has been associated with increased cortical thickness in certain areas of the brain, including regions involved in attention and introspection (336). Prayer can also reduce activity in the default mode network (DMN). Regular prayer can enhance connectivity between brain regions, improving overall cognitive function and mental well-being.

In summary, praying can stimulate brain regions involved in spiritual experiences and personal growth, contributing to a sense of purpose and meaning. Praying impacts neuroplasticity by engaging and enhancing cognitive functions, emotional regulation, stress reduction, mindfulness, social cognition, and structural and functional aspects of the brain. These changes can improve cognitive function, flexibility, emotional well-being, and overall brain health.

Open Up to Your Dreams

You may find it curious that I would put dreams as a plasticity ritual. Still, evidence suggests that we can nurture our dreams to maximize neuroplastic benefits. Dr. Matthew Walker and Dr. Robert Stickgold proposed that sleep and periods associated with dreaming facilitate memory consolidation through neuroplastic processes. Their review suggests that sleep after learning encourages the neural reorganization necessary for memory consolidation, hinting at the role of dreams in this context (337). In their model of dream production, Nielsen and Lara-Carrasco proposed that dreaming reflects brain plasticity since it is actively involved in reorganizing emotional memory. They believe dreams may participate in the processing and integrating emotional experiences, supporting adaptive neuroplastic changes in emotional brain circuits (338).

Meanwhile, a study by Dr. Masako Tamaki and his colleagues found that during sleep, the brain reactivates and consolidates memories of motor skills learned while awake, a process essential for neuroplasticity. Though not exclusively focused on dreams, this study highlights the critical role of sleep, a state closely associated with dreaming, in facilitating neuroplastic changes related to learning and memory (339). Finally, other researchers

Figure 112: Open Up to Your Dreams (AI)

suggest that dreams may play a role in cognitive and emotional processing that indirectly supports neuroplasticity by reinforcing or modifying neural connections associated with learning and memory (340).

Clearly, nurturing your dreams is crucial to benefit from all the properties I discussed. This involves creating an environment and adopting practices that promote healthy sleep and potentially more meaningful or vivid dreaming. There are essential strategies to help nurture your dreams. First, stick to a regular sleep routine by going to bed and waking up at the same time every day, even on weekends. It helps regulate your body's internal clock and can lead to more consistent dream patterns. Ensure your sleeping environment is comfortable, quiet, dark, and relaxed. Consider using blackout curtains, eye masks, and earplugs if needed. Invest in a comfortable mattress and pillows.

Then, avoid screens (phones, tablets, computers, TV) at least an hour before bedtime, as the blue light emitted by digital screens can disrupt your body's ability to produce melatonin, a crucial hormone that regulates sleep. Avoid heavy meals, caffeine, and alcohol close to bedtime, as they can disrupt sleep. A light snack before bed, especially one that includes ingredients known to be sleep-inducing, like dairy or nuts, might be beneficial. Engage in relaxing activities before bed, such as reading, taking a warm bath, or practicing relaxation techniques like deep breathing, meditation, or gentle yoga.

Upon waking, write down whatever you remember from your dreams. This practice can enhance your recall of dreams and may lead to more detailed and vivid dreams over time. Reflect on your dreams and consider what they might mean to you. *Dream tending* is one method of working with dreams that emphasizes engaging with the images of dreams as if they were alive and capable of interaction. Developed by Dr. Stephen Aizenstat, this approach is based on the understanding that dreams are not just random neural firings in the brain but are meaningful experiences filled with vital information about our lives, relationships, and the deeper aspects of our psyche (341).

Unlike traditional dream analysis, which often involves interpreting a dream to find its symbolic meaning related to one's life, *Dream Tending* encourages individuals to interact with their dream images actively. This consists in seeing dream figures not merely as parts of oneself but as beings with the right to exist. The process allows for a dynamic engagement with these figures, fostering a dialogue that can lead to deeper insights and emotional healing.

Dream tending incorporates principles from depth psychology, particularly Jungian and archetypal psychology, and integrates aspects of phenomenology. The practice can lead to a profound connection with the unconscious, offering new perspectives on personal issues, creative blocks, and life transitions. It is used therapeutically to enhance psychological growth, creativity, and well-being and can make the experience of dreaming more meaningful and insightful.

Since stress and anxiety can negatively impact sleep and dreams, you can always look to effective ways to manage stress through other habits I have discussed to protect your sleep. Exploring lucid dreaming through techniques like reality checks and dream visualization can lead to a different level of interaction with dreams (342). *Lucid dreaming* is a phenomenon where a person becomes aware that they are dreaming while still in the dream. This state of consciousness allows the dreamer to control their actions within the dream or even the dream's narrative. Lucid dreams can occur spontaneously or be induced through various techniques. Remember, while these practices can enhance sleep quality and dream experiences, the nature of dreams can vary significantly from person to person. What works for one individual may not work for another, and finding a routine that suits you best is essential.

Indeed, dreams can positively impact neuroplasticity, the brain's ability to reorganize and form new neural connections. Dreams often involve emotional content, which can help process and integrate emotional experiences, an essential aspect of memory consolidation. For many, dreams can have a therapeutic effect, allowing the mind to work through and reduce stress and anxiety, creating a better neuroplasticity environment. Dreams' unique and often illogical nature can encourage creative, lateral, and unconstrained thinking. This can lead to novel insights or problem-solving approaches that may not occur during wakeful, linear thought processes. Dreaming can provide a 'safe space' for the brain to process complex emotions and experiences, contributing to emotional regulation and mental health. Good sleep quality, which includes healthy dreaming, is essential for overall brain health and cognitive function. Poor sleep can impair cognitive abilities and reduce neuroplasticity.

Finally, dreams might play a role in synaptic pruning, a process where extra neurons and synaptic connections are eliminated to increase the efficiency of neural transmission (343). During REM sleep, various brain regions are activated in a pattern like wakefulness, which may help maintain and strengthen neural circuits.

In summary, dreams, as a component of sleep, are integral to various cognitive and emotional processes that contribute to neuroplasticity. They play roles in memory consolidation, emotional regulation, stress reduction, creativity, learning, and overall brain health. Understanding the full impact of dreams on neuroplasticity is an ongoing area of scientific research.

Open Up to Stargazing

Stargazing, or the act of observing the night sky, can potentially have a positive impact on neuroplasticity. However, this area has not been extensively studied in the context of astronomy. The benefits are likely indirect and associated broader cognitive, emotional, and psychological effects of such an activity.

Figure 113: Open Up to Stargazing (AI)

Stargazing stimulates learning and curiosity about the universe, leading to cognitive engagement and acquiring new knowledge, thereby promoting neuroplasticity. Understanding celestial patterns and spatial relationships in the night sky can enhance spatial reasoning skills.

Dr Jennifer Stellar (what a name to cite for the benefits of stargazing!) and her colleagues explored the emotion of awe and its effects on well-being and cognitive function. Awe, which might be elicited by stargazing, is associated

347

with decreased self-focus, increased social connection, and a sense of being part of something larger than oneself. These psychological effects could contribute to neural changes over time, although this study did not examine direct links to neuroplasticity (344).

Observing the vastness of the night sky can be a tranquil and meditative experience. It reduces stress and promotes a state of mindfulness, both of which are beneficial for neuroplasticity. Experiencing awe while observing the stars can have a profound emotional impact, potentially leading to increased connectedness and well-being. Such emotional experiences can contribute to neural changes.

The beauty and mystery of the night sky can inspire creative thinking and imagination, stimulating parts of the brain involved in creative processes. Being outdoors and exposed to natural light patterns, including the darkness of the night sky, can help regulate circadian rhythms, improving sleep quality. As noted earlier, good sleep is crucial for brain health and neuroplasticity. Stargazing can be a social activity, often done in groups or with astronomy clubs, fostering social connections and discussions that can be cognitively stimulating.

While the direct effects of stargazing on neuroplasticity might not be as pronounced or well-studied as other activities like learning a new language or playing a musical instrument, it likely contributes to brain health and cognitive function in more subtle ways through relaxation, inspiration, cognitive engagement, and emotional experiences. As with many activities that enrich our lives, the benefits of stargazing may extend beyond what can be easily measured or observed in traditional neuroscientific research.

Open Up to Entheogens

The use of psychedelics for "journeying" or experiencing altered states of consciousness has garnered significant interest for its potential impact on neuroplasticity and mental health. As we discussed earlier, many studies have shown that psychedelics promote structural and functional plasticity while potentially providing significant therapeutic benefits (345, 346).

Since psychedelics can induce profound changes in perception, emotion, and cognition, they can promote the growth of dendrites, dendritic spines, and synapses, effectively increasing neural connectivity. There is evidence suggesting psychedelics may stimulate the development of new neurons, particularly in areas of the brain related to emotion and memory.

Psychedelics can disrupt entrenched patterns of thought, potentially aiding in breaking rigid cognitive patterns associated with SADAT conditions like depression, PTSD, or addiction. The altered states of consciousness induced by psychedelics can lead to increased creativity and novel problem-solving approaches. Many psychedelics act on serotonin receptors, influencing mood, cognition, and perception. This can lead to improved emotional processing and mood regulation. In controlled settings, psychedelics

Figure 114: Open Up to Entheogens (AI)

have shown promise as therapeutic agents, potentially aiding in the treatment of mental health disorders by facilitating deeply introspective and transformative experiences that can alleviate symptoms of SADAT.

Finally, some psychedelics can decrease the activity of the default mode network, leading to a sense of oneness and connectedness with the world (347). They induce the release of neurotransmitters, leading to various emotional and cognitive experiences. For more information on the specific role of entheogens on neuroplasticity, visit the APPENDIX section.

Open Up to Mantras

Mantras, originating from ancient spiritual traditions, particularly in Eastern cultures, have been used for millennia as a powerful tool for meditation and mindfulness. "Mantra," derived from Sanskrit, translates to "*an instrument of thought*." These vocalized phrases are often repeated for those who meditate regularly; they serve as a method to focus the mind and bring about a calming effect through the repetitive vocalization of specific sounds, words, or sentences. The

Figure 115: Open Up to Mantras (AI)

neuroscience behind the impact of mantras on the brain is a relatively new field of study and presents unique research challenges. A notable study by Berkovich and colleagues utilized functional Magnetic Resonance Imaging (fMRI) to explore the effects of mantra repetition on brain activity (348). They examined 23 participants, none of whom were regular meditators, as they repeated a single word and compared this activity to their resting-state brain function. The results were enlightening; mantra repetition decreased activation in various cortical networks, particularly the default mode network (DMN). As you recall, the DMN is involved in many internal thought processes, including self-referential thinking, planning, predicting future events, and mind-wandering.

For many spiritual practitioners, reducing activity in the DMN is crucial, as mind-wandering can lead to increased stress, anxiety, and depression. Therefore, the simple practice of mantra repetition can significantly impact the brain, specifically in diminishing DMN activity, and is a valuable technique in fostering mental wellness. Findings also suggest that repetitive chanting mantras can influence neural networks, potentially contributing to neuroplasticity and the brain's ability to reorganize and form new neural connections throughout life.

Meanwhile, reciting mantras requires focused attention, which can strengthen the brain's attention networks. Regular practice can make these networks more efficient, enhancing overall cognitive function. Mantra recitation can be calming and stress-relieving. Lower stress levels benefit neuroplasticity, as chronic stress can impair the brain's ability to form new connections. The rhythmic and repetitive nature of mantras can have a soothing effect on the limbic system, the brain's emotional center, aiding emotional regulation.

Mantras can promote a state of mindfulness, helping individuals to be present in the moment. This mindfulness is associated with changes in brain regions related to self-awareness and introspection (349). Mantra chanting can induce altered brainwave states, remarkably increasing theta wave activity, and is associated with deep relaxation and meditative states. The reflective aspect of mantra practice, especially when the mantras have specific meanings or intentions, can engage the prefrontal cortex, aiding in cognitive reappraisal and positive thinking.

Additionally, the rhythmic aspect of mantra chanting can lead to neural synchronization, where different regions of the brain exhibit coordinated activity, potentially enhancing cognitive efficiency and mental clarity. Regular meditation practices, including mantra chanting, have been linked to increased gray matter density in various brain parts, a sign of enhanced neuroplasticity.

In summary, the practice of mantras contributes to neuroplasticity by improving focus and attention, reducing stress, aiding emotional regulation, enhancing mindfulness, altering brainwave states, engaging in cognitive reappraisal, modulating the DMN, synchronizing brain activity, and even causing structural changes in the brain. These effects collectively support the brain's ability to adapt and develop new neural pathways. To incorporate this practice effectively, here are examples of self-love mantras I recommend:

- I love myself.
- Every day is a new beginning.
- I accept all that is good in my life.
- I acknowledge my self-worth.
- I am blessed.
- I am confident in my decisions.
- I am evolving positively.
- I make positive changes.
- My emotions do not define me.
- I am grateful.
- I deserve the positivity in my life.
- I believe in my abilities.
- I trust myself.
- I choose positive thoughts.
- Self-love is my choice.
- Happiness is within my grasp.
- I am worthy of a fulfilling life.
- I am deserving of love.
- I have all I need within me.
- I can overcome any challenge.
- My ideas have value.
- I release self-criticism.
- I embrace the person I am becoming.
- I contribute positively.
- I attract what I desire.
- I observe myself without judgment.
- I support myself.
- I remain true to my authentic self.

To prime your brain for self-love mantras, especially if you are experiencing high levels of SADAT, gradual steps involved in selecting and performing them are essential. Narrative transportation and mirror neuron engagement can aid this process. The approach involves observing others practicing mantras through short video clips, gradually incorporating the habit into daily practice, and eventually integrating it into a daily or weekly ritual. This systematic approach allows the brain to adapt and embrace these habits effectively, fostering a journey toward greater mental wellness and self-love.

Engaging in *OPEN* rituals is the key to any successful shift in your patterns of behaviors and thoughts. I have identified three categories of rituals. The key to your ability to sustain an *OPEN* life is to create a routine that can raise consciousness and promote new neuronal growth.

PRIMAL RITUALS (8) are designed to release the toxic loops activated by elevating levels of stress and anxiety, characteristic of the bottom-up effect.

- *Open up to your body*, like practicing yoga, dancing, or running, significantly maintains neuroplasticity. These activities promote the release of endorphins to alleviate pain, reduce stress, and enhance neurogenesis.
- *Open up to your breath* can help balance your autonomic nervous system's sympathetic (fight-or-flight) and parasympathetic (rest-and-digest) branches. Controlled breathing exercises can decrease activation in the amygdala, the brain region involved in emotional processing, particularly fear and anxiety responses.
- *Open up to nature* positively impacts the brain, enhancing cognitive function, emotional well-being, and neural connectivity. Spending time in nature can lower cortisol levels, reduce stress, and improve both cognitive functioning and mental clarity.
- *Open up to your food* provides essential nutrients that support brain health, cognitive function, and neuroplasticity.
- *Open up to your sleep* is essential for restoring and rejuvenating your brain, especially to waste products that accumulate during waking hours.
- *Open up to your pets* involves physical activity like walking, which can improve overall health, including brain health. Spending time with your pets can promote a state of mindfulness and presence, which is beneficial for mental clarity and neuroplasticity.

RATIONAL RITUALS (11) lower the Default Mode Network (DMN) activity while providing flow and comfort. These rituals trigger neurotransmitters that relax and stimulate the most evolved part of the brain while opening critical periods of neuroplasticity.

- *Open up to evocative music* activates the auditory cortex and stimulates complex neural networks that process sound, rhythm, melody, and harmony.

It also engages memory, attention, emotion, and motor functions, enhancing the brain's cross-modal plasticity.

- ***Open up to an instrument*** can engage the brain's emotional centers, like the limbic system, enhancing emotional processing and empathy. Engaging in complex activities like playing an instrument can stimulate the growth of new neurons, particularly in the hippocampus, which is essential for memory and learning.

- ***Open up to laughter*** reduces cortisol, the body's primary stress hormone. Lower cortisol levels are beneficial for cognitive functioning and neuroplasticity, as chronic stress can inhibit the formation of new neural connections.

- ***Open up to uplifting movies*** can stimulate the production of Brain-Derived Neurotrophic Factor (BDNF), a protein that supports neuron survival, differentiation, and synaptic plasticity.

- ***Open up to gardening*** engages multiple brain functions and regions. It involves sensory stimulation, cognitive and motor skills, stress reduction, emotional regulation, learning, social interaction, and a connection with nature. These aspects support the brain's ability to adapt and develop new neural pathways.

- ***Open up to cooking*** enhances various brain functions, including sensory integration, cognitive abilities, motor skills, creativity, emotional well-being, and social interaction, all contributing to neuroplasticity.

- ***Open up to board games*** involves complex cognitive processes, including executive functions, problem-solving, memory, concentration, and strategic thinking. These activities stimulate and challenge the brain, leading to increased neuroplasticity by developing new neural connections, enhancing cognitive skills, and improving overall brain health.

- ***Open up to non-violent video games*** can enhance neuroplasticity, enabling the brain to reorganize itself by forming new neural connections. Studies suggest that certain types of video games, especially those requiring strategy, problem-solving, and spatial navigation, can improve cognitive functions like attention, memory, and problem-solving skills

- ***Open up to learning*** enhances neuroplasticity through synaptic plasticity, neurogenesis, remodeling of neural circuits, neurotransmitter modulation, improvement in cognitive functions, reduction in cognitive decline, emotional regulation, and increased brain connectivity. These processes contribute to the brain's adaptability, resilience, and capacity for continued growth and development.

- ***Open up to traveling*** can enhance neuroplasticity through exposure to novel experiences, cognitive challenges, rich sensory stimulation, emotional engagement, social interactions, physical activity, mindfulness, and creative inspiration. These varied and dynamic experiences encourage the brain to adapt, reorganize, and strengthen neural connections, improving cognitive function, flexibility, and overall brain health.

- ***Open up to art*** enhances neuroplasticity through multisensory stimulation, cognitive enhancement, motor skill development, creativity and problem-solving, emotional regulation, neurotransmitter modulation, reflective thinking, and synaptic plasticity. These processes collectively stimulate the brain, contributing to its adaptability and developing new neural pathways.

SPIRITUAL RITUALS (6) may have evolved to counterbalance the PRIMAL brain's overactivity, which is constantly geared toward survival. Spiritual practices can impact various brain areas associated with attention, emotion, and stress response. They can increase relaxation, reduce stress, and enhance mood by affecting the autonomic nervous system, alter neurotransmitter levels, and improve neural connectivity.

- ***Open up to meditation*** significantly impacts neuroplasticity through structural and functional changes in the brain. These changes contribute to improved cognitive functions, emotional regulation, stress reduction, and overall brain health, demonstrating the powerful influence of meditation on the brain's adaptability and resilience.
- ***Open up to prayers*** can stimulate brain regions involved in spiritual experiences and personal growth, contributing to a sense of purpose and meaning. Praying impacts neuroplasticity by engaging and enhancing cognitive functions, emotional regulation, stress reduction, mindfulness, social cognition, and structural and functional aspects of the brain. These changes can improve cognitive function, flexibility, emotional well-being, and overall brain health.
- ***Open up to your dreams*** is integral to various cognitive and emotional processes contributing to neuroplasticity. Dreams play key roles in memory consolidation, emotional regulation, stress reduction, creativity, learning, and overall brain health. Understanding the full impact of dreams on neuroplasticity is an ongoing area of scientific research.
- ***Open up to stargazing*** contributes subtly to brain health and cognitive function through relaxation, inspiration, cognitive engagement, and emotional experiences. The benefits of stargazing may extend beyond what can be easily measured or observed in traditional neuroscientific research.
- ***Open up to entheogens*** can induce profound changes in perception, emotion, and cognition. As a result, they can promote the growth of

dendrites, dendritic spines, and synapses, effectively increasing neural connectivity.

- ***Open up to mantras*** contributes to neuroplasticity by improving focus and attention, reducing stress, aiding emotional regulation, enhancing mindfulness, altering brainwave states, engaging in cognitive reappraisal, modulating the DMN, synchronizing brain activity, and even causing structural changes in the brain. These effects collectively support the brain's ability to adapt and develop new neural pathways

CONCLUSION

Flow into Your OPEN Future

"The measure of intelligence is the ability to change."

--Albert Einstein, Physicist

I made the bold and unprecedented move of being transparent and vulnerable so that my story could inspire you. I shared multiple perspectives and proposed a neurospiritual model of consciousness through which you can now re-imagine your daily activities. From the mundane to the extraordinary, you have the power to use neurobiology to enhance your mental wellness.

OPEN aims to motivate you to continue exploring your brain's remarkable capacity for adaptation and growth. *OPEN* rituals provide intellectual stimulation and can contribute to your emotional and spiritual well-being, emphasizing the interconnectedness of your mental, physical, and emotional health. *OPEN* rituals involving your body, art, and cooking can enhance your cognitive functions, emotional regulation, and overall mental well-being. The profound impact of playing musical instruments, learning new languages, and engaging in physical exercise to enhance your brain's flexibility and ability to form new connections is well documented. So is the importance of embracing new experiences and challenges, whether through travel, playing games, or even engaging with the simple act of stargazing, in nurturing your brain's inherent capacity to grow and evolve. Finally, when used responsibly, entheogens may raise your consciousness and neuroplasticity, albeit with necessary caution and within the bounds of legal and ethical considerations.

In essence, the journey of each ritual is a testament to your brain's incredible adaptability and resilience to achieve *flow*. Flow is a fascinating concept introduced by Mihaly Csikszentmihalyi several decades ago (350). Flow describes a mental state of deep immersion and optimal experience that leads to effortless productivity and heightened performance. From a neuroscientific perspective, flow engages specific brain mechanisms that balance automaticity and conscious control. This has profound implications for neuroplasticity and the formation of habits. Understanding the neurobiology behind flow and its links to consciousness and neuroplasticity is crucial for recognizing how to cultivate habits that enhance cognitive flexibility and a more profound sense of awareness in your daily lives.

Consciousness in Flow

Flow is often described as a state of effortless consciousness in which you are absorbed in the present moment without the intrusive thoughts and self-monitoring typical of everyday consciousness. You have likely experienced flow by losing yourself daydreaming, playing video games, or simply cooking or gardening. Remarkably, this state is marked by decreased activity in the prefrontal cortex, particularly the dorsolateral prefrontal cortex, which is responsible for executive functions such as planning, decision-making, and self-reflection. This phenomenon, termed *transient hypofrontality,* allows your brain to bypass the usual conscious control mechanisms and tap into more intuitive and automatic processes (351).

In this state, your sense of self becomes temporarily muted, and you may feel a merging of action and awareness, losing track of time and external distractions. In flow, consciousness is not the deliberate, effortful form of awareness often associated with tasks like problem-solving or introspection. Instead, it is a more fluid, open state of consciousness where automaticity and subconscious processes are allowed to take center stage. This links directly to the notion that elevated states of consciousness are not necessarily defined by higher cognitive effort but rather by the ability to remain present and fully engaged without the interference of self-criticism or over-analysis.

Importantly, while flow might appear to reduce conscious control, it paradoxically requires high skill and awareness to enter this state. The balance between skill and challenge is vital for inducing flow; too little challenge leads to boredom, and too much overwhelms cognitive resources. This balancing act indicates that consciousness plays a crucial role in flow by helping us adjust our focus, attention, and effort based on the task's complexity. This adaptive form of consciousness is crucial for sustained engagement and performance, reinforcing your exploration of how consciousness can be trained to achieve more profound mental states.

Neuroplasticity in Flow

Engaging in flow rituals can significantly impact your neuroplasticity, primarily because flow states involve repeated practice and engagement in challenging activities that push your brain to adapt and grow. When you repeatedly enter flow during *OPEN* rituals, such as learning a new skill or mastering a demanding physical activity, your brain undergoes structural changes that strengthen the neural circuits involved in those activities.

As discussed, plasticity is driven by long-term potentiation, where synaptic connections between neurons are strengthened, and even neurogenesis, the formation of new neurons. Flow states, which constantly balance challenge and skill, activate these processes by providing the brain with novel stimuli and encouraging it to find efficient solutions. When you regularly enter flow, your brain adapts, making it easier for you to achieve flow in the future.

This relationship between neuroplasticity and flow is especially relevant to the rituals that sustain both. Flow requires a momentary engagement with a challenging task and sustained effort over time. As neuroplasticity strengthens the brain's ability to enter flow more readily, it reinforces behaviors that promote continuous improvement and mastery. In our discussions on neuroplasticity and its role in habit formation, this connection highlights the importance of creating environments and routines that support flow-inducing activities. Whether in professional, athletic, or creative settings, regular engagement in tasks that challenge your brain fosters neuroplasticity and makes it easier to access flow states, promoting long-term cognitive growth and flexibility.

The neuroscience of flow reveals its deep connection to consciousness and neuroplasticity. Flow is like friction-free consciousness, where you are fully absorbed in the present moment without intrusive thoughts and the need to self-monitor your mental activity. While playing music, engaging in deep meditation, or losing yourself in creative work, flow emerges as a dynamic state where the mind and body synchronize seamlessly. This remarkable relationship suggests that consciousness and energy efficiency are critical conditions for achieving higher states of cognitive functioning.

Dr. Karl Friston, a renowned neuroscientist, theoretical biologist, and Professor of Neuroscience at University College London, proposed a theory that explains how the brain optimizes perception, action, and consciousness by minimizing uncertainty. Friston's revolutionary *Free Energy Model* is a robust framework to decode neuroplasticity, flow, and self-transformation but also offers a compelling rationale to understand flow as a state of minimal cognitive resistance and optimal efficiency (52). The term "free energy" originates from statistical physics and information theory, where free energy measures disorder or uncertainty in a system. According to Friston, the brain constantly makes predictions about the world and updates them to minimize free energy or uncertainty.

Thus, flow represents a state where these predictions are finely tuned— the brain no longer struggles to resolve errors, allowing thoughts, actions, and awareness to merge effortlessly. This is why highly trained skills (e.g., playing an instrument, driving) require less mental effort over time—the brain's models become more precise. A study by Dr. Yan-Ling Pi and his colleagues confirmed that performance athletes' brain networks exhibited shorter path lengths and higher global efficiency, indicating faster neuronal transmission when comparing 21 high-level basketball players to 25 non-athlete controls using diffusion-tensor imaging and graph theory analysis (352). Regionally, increased nodal parameters were observed in areas associated with visual, default mode, and attention networks. These changes correlated significantly with the athletes' years of training, suggesting that extensive practice leads to structural brain adaptations that enhance performance.

From Friston's perspective, flow occurs when the brain achieves an optimal balance between prediction and adaptation—when it no longer expends excess energy to correct mismatches between expectation and reality. This explains why flow is most easily achieved when there is a perfect balance between challenge and skill:

- Too little challenge → The brain's predictions become too rigid, leading to boredom.
- Too much challenge → The brain struggles with uncertainty, leading to stress.
- Perfect balance → The brain efficiently updates its models in real-time, reducing cognitive resistance and producing effortless action.

Meanwhile, rituals invite surrender—letting go of excessive control and allowing the experience to unfold naturally. Many mystical traditions describe this state as one where the ego dissolves, and a deeper connection to the present moment emerges. Therefore, flow is not just a neurological event but a gateway to expanded consciousness, where presence, intuition, and action become one.

Meanwhile, there is also a strong relationship between the Free Energy Principle and the default mode network (DMN), particularly in how the brain balances self-referential thinking, uncertainty, and cognitive control. As we have discussed in multiple sections of the book, the default mode network is active when we are engaged in self-reflection, daydreaming, and processing past experiences.

From a predictive perspective, the DMN creates high-level predictions about who we are and how the world works. If the default mode network is overactive, the brain can become trapped in repetitive self-narratives (e.g., depression, rumination, trauma), leading to excess free energy because the brain struggles to update its self-model with new experiences. However, when the DMN quiets down, the brain becomes more open to revising its models—this is often linked to psychedelic states, deep meditation, and flow.

To conclude, cultivating daily *OPEN rituals* that encourage flow—such as meditation, physical activity, deliberate practice, and adequate rest—can enhance cognitive flexibility and awareness, forming a foundation for sustained personal and professional growth. These connections underscore the importance of designing a life where flow, consciousness, and neuroplasticity are interconnected and actively nurtured, supporting long-term well-being and mental resilience. As we close, the journey of learning and self-discovery on flow is ongoing, with each new experience offering growth and transformation opportunities. The brain's plasticity is not just a scientific fact but a beacon of hope and a reminder of our limitless potential for change and development. *OPEN* can serve as a comprehensive guide to enhance your neuroplasticity while protecting you from the toxic effects of stress, anxiety, depression, addictions, and trauma.

Key Takeaways from Each Chapter.

Chapter ONE: Living Fully OPEN

Chapter ONE provided an example of my *OPEN* life journey marked by a struggle with low plasticity, leading to stress, anxiety, and depression. It illustrated how embracing plasticity transformed my life, enabling a shift to higher consciousness and better mental flexibility. The chapter shared my intimate and productive dialogue with Ayahuasca during my month-long shamanic initiation in Peru. During this trip, I had the urge to start this book, receiving many insights that are no less than fundamental building blocks of the book.

Chapter TWO: The Self-healing Power of Consciousness and Neuroplasticity

Chapter TWO delved into the neuroscience of plasticity and consciousness. I laid the scientific groundwork for understanding the effect of stress, anxiety, depression, addiction, and trauma (SADAT) on consciousness and health, elucidating how enhancing plasticity is crucial for overcoming these conditions. While this chapter was dense and possibly challenging, it gave you the knowledge to appreciate the incredible biological symphony of our nervous system and the degree to which you can regulate it. The chapter simplified complex concepts with multiple charts and visual aids to make the science behind neuroplasticity and the value of accessing meta-consciousness accessible and engaging.

Chapter THREE: OPEN Your Mind

The focus of Chapter THREE shifted to elevating consciousness as a critical step in enhancing the role of neuroplasticity. While this section made it clear that raising our awareness is front and center of our mental health, the chapter also introduced the role of multiple barriers of consciousness that can compromise, delay, or sabotage your ability to see and understand the degree to which you may be affected by maladaptive responses to SADAT. The chapter discussed how higher consciousness levels can be achieved through natural modalities, psychedelic experiences, and ancient healing practices, providing fresh perspectives on tackling complex mental health challenges.

Chapter FOUR: OPEN Your Self

In Chapter FOUR, I focused on personal stories and the intricacies of our personality to help you gain the awareness needed to rewire old patterns and create new, healthier narratives. This rewiring process isn't just about altering behavior; it's about transforming the stories that shape your identity. Reframing your personal narrative is central to your healing process to eliminate old patterns that no longer serve you. You cannot pass on this gift of nature: it is your ability to discharge your nervous system from the agony of worries, stress, regrets, shame, anger, and so much more. I explored the value of powerful personality models such as the Big Five and the Enneagram, which provide unique insights into your character traits' blueprint and imprint. Meanwhile, I also explained the science of narrative transportation to help you identify and alter detrimental narratives contributing to SADAT, thereby enabling a positive reconstruction of your personal story.

Chapter FIVE: OPEN Your Life

Chapter FIVE offered practical advice on *OPEN* rituals reinforcing the lifelong journey toward cognitive flexibility and well-being. I reviewed activities targeting the PRIMAL, RATIONAL, and SPIRITUAL systems through which varying levels of consciousness can navigate. I presented how *OPEN* rituals have scientifically backed benefits on neuroplasticity, confirming the role of habit formation in sustaining mental health improvements.

I sincerely hope his book has enabled you to integrate cutting-edge scientific findings with traditional healing wisdom to form a unique narrative about the power of consciousness and plasticity. For me, at least, *OPEN* is more than just a text; it is a testament to the power of overcoming the challenges of SADAT and living a soul-centered and joyful life.

Figure 116: In Front of My Tambo

I started writing OPEN on November 8, 2023, in Iquitos, Peru, and finished the first draft 94 days later, on February 10, 2024, in Kona, Hawaii. By October 30 2024, I had completed a thorough revision and rewriting process, adding over 300 scientific citations and well over 100 images to the manuscript, many generated by AI under my prompts.

KEY NUMBERS ABOUT OPEN

- **134,204** **Words:** It took exactly 14 months to write *OPEN*.

- **364** **Pages:** The length of *OPEN* without citations and tables.

- **58%** **Comorbidity:** The percentage of individuals who experience both anxiety and depression.

- **26** **Days of Healing:** length of my shamanic retreat in the Amazon, during which I started to write *OPEN*.

- **5** **Conditions (SADAT):** The five major mental health conditions that a neurospiritual approach can address better than most conventional treatment plans.

- **3**-**Step Model:** *OPEN* is structured around only three steps: OPEN your Mind, OPEN your Self, and OPEN your Life.

- **2** **Decades of Experience:** I have studied neuroplasticity and consciousness for 20 years to overcome SADAT. You don't need to spend all those years if you read *OPEN*!

- **1** **Simple Conclusion:** Ultimately, *OPEN* points to **one simple truth**:

  Your brain can self-heal. By embracing neuroplasticity, consciousness practices, and *OPEN* rituals, you can reprogram your mind for greater happiness and flow.

ABOUT THE AUTHOR

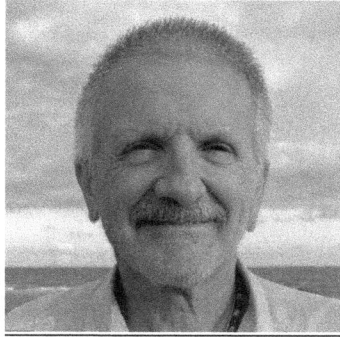

Dr. Christophe Morin holds an MA and a Ph.D. in Media Psychology from Fielding Graduate University and an MBA from Bowling Green State University.

Dr. Morin is a distinguished lecturer at Johns Hopkins Whiting School of Engineering and an Adjunct Faculty at Fielding Graduate University, where he teaches several postgraduate courses on Prompt Engineering, Generative AI, the neuroscience of media effect, and neuromarketing. He actively volunteers his expertise for many non-profit organizations seeking to use cutting-edge science to raise awareness and money.

Dr. Morin has over 30 years of experience in consumer, advertising, and public health research using cognitive psychology, media psychology, personality psychology, and neuroscience. He has authored three best-sellers on neuromarketing and scientific persuasion, currently available in multiple languages. He earned many prestigious speaking, research, and publication awards during his remarkable career. During the COVID crisis, Dr. Morin pivoted his research and writing to the neurospiritual basis of mental health. In 2020, he published *The Serenity Code*, proposing a scientific formula for achieving more serenity through self-love habits.

Since then, Dr. Morin has delved deeper into neuroplasticity's materialistic and non-materialistic foundations, exploring its powerful connection to transformation and self-healing. In November 2023, he embarked on a month-long shamanic initiation retreat in Peru with the Shipibo tribe, an experience that inspired him to begin writing *OPEN*—his most personal work to date. The book combines cutting-edge research on the neuroscience of self-healing with ancient traditions and the responsible, safe use of entheogens to help alleviate or eliminate these conditions.

APPENDIX

The information about entheogens's relationship to neuroplasticity is only for educational and informational purposes. Psychedelics and entheogens are powerful substances that can have profound effects on consciousness and well-being but also carry potential risks. These substances are currently regulated or illegal in many jurisdictions, and their use should be approached with caution and respect. If you are considering using psychedelics or entheogens for therapeutic or spiritual purposes, it is essential to do so under the guidance of a trained professional in a safe and controlled setting, where legally permitted. Note that psychedelic experiences are not suitable for everyone and may not be appropriate for individuals with certain medical or mental health conditions. This information is not a substitute for medical, legal, or psychological advice, and you should consult qualified professionals for personal guidance.

Ayahuasca

Recent scientific research has begun exploring Ayahuasca's potential impact on neuroplasticity, which refers to the brain's ability to adapt and reorganize by forming new neural connections. Some studies suggest that Ayahuasca may promote neurogenesis. Research indicates that Ayahuasca may increase the levels of brain-derived neurotrophic factors (BDNF) in the brain, which are critical proteins that can facilitate neuroplasticity (353). Higher BDNF levels are associated with improved cognitive functioning and resilience against neurodegenerative diseases.

Meanwhile, Ayahuasca's active component, DMT, acts as a potent agonist at serotonin receptors, particularly the 5-HT2A receptor. The activation of these receptors is closely associated with increased neurogenesis synaptic formation and also plays a role in the emotional and introspective effects Ayahuasca users report, as well as in the potential therapeutic effects for conditions such as depression and PTSD (354).

So Ayahuasca has been shown to promote cortical plasticity, making the brain more adaptable and enhancing the efficiency of neural networks. This is thought to improve cognitive flexibility and problem-solving skills (355). The complex psychological effects of Ayahuasca and its varied legal status continue to pose challenges for research and therapeutic use. Further studies are needed to fully understand and harness Ayahuasca's effects on neuroplasticity in clinical settings.

Finally, there is strong evidence indicating that Ayahuasca consumption affects the Default Mode Network (DMN). A study published in *PLOS ONE* utilized functional magnetic resonance imaging (fMRI) to observe the DMN activity of ten experienced ayahuasca users (356). The findings revealed a significant decrease in activity across most parts of the DMN, including key regions such as the posterior cingulate cortex (PCC)/precuneus and the medial prefrontal cortex (mPFC). Additionally, functional connectivity within the PCC/precuneus decreased after Ayahuasca intake. These results suggest that the altered state of consciousness induced by ayahuasca is associated with modulation of both the activity and connectivity of the DMN.

Psilocybin

Psilocybin disrupts the brain's regular communication networks. It also decreases activity in the brain's default mode network (DMN), which is active during rest and involved in introspection and self-referential thought processes.

This disruption contributes to the experience of ego dissolution—a loss of the sense of self. Recent research suggests that psilocybin enhances neuroplasticity. This could explain improvements in mood and cognitive flexibility observed after psilocybin use.

Indeed, psilocybin has been shown to increase the expression of genes associated with neuroplasticity, particularly those linked to the growth and formation of synaptic connections. Research indicates that psilocybin promotes the development of ***dendritic spines,*** which are small protrusions on nerve cells where synapses are located. Dendritic spines are critical for synaptic plasticity and overall brain health (357). For instance, research found that a single dose of psychedelics, including psilocybin, can increase the density of dendritic spines and promote structural and functional neuroplasticity in the prefrontal cortex of mice.

Meanwhile, studies have confirmed that psilocybin leads to increased BDNF production, which could underlie enhanced neurogenesis and synaptic plasticity (358). Psilocybin is also known to re-open critical periods of neuroplasticity (250).

The research into psilocybin and neuroplasticity is still evolving, but the evidence points to its significant potential to enhance brain function and treat neuropsychiatric disorders. As regulatory barriers are addressed and more research is conducted, psilocybin may become a key component of future therapeutic interventions that harness the power of neuroplasticity.

Peyote

Mescaline, the active compound in peyote, is known for inducing altered states of consciousness characterized by visual hallucinations, altered perceptions, and profound introspective experiences. Many users report a heightened sense of connection to the natural world, feelings of transcendence, and profound existential insight. Similar to other psychedelics, peyote can induce experiences of ego dissolution, leading to a temporary loss of the sense of self, which can be transformative and lead to long-lasting changes in perspective. By promoting neural flexibility, mescaline can break rigid patterns that often characterize SADAT conditions and foster psychological resilience.

Mescaline can reduce DMN activity, disrupting the typical pattern of self-referential thought and introspection that the DMN supports. While the activity in the DMN decreases, global brain connectivity increases. This means that regions of the brain that do not typically communicate directly with each other start to interact more. Such increased connectivity can lead to the unique cognitive and perceptual shifts experienced during a psychedelic state.

Like many classic psychedelics, mescaline functions primarily through its action on serotonin receptors, particularly the 5-HT2A receptor. Activation of this receptor is believed to play a key role in promoting neuroplasticity. This activation can lead to increased synaptic formation, enhanced dendritic growth, and greater brain network connectivity (359).

Wachuma

Also known as the San Pedro cactus (Echinopsis pachanoi), Wachuma has been used for thousands of years in traditional Andean medicine primarily for its psychoactive properties due to the alkaloid mescaline. Similar to other psychedelic substances, mescaline's influence on neuroplasticity is a subject of interest. Specific studies on Wachuma are less prevalent compared to more commonly researched psychedelics like LSD or psilocybin. However, drawing on what is known about mescaline can provide insights into how Wachuma might affect neuroplasticity.

Wachuma's active compound, mescaline, functions as a serotonin receptor agonist, particularly targeting the 5-HT2A receptor. This receptor's activation is critical for many of the effects of classic psychedelics, including their potential to induce neuroplastic changes. Activation of 5-HT2A facilitates increased synaptic plasticity, meaning it helps the brain form new neural connections more readily.

While direct studies on Wachuma are limited, research on similar psychedelics suggests that mescaline could enhance the expression of BDNF. This protein is vital for maintaining and growing new neurons, thus supporting overall brain health and plasticity.

The potential for mescaline and, by extension, Wachuma to enhance neuroplasticity may underlie its therapeutic benefits, such as in the treatment of conditions like depression, anxiety, and trauma-related disorders. By fostering neuroplasticity, Wachuma could help reset entrenched neural patterns associated with these disorders (360).

Iboga

Research into the specific effects of ibogaine on neuroplasticity is still in the early stages. However, preliminary studies and theoretical models suggest that ibogaine may potentially enhance the brain's ability to form new neural connections and adapt to new experiences. Some research indicates that ibogaine may stimulate the release of growth factors in the brain, such as Brain-

371

Derived Neurotrophic Factor (BDNF) (361). BDNF plays a crucial role in neuroplasticity, promoting the survival of neurons and encouraging the growth of new synaptic connections.

While specific studies on ibogaine's effect on the default mode network (DMN) are limited, its psychedelic properties suggest that it may disrupt the DMN, similar to other psychedelics, thereby affecting self-referential thought processes and potentially leading to ego dissolution. Similar to other psychedelics, ibogaine also interacts with serotonin receptors, particularly the 5-HT2A receptor, which is known to play a role in neuroplasticity. This interaction can facilitate the remodeling and strengthening of neural pathways (362).

One of the most prominent applications of ibogaine is in the treatment of substance addiction. Ibogaine has been reported to reduce withdrawal symptoms and decrease the cravings for drugs such as opioids, alcohol, and stimulants. The neuroplastic changes induced by ibogaine might help to reset addicted brain circuits, offering a new therapeutic pathway for recovery (363).

Salvia

Current scientific understanding of Salvia's effects on neuroplasticity is limited. More research is needed to determine whether salvinorin A significantly impacts neuroplasticity. Salvinorin A's primary action is as a kappa-opioid receptor agonist. The activation of these receptors has been linked to various effects on perception and consciousness, but their role in neuroplasticity is complex and not fully understood. There is some evidence to suggest that kappa-opioid receptor activity can lead to different neuroplastic outcomes compared to the serotonin-based mechanisms typically associated with psychedelics (364).

While direct studies on salvia and neuroplasticity are limited, kappa-opioid receptors are known to play a role in modulating synaptic plasticity and neuronal connectivity in specific brain areas (365). Research in this area is still emerging, and the exact implications of salvia's kappa-opioid activity on long-term neuroplastic changes are unclear.

Meanwhile, research into the specific effects of Salvia divinorum, particularly its active compound salvinorin A, on the brain's default mode network (DMN) is still developing and not as extensive as research on other psychedelics like LSD or psilocybin. However, given Salvia's potent psychoactive effects, it is plausible that it could impact the DMN.

So, the role of salvia in promoting neuroplasticity is an area of research still in its infancy. The activation of kappa-opioid receptors by salvinorin A suggests a complex interplay with brain mechanisms that could impact neuroplasticity in ways that are distinct from the mechanisms employed by classical psychedelics. Further research is necessary to elucidate these effects and to determine whether salvia has practical applications in therapeutic settings or in enhancing cognitive or mental health. Salvia remains a substance of interest primarily for its unique pharmacological properties and its traditional use in shamanic contexts rather than for its neuroplastic potential.

Cannabis

Research suggests that cannabis might affect neuroplasticity both positively and negatively. THC, a component of cannabis, has been shown to affect brain regions involved in learning and memory, which could impair neuroplasticity with long-term use. However, other studies indicate that cannabinoids can promote neurogenesis (the growth of new neurons) and might have neuroprotective effects. Therefore, the impact of cannabis on neuroplasticity is complex and depends on various factors, including dosage, frequency of use, and individual biology.

Cannabis acts primarily through the endocannabinoid system, which plays a crucial role in modulating neuroplasticity. This system includes cannabinoid receptors (CB1 and CB2), endogenous cannabinoids (endocannabinoids), and the enzymes involved in their synthesis and degradation (366). THC, the psychoactive component of cannabis, has been shown to affect synaptic plasticity in various brain regions. Studies have shown both enhancement and reduction in synaptic plasticity, suggesting dose-dependent effects (367).

Cannabis can be used for journeying, which means engaging deeply with one's inner landscape or external environments to stimulate cognitive and emotional growth. While direct studies on "journeying" per se and neuroplasticity are sparse, the underlying principles of new experiences, learning, and emotional engagement are well-documented to contribute to neuroplastic changes. Engaging in new, challenging, and enriching experiences can enhance cognitive flexibility, encourage the development of new neural connections, and support cognitive and emotional resilience. This process can be akin to the mental and emotional journeys one might undertake through meditation, mindfulness practices, or immersive learning experiences, all known to foster neuroplasticity.

To conclude, cannabis has complex, bidirectional effects on neuroplasticity, with the potential to both enhance and impair neural adaptability. This complexity underscores the need for cautious and well-regulated medicinal use of cannabis, especially considering factors like age, dosage, and the ratio of THC to CBD. More research is needed to understand these dynamics fully and to harness cannabis' potential benefits while mitigating its risks.

Bufo and 5-Meo-DMT

5-MeO-DMT is a potent psychedelic found in various plant species and the venom of the Bufo alvarius toad. It is known for its intense and rapid psychedelic effects, which are markedly different from those of other classic psychedelics such as LSD or psilocybin. The role of 5-MeO-DMT on neuroplasticity is a burgeoning area of research, with interest in how its profound psychoactive effects might translate into long-term changes in brain structure and function.

Similar to other psychedelics, 5-MeO-DMT acts primarily as an agonist at serotonin receptors, particularly the 5-HT1A and 5-HT2A receptors. Activation of these receptors is known to play a significant role in modulating neuroplasticity, potentially leading to increased synaptic plasticity and neurogenesis (368). Szabo's study found that 5-MeO-DMT induced changes in protein expression related to neuronal development and synaptic plasticity in cerebral organoids.

Research has also suggested that 5-MeO-DMT may promote the expression of BDNF, a crucial factor in neuroplasticity that supports the survival and differentiation of neurons and the growth of new synapses (369).

Meanwhile, the ability of 5-MeO-DMT to induce rapid and profound changes in consciousness and potential long-term neuroplasticity might help treat conditions such as depression, PTSD, and anxiety. Its effects on neuroplasticity suggest it could help 'reset' maladaptive neural circuits.

Thus, 5-MeO-DMT appears to influence neuroplasticity through mechanisms similar to other psychedelics, potentially enhancing neuroplasticity and promoting psychological well-being through its psychoactive effects. However, the exact pathways and long-term impacts of these changes remain to be fully elucidated in rigorous scientific studies. As research continues, the therapeutic potentials of 5-MeO-DMT could become more apparent, offering new avenues for treating a range of psychiatric and neurological conditions.

LSD

LSD (lysergic acid diethylamide) is a well-known psychedelic compound that has been extensively studied for its profound effects on the mind and brain. Over the past few decades, emerging research has explored LSD's impact on neuroplasticity. LSD is known primarily for its potent agonist activity at serotonin 5-HT2A receptors, which are heavily implicated in cognitive processes, mood regulation, and perception. Activation of these receptors is central to LSD's ability to enhance neuroplasticity. This includes promoting the growth and survival of neurons and facilitating synaptic plasticity, essential for learning and memory (370).

LSD has been shown to influence the expression of BDNF, thus facilitating neuroplastic processes. BDNF is a protein that plays a critical role in neuroplasticity, helping to support the survival of existing neurons and encourage the growth of new neurons and synapses (371).
The potential of LSD to enhance neuroplasticity offers promising avenues for the treatment of various mental health disorders, such as depression, anxiety, and PTSD. By promoting neuroplastic changes, LSD could help to break the rigid, maladaptive neural circuits associated with these conditions (372).
LSD appears to influence neuroplasticity through several mechanisms, including serotonin receptor activation, BDNF enhancement, and increased cortical connectivity.

These effects could underpin the profound cognitive, emotional, and perceptual changes induced by LSD, offering potential therapeutic benefits for mental health disorders. However, further research is essential to optimize safe and effective use in therapeutic settings.

Ketamine

Ketamine, initially developed as an anesthetic, has gained significant attention in recent years for its rapid-acting antidepressant effects. Research into ketamine's impact on neuroplasticity has provided insights into its potential mechanisms of action in treating mood disorders, particularly treatment-resistant depression.

Ketamine is primarily known for its antagonistic effects on the brain's N-methyl-D-aspartate (NMDA) receptors, critical for synaptic transmission in the central nervous system. By blocking these receptors, ketamine is thought to cause a cascade of biochemical reactions that promote neuroplasticity. Studies show the role of NMDA receptor antagonism in reversing the synaptic deficits observed in mood disorders, which is central to ketamine's mechanism. (373).

Following ketamine administration, there is an increase in the levels of BDNF, a crucial mediator of neuroplasticity that promotes the growth and differentiation of new neurons and synapses (374). Meanwhile, ketamine activates protein synthesis and synaptic formation, which may explain its rapid antidepressant effects (375).

Ketamine's influence on neuroplasticity offers a powerful insight into its potential as a novel treatment for depression and possibly other psychiatric disorders. By antagonizing NMDA receptors, enhancing BDNF levels, and activating the mTOR pathway, ketamine can induce rapid synaptic connectivity and function changes. These mechanisms contribute to its rapid antidepressant actions, providing a critical tool for cases where traditional treatments fail. However, further research is necessary to harness its therapeutic potential while minimizing risks.

MDMA

MDMA (3,4-methylenedioxymethamphetamine), commonly known as ecstasy, is a psychoactive drug that has gained attention in recent years for its potential therapeutic benefits, particularly in the treatment of PTSD and other mental health disorders. Emerging research into MDMA's impact on neuroplasticity suggests it may facilitate significant changes in brain structure and function, contributing to its therapeutic effects.

MDMA primarily exerts its effects by increasing the release of serotonin, a neurotransmitter crucial for mood regulation, social behavior, and cognitive functions. The surge in serotonin levels following MDMA administration can lead to enhanced mood and empathy, as well as increased neuroplasticity (376).

Similar to other psychotropic drugs, MDMA has been shown to influence the levels of BDNF. This protein is essential for the growth, maintenance, and survival of neurons, and its increase is associated with improved neuroplasticity (377).

Meanwhile, studies using neuroimaging have shown that MDMA can decrease functional connectivity in the amygdala (the brain region involved in processing emotional responses) and increase connectivity in the prefrontal cortex (involved in higher cognitive functions). These changes suggest a potential mechanism by which MDMA could enhance emotional regulation and stress resilience, contributing to neuroplasticity (378).

To conclude, MDMA's influence on neuroplasticity, mainly through its effects on serotonin, BDNF, and brain connectivity, provides a compelling basis for its potential therapeutic benefits. These mechanisms contribute to its ability to alter emotional processing and enhance cognitive functions, making it a valuable candidate for treating disorders like PTSD. However, ongoing research is essential to fully harness MDMA's neuroplastic potential while ensuring patient safety in therapeutic settings.

REFERENCES

1.	Jung CG. The structure and dynamics of the psyche: Routledge; 2014.
2.	Freud S. Beyond the pleasure principle. London, Vienna1922.
3.	Kessler RC, Berglund P, Demler O, Jin R, Merikangas KR, Walters EE. Lifetime prevalence and age-of-onset distributions of DSM-IV disorders in the National Comorbidity Survey Replication. Archives of general psychiatry. 2005;62(6):593-602.
4.	Brady KT, Killeen TK, Brewerton T, Lucerini S. Comorbidity of psychiatric disorders and posttraumatic stress disorder. Journal of clinical psychiatry. 2000;61:22-32.
5.	Swendsen J, Conway KP, Degenhardt L, Glantz M, Jin R, Merikangas KR, et al. Mental disorders as risk factors for substance use, abuse and dependence: results from the 10-year follow-up of the National Comorbidity Survey. Addiction. 2010;105(6):1117-28.
6.	Hadar A, David J, Shalit N, Roseman L, Gross R, Sessa B, et al. The psychedelic renaissance in clinical research: a bibliometric analysis of three decades of human studies with psychedelics. Journal of psychoactive drugs. 2023;55(1):1-10.
7.	Tafur J. The Fellowship of the River: A Medical Doctor's Exploration into Traditional Amazonian Plant Medicine: Espiritu Books; 2017.
8.	Grawe K. Neuropsychotherapy: How the neurosciences inform effective psychotherapy: Routledge; 2017.
9.	Garakani A, Murrough JW, Freire RC, Thom RP, Larkin K, Buono FD, et al. Pharmacotherapy of Anxiety Disorders: Current and Emerging Treatment Options. Frontiers in Psychiatry. 2020;11.
10.	Shultes R, Hofman A. Plants of the gods : origins of hallucinogenic use. New York: A. van der Marck Editions; 1987.
11.	Gonzalez D, Cantillo J, Perez I, Carvalho M, Aronovich A, Farre M, et al. The Shipibo Ceremonial Use of Ayahuasca to Promote Well-Being: An Observational Study. Frontiers in Pharmacology. 2021;12.
12.	Mehl-Madrona L, Mainguy B. Remapping your mind: the neuroscience of self-transformation through story: Bear & Company; 2015.
13.	Seth A. Being You: A New Science of Consciousness: Dutton; 2021.
14.	Dehaene S. Consciousness and the Brain: Deciphering How the Brain Codes Our Thoughts New York: Penguin Books; 2014.
15.	Beauregard M, O'Leary D. The spiritual brain: A neuroscientist's case for the existence of the soul. San Francisco, CA, US: HarperOne/HarperCollins; 2007. xvi, 368-xvi, p.
16.	Lightman A. The transcendent brain : spirituality in the age of science. NY: Pantheon Books; 2023.

17.   Maté G, Maté D. The Myth of Normal: Trauma, Illness, and Healing in a Toxic Culture2022.

18.   Siegel D. Aware: The Science and Practice of Presence--The Groundbreaking Meditation Practice: TarcherPerigee; 2020.

19.   Chopra D. Quantum Healing (Revised and Updated): Exploring the Frontiers of Mind/Body Medicine: Random House Publishing Group; 2015.

20.   Spriggs MJ, Murphy-Beiner A, Murphy R, Bornemann J, Thurgur H, Schlag AK. ARC: a framework for access, reciprocity and conduct in psychedelic therapies. Frontiers in psychology. 2023;14:1119115.

21.   FDA gives thumbs down to MDMA for now, demanding further research [press release]. NPR2024.

22.   Tang YY, Ma Y, Wang J, Fan Y, Feng S, Lu Q, et al. Short-term meditation training improves attention and self-regulation. Proc Natl Acad Sci U S A. 2007;104(43):17152-6.

23.   Morin C. The Serenity Code: Depth Insights; 2020.

24.   Perkins D, Schubert V, Simonová H, Tófoli LF, Bouso JC, Horák M, et al. Influence of Context and Setting on the Mental Health and Wellbeing Outcomes of Ayahuasca Drinkers: Results of a Large International Survey. Front Pharmacol. 2021;12:623979.

25.   Nakajima Y, Tanaka N, Mima T, Izumi S-I. Stress recovery effects of high-and low-frequency amplified music on heart rate variability. Behavioural Neurology. 2016;2016.

26.   Bustos S. The healing power of the icaros: A phenomenological study of ayahuasca experiences. In: CIIS, editor. 2008.

27.   Callicott C. nterspecies communicationin the Western Amazon: Music as a form ofconversation between plants and people. European Journal of Ecopsychology. 2013;4(23):32-53.

28.   Smith AM, Macheski A. Sensing Shipibo Aesthetics Beyond the Peruvian Amazon: Kené Design in Icaros: A Vision (2016). Journal of Latin American Cultural Studies. 2023;32:333 - 60.

29.   Hook JN, Hall TW, Davis DE, Van Tongeren DR, Conner M. The Enneagram: A systematic review of the literature and directions for future research. Journal of clinical psychology. 2021;77(4):865-83.

30.   Universe Within Podcast Ep31 - Imika Tarirú - Tubú, Ancestral Wisdom, Plant Medicine & The Word. 2021.

31.   AlphaGo2016.

32.   Chalmers DJ. Facing Up to the Problem of Consciousness. The Character of Consciousness: Oxford University Press; 2010. p. 0.

33.   Kahneman D. Thinking, fast and slow. New York, NY: Farrar, Straus and Giroux; 2011.

34.   Hebb DO. The organization of behavior. New York, NY: Wiley; 1949.

35.   Zheng J, Meister M. The unbearable slowness of being: Why do we live at 10 bits/s? Neuron. 2024.

36.   Norretranders T. The user illusion: Cutting consciousness down to size: Penguin; 1999.

37.   Wong-Riley MT. Energy metabolism of the visual system. Eye Brain. 2010;2:99-116.

38.   Gottfried JA. Neurobiology of sensation and reward. 2011.

39.   Radtke EL, Düsing R, Kuhl J, Tops M, Quirin M. Personality, Stress, and Intuition: Emotion Regulation Abilities Moderate the Effect of Stress-Dependent Cortisol Increase on Coherence Judgments. Frontiers in psychology. 2020;11.

40.   Nima AA, Rosenberg P, Archer T, Garcia D. Anxiety, Affect, Self-Esteem, and Stress: Mediation and Moderation Effects on Depression. PLOS ONE. 2013;8(9):e73265.

41.   Avetisyan M, Schill EM, Heuckeroth RO. Building a second brain in the bowel. The Journal of clinical investigation. 2015;125(3):899-907.

42.   Kane L, Kinzel J. The effects of probiotics on mood and emotion. JAAPA. 2018;31(5).

43.   Antonenko LM. Psychogenic dizziness. Nevrologiĭa, neĭropsikhiatriĭa, psikhosomatika. 2016;8(2):50-4.

44.   Vickery S, Patil KR, Dahnke R, Hopkins WD, Sherwood CC, Caspers S, et al. The uniqueness of human vulnerability to brain aging in great ape evolution. Science Advances. 2024;10(35):eado2733.

45.   Damasio AR. The feeling of what happens. 1999.

46.   Davidson RJ, Schwartz, G. E., Saron, C., Bennett, J., Goleman, D. J. Frontal versus parietal EEG asymmetry during positive and negative affect. Psychophysiology. 1979;16:202-3.

47.   Darwin C. The expression of the emotions in man and animals. London, England: John Murray; 1872.

48.   Ekman P, Friesen, W. V. Constants Across Cultures in the Face and Emotion. Journal of Personality and Social Psychology. 1971;17:124-9.

49.   Plutchik R, Kellerman H. Emotion: Theory, research and experience. London, UK: Academic Press; 1980.

50.   Ledoux JE. Anxious: Using the brain to understand and treat fear and anxiety. New York, NY: Penguin Books; 2016.

51.   Barrett LF. How emotions are made: The secret life of the brain. Pan Macmillan. 2017.

52.   Friston K. The free-energy principle: a unified brain theory? Nature reviews neuroscience. 2010;11(2):127-38.

53.   Dehaene S, Changeux JP. Ongoing spontaneous activity controls access to consciousness: A neuronal model of inattentional blindness. PLos ONE. 2005(3):e141.

54.   Freud S. Inhibitions, Symptoms and Anxiety. The Psychoanalytic Quarterly. 1936;5(1):1-28.

55.   McGaugh JL. Making lasting memories: Remembering the significant. Proceedings of the National Academy of Sciences. 2013;110(Supplement 2):10402-7.

56.   Gillan CM, Poldrack R, Desmond J. The role of left prefrontal cortex in language and memory. Proceedings of the National Academy of Science. 1998;95:906-13.

57.   Apperly I. Mindreaders: The Cognitive Basis of "Theory of Mind". 1 ed. Hove: Psychology Press; 2010.

58.   Soares JM, Marques P, Magalhaes R, Santos NC, Sousa N. The association between stress and mood across the adult lifespan on default mode network. Brain Structure and Function. 2017;222:101-12.

59.   Padmanabhan A, Lynch CJ, Schaer M, Menon V. The default mode network in autism. Biological Psychiatry: Cognitive Neuroscience and Neuroimaging. 2017;2(6):476-86.

60.   Mevel K, Chételat G, Eustache F, Desgranges B. The default mode network in healthy aging and Alzheimer's disease. International journal of Alzheimer's disease. 2011;2011.

61.   Damasio AR. Descartes Error Revisited. Journal of the history of the neurosciences. 2001;10(2):192-4.

62.   Koch C. Review. What is the function of the claustrum? 2005. p. 1271-9.

63.   Gazzaniga MS, Ivry RB, Mangun GR. Cognitive neuroscience: the biology of the mind. Third Edition ed. New York, NY: W. W. Norton & Company; 2009.

64.   Gusnard DA, Akbudak E, Shulman GL, Raichle ME. Medial prefrontal cortex and self-referential mental activity: relation to a default mode of brain function. Proc Natl Acad Sci U S A. 2001;98(7):4259-64.

65.   Dennett DC. Quining qualia. In: Marcel A, Bisiach E, editors. Consciousness in modern science. Oxford: Oxford University Press; 1985.

66.   Libet B. The timing of brain events: Reply to the "Special Section" in this journal of September 2004, edited by Susan Pockett. Conscious Cogn. 2006;15(3):540-7.

67.   Dehaene S, Changeux J, Naccache L, Sackur J, Sergent C. Conscious, preconscious, and subliminal processing: A testable taxonomy. TRENDS IN COGNITIVE SCIENCES. 2006;10(5):204-11.

68.   Paus T. Functional anatomy of arousal and attention systems in the human brain. Progress in Brain Research. 2000;126:65-77.

69.   Lamme VA. Why visual attention. TRENDS IN COGNITIVE SCIENCES. 2003;7:12-8.

70.   Zeki S. The disunity of consciouness. TRENDS IN COGNITIVE SCIENCES. 2003;7:214-8.

71.   Melloni L, Mudrik L, Pitts M, Bendtz K, Ferrante O, Gorska U, et al. An adversarial collaboration protocol for testing contrasting predictions of global neuronal workspace and integrated information theory. PLoS One. 2023;18(2):e0268577.

72.   Finkel E. This Contest Put Theories of Consciousness to the Test. Here's What It Really Proved: Wired; 2023 [Available from: https://www.wired.com/story/this-contest-put-theories-of-consciousness-to-the-test-heres-what-it-really-proved/.

73.   Koch C, Crick FC. Towards a Neurobiological Theory of Consciousness. The Academic Press. 1990.

74.   Feinberg TE, Mallatt JM. The Ancient Origins of Consciousness: How the Brain Created Experience: MIT Press; 2016.

75.   Maslow AH. Religions Values and Peak-Experiences: Rare Treasure Editions; 2021.

76.   Grof S, Grof C, Kornfield J. Holotropic Breathwork : A New Approach to Self-Exploration and Therapy. Albany, UNITED STATES: State University of New York Press; 2010.

77.   Wilber K. An integral theory of consciousness. Journal of consciousness studies. 1997;4(1):71-92.

78.   Hodgkinson B. The essence of Vedanta: Arcturus Publishing; 2006.

79.   Waldron WS. Making Sense of Mind Only: Why Yogacara Buddhism Matters: Wisdom Publications; 2023.

80.   Stutley M. Shamanism: an introduction: Routledge; 2003.

81.   Jones PN. Shamanism: An Inquiry into the History of the Scholarly Use of the Term in English-Speaking North America. Anthropology of Consciousness. 2006;17(2):4-32.

82.   Rains WL. Journey of the Goddess: Second wave feminism and the evolution of American Wicca: ProQuest Dissertations Publishing; 2013.

83.   Alper B, Rotolo M, Tevington P, Nortey J, Kallo A. Spirituality Among Americans. Pew Research Center; 2023.

84.   Pelczar M. The case for panpsychism: a critical assessment. Synthese (Dordrecht). 2022;200(4).

85.   Cook G. Does Consciousness Pervade the Universe? Scientific American. 2020.

86.   Hoffman D. The Case Against Reality: Why Evolution Hid the Truth from Our Eyes: W. W. Norton; 2019.

87.   Hameroff S, Penrose R. Conscious Events as Orchestrated Space-Time Selections. NeuroQuantology. 2007;1(1).

88.   Khan S, Huang Y, Timuçin D, Bailey S, Lee S, Lopes J, et al. Microtubule-Stabilizer Epothilone B Delays Anesthetic-Induced Unconsciousness in Rats. eneuro. 2024;11(8):ENEURO.0291-24.2024.

89.   Bond E. The contribution of coherence field theory to a model of consciousness: electric currents, EM fields, and EM radiation in the brain. Front Hum Neurosci. 2022;16:1020105.

90.   Hunt T, Jones M, McFadden J, Delorme A, Hales CG, Ericson M, et al. Editorial: Electromagnetic field theories of consciousness: opportunities and obstacles. Frontiers in Human Neuroscience. 2024;17.

91.   Hassani H, Silva ES, Unger S, TajMazinani M, Mac Feely S. Artificial intelligence (AI) or intelligence augmentation (IA): what is the future? Ai. 2020;1(2):8.

92.   Radholm K, Af Geijerstam P, Woodward M, Chalmers J, Hellgren M, Jansson S, et al. Dog ownership, glycaemic control and all-cause death in patients with newly diagnosed type 2 diabetes: a national cohort study. Front Public Health. 2023;11:1265645.

93.	Gonzalez-Rodriguez D, Perez-Carmona M. Psychedelics and Artificial Intelligence: Integrating AI-Powered Analysis in Phenomenological Mental Health Studies. 2023.

94.	APA. Stress in America™ 2023: A nation grappling with psychological impacts of collective trauma. 2023 2023.

95.	Baradell JG, Klein K. Relationship of life stress and body consciousness to hypervigilant decision making. Journal of Personality and Social Psychology. 1993;64(2):267.

96.	Fumagalli F, Molteni R, Racagni G, Riva MA. Stress during development: Impact on neuroplasticity and relevance to psychopathology. Progress in neurobiology. 2007;81(4):197-217.

97.	Kolassa I-T, Elbert T. Structural and functional neuroplasticity in relation to traumatic stress. Current directions in psychological science. 2007;16(6):321-5.

98.	Julia N. Anxiety Statistics & Facts: How Many People Have Anxiety? : CFHA; 2023 [Available from: https://cfah.org/anxiety-statistics/#Anxiety_Statistics_How_Many_People_Have_Anxiety.

99.	McEwen BS. Physiology and neurobiology of stress and adaptation: central role of the brain. Physiological reviews. 2007;87(3):873-904.

100.	Wilson RS, Arnold SE, Schneider JA, Kelly JF, Tang Y, Bennett DA. Chronic psychological distress and risk of Alzheimer's disease in old age. Neuroepidemiology. 2006;27(3):143-53.

101.	Lupien SJ, McEwen BS, Gunnar MR, Heim C. Effects of stress throughout the lifespan on the brain, behaviour and cognition. Nature reviews neuroscience. 2009;10(6):434-45.

102.	Shields GS, Sazma MA, Yonelinas AP. The effects of acute stress on core executive functions: A meta-analysis and comparison with cortisol. Neuroscience & Biobehavioral Reviews. 2016;68:651-68.

103.	Nir Y, Tononi G. Dreaming and the brain: from phenomenology to neurophysiology. TRENDS IN COGNITIVE SCIENCES. 2010;14(2):88-100.

104.	Graves L, Pack A, Abel T. Sleep and memory: a molecular perspective. Trends in neurosciences. 2001;24(4):237-43.

105.	Maquet P. The role of sleep in learning and memory. science. 2001;294(5544):1048-52.

106.	Misquitta KA, Miles A, Prevot TD, Knoch JK, Fee C, Newton DF, et al. Reduced anterior cingulate cortex volume induced by chronic stress correlates with increased behavioral emotionality and decreased synaptic puncta density. Neuropharmacology. 2021;190:108562.

107.	Boissy A. Fear and fearfulness in animals. The quarterly review of biology. 1995;70(2):165-91.

108.	Grillon C. Associative learning deficits increase symptoms of anxiety in humans. Biological psychiatry. 2002;51(11):851-8.

109.	Krystal JH, Tolin DF, Sanacora G, Castner SA, Williams GV, Aikins DE, et al. Neuroplasticity as a target for the pharmacotherapy of anxiety disorders, mood disorders, and schizophrenia. Drug discovery today. 2009;14(13-14):690-7.

110.   van Tol M-J, van der Wee NJ, van den Heuvel OA, Nielen MM, Demenescu LR, Aleman A, et al. Regional brain volume in depression and anxiety disorders. Archives of general psychiatry. 2010;67(10):1002-11.

111.   NIH. Statistics for Major Depression 2024 [Available from: https://www.nimh.nih.gov/health/statistics/major-depression.

112.   Palazidou E. The neurobiology of depression. British medical bulletin. 2012;101(1):127-45.

113.   Whiteley CM. Depression as a disorder of consciousness. 2021.

114.   Roy D, Peters ME, Everett A, Leoutsakos J-M, Yan H, Rao V, et al. Loss of consciousness and altered mental state predicting depressive and post-concussive symptoms after mild traumatic brain injury. Brain injury. 2019;33(8):1064-9.

115.   Fuchs E, Czéh B, Kole MH, Michaelis T, Lucassen PJ. Alterations of neuroplasticity in depression: the hippocampus and beyond. European Neuropsychopharmacology. 2004;14:S481-S90.

116.   Hirschfeld RM. The comorbidity of major depression and anxiety disorders: recognition and management in primary care. Primary care companion to the Journal of clinical psychiatry. 2001;3(6):244.

117.   SAMHSA. 2022 National Survey on Drug Use and Health 2024 [Available from: https://www.samhsa.gov/data/release/2022-national-survey-drug-use-and-health-nsduh-releases#infographic.

118.   CDC. New Report Confirms U.S. Life Expectancy has Declined to Lowest Level Since 1996. 2022.

119.   Mann B. In 2023 fentanyl overdoses ravaged the US and fueled a new culture war fight: NPR; 2023 [Available from: https://www.opb.org/article/2023/12/28/fentanyl-crisis-addiction-overdose/.

120.   Potenza MN, Sofuoglu M, Carroll KM, Rounsaville BJ. Neuroscience of behavioral and pharmacological treatments for addictions. Neuron. 2011;69(4):695-712.

121.   Alter A. Irresistible: The rise of addictive technology and the business of keeping us hooked: Penguin; 2017.

122.   Goeders NE. The impact of stress on addiction. European Neuropsychopharmacology. 2003;13(6):435-41.

123.   Hirschman EC. The consciousness of addiction: Toward a general theory of compulsive consumption. Journal of Consumer Research. 1992;19(2):155-79.

124.   Muskiewicz DE, Uhl GR, Hall FS. The role of cell adhesion molecule genes regulating neuroplasticity in addiction. Neural Plasticity. 2018;2018.

125.   CNTR. Coalition for National Trauma Research 2024 [Available from: https://www.nattrauma.org/.

126.   Bramley EV. The trauma doctor: Gabor Maté on happiness, hope and how to heal our deepest wounds. 2023.

127.   Bernstein JS. Living in the borderland: The evolution of consciousness and the challenge of healing trauma: Routledge; 2006.

128.   Giotakos O. Neurobiology of emotional trauma. Psychiatriki. 2020;31(2):162-71.

129.   Michaels TI, Stone E, Singal S, Novakovic V, Barkin RL, Barkin S. Brain reward circuitry: the overlapping neurobiology of trauma and substance use disorders. World journal of psychiatry. 2021;11(6):222.

130.   de Ribaupierre S. Trauma and impaired consciousness. Neurologic clinics. 2011;29(4):883-902.

131.   Frewen PA, Brown MF, Lanius RA. Trauma-related altered states of consciousness (TRASC) in an online community sample: Further support for the 4-D model of trauma-related dissociation. Psychology of Consciousness: Theory, Research, and Practice. 2017;4(1):92.

132.   Lanius RA. Trauma-related dissociation and altered states of consciousness: a call for clinical, treatment, and neuroscience research. European Journal of Psychotraumatology. 2015;6(1):27905.

133.   Sasmita AO, Kuruvilla J, Ling APK. Harnessing neuroplasticity: modern approaches and clinical future. International Journal of Neuroscience. 2018;128(11):1061-77.

134.   Knudsen EI. Sensitive periods in the development of the brain and behavior. Journal of cognitive neuroscience. 2004;16(8):1412-25.

135.   Hensch TK, Bilimoria PM, editors. Re-opening windows: manipulating critical periods for brain development. Cerebrum: the Dana forum on brain science; 2012: Dana Foundation.

136.   Park C, Rosenblat JD, Brietzke E, Pan Z, Lee Y, Cao B, et al. Stress, epigenetics and depression: A systematic review. Neurosci Biobehav Rev. 2019;102:139-52.

137.   Jung CG. The practice of psychotherapy: essays on the psychology of the transference and other subjects.(Bollingen Series 20.). 1954.

138.   Maxwell ML. Rage and social media: The effect of social media on perceptions of racism, stress appraisal, and anger expression among young African American adults: Virginia Commonwealth University; 2016.

139.   McIntosh D. The ego and the self in the thought of Sigmund Freud. The International Journal of Psycho-Analysis. 1986;67:429.

140.   Sharp M-L, Fear NT, Rona RJ, Wessely S, Greenberg N, Jones N, et al. Stigma as a barrier to seeking health care among military personnel with mental health problems. Epidemiologic reviews. 2015;37(1):144-62.

141.   Erdelyi MH. The unified theory of repression. Behavioral and Brain Sciences. 2006;29(5):499-511.

142.   Bernstein JS. Spiritual Redemption or Spiritual Bypass. Living in the Borderland: Routledge; 2005. p. 187-92.

143.   Fisher HE. Lust, attraction, and attachment in mammalian reproduction. Human nature. 1998;9:23-52.

144.   Hermann AD, Leonardelli GJ, Arkin RM. Self-doubt and self-esteem: A threat from within. Pers Soc Psychol Bull. 2002;28(3):395-408.

145.   Sakulku J. The impostor phenomenon. The Journal of Behavioral Science. 2011;6(1):75-97.

146.   Das S, Kramer A, editors. Self-censorship on Facebook. Proceedings of the International AAAI Conference on Web and Social Media; 2013.

147.   Hardy B. Personality Isn't Permanent: Break Free from Self-Limiting Beliefs and Rewrite Your Story. New York: Penguin Random House; 2020.

148.   Jacobson E. Denial and repression. Journal of the American Psychoanalytic Association. 1957;5(1):61-92.

149.   Von Hippel W, Trivers R. The evolution and psychology of self-deception. Behavioral and brain sciences. 2011;34(1):1-16.

150.   Mark G. Attention Span: A Groundbreaking Way to Restore Balance, Happiness and Productivity: Hanover Square Press; 2023.

151.   Wright P. The harassed decision maker: Time pressures, distractions, and the use of evidence. Journal of applied psychology. 1974;59(5):555.

152.   Koessmeier C, Büttner OB. Why are we distracted by social media? Distraction situations and strategies, reasons for distraction, and individual differences. Frontiers in psychology. 2021;12:711416.

153.   Acciarini C, Brunetta F, Boccardelli P. Cognitive biases and decision-making strategies in times of change: a systematic literature review. Management Decision. 2021;59(3):638-52.

154.   Benson B. Cognitive Bias Codex 2016 [Available from: https://medium.com/thinking-is-hard/4-conundrums-of-intelligence-2ab78d90740f.

155.   Kahneman D, Tversky A. Prospect Theory: An Analysis of Decision under Risk. Econometrica. 1979;47(2):263-91.

156.   Peters U. What is the function of confirmation bias? Erkenntnis. 2022;87(3):1351-76.

157.   Nisbett RE, Wilson TD. The halo effect: Evidence for unconscious alteration of judgments. Journal of Personality and Social Psychology. 1977;35(4):250.

158.   Janov A. The biology of love: Prometheus Books; 2010.

159.   Fotiou E, Gearin AK. Purging and the body in the therapeutic use of ayahuasca. Social Science & Medicine. 2019;239:112532.

160.   Levine PA. Healing Traum: A Pioneering Program for Restoring the Wisdom of Your Body: ReadHowYouWant.com, Limited; 2010.

161.   van der Kolk BA. The Body Keeps the Score: Memory and the Evolving Psychobiology of Posttraumatic Stress. Harvard Review of Psychiatry. 1994;1(5):253-65.

162.   Bragg PC. The miracle of fasting: Health Science Publications, Inc.; 2004.

163.   Berthelot E, Etchecopar-Etchart D, Thellier D, Lancon C, Boyer L, Fond G. Fasting interventions for stress, anxiety and depressive symptoms: a systematic review and meta-analysis. Nutrients. 2021;13(11):3947.

164.   Hoffman MD, Hoffman DR. Exercisers Achieve Greater Acute Exercise-Induced Mood Enhancement Than Nonexercisers. Archives of Physical Medicine and Rehabilitation. 2008;89(2):358-63.

165.   Grof S, Grof C. Holotropic Breathwork, Second Edition: A New Approach to Self-Exploration and Therapy: State University of New York Press; 2023.

166.	Miller T, Nielsen L. Measure of Significance of Holotropic Breathwork in the Development of Self-Awareness. The Journal of Alternative and Complementary Medicine. 2015;21(12):796-803.

167.	Rock AJ, Denning NC, Harris KP, Clark GI, Misso D. Exploring holotropic breathwork: An empirical evaluation of altered states of awareness and patterns of phenomenological subsystems with reference to transliminality. Journal of Transpersonal Psychology. 2015;47(1).

168.	Sarbacker SR, Samuel G, Patton LL, Bronkhorst J, Chapple CK, Wallace V. Contextualizing the History of Yoga in Geoffrey Samuel's "The Origins of Yoga and Tantra": A Review Symposium. Dordrecht: Springer; 2011. p. 303-57.

169.	Li Q, Kobayashi M, Kumeda S, Ochiai T, Miura T, Kagawa T, et al. Effects of Forest Bathing on Cardiovascular and Metabolic Parameters in Middle-Aged Males. Evidence based Complementary and Alternative Medicine. 2016.

170.	Keniger LE, Gaston KJ, Irvine KN, Fuller RA. What are the benefits of interacting with nature? International journal of environmental research and public health. 2013;10(3):913-35.

171.	Payne M, Delphinus E. A review of the current evidence for the health benefits derived from forest bathing. The International Journal of Health, Wellness and Society. 2018;9(1):19.

172.	Capaldi CA, Dopko RL, Zelenski JM. The relationship between nature connectedness and happiness: A meta-analysis. Frontiers in psychology. 2014;5:92737.

173.	Brewer JA, Worhunsky PD, Gray JR, Tang Y-Y, Weber J, Kober H. Meditation experience is associated with differences in default mode network activity and connectivity. Proc Natl Acad Sci U S A. 2011;108(50):20254-9.

174.	Newberg A, Iversen J. The neural basis of the complex mental task of meditation: neurotransmitter and neurochemical considerations. Med Hypotheses. 2003;61(2):282-91.

175.	Zeidan F, Grant JA, Brown CA, McHaffie JG, Coghill RC. Mindfulness meditation-related pain relief: evidence for unique brain mechanisms in the regulation of pain. Neuroscience letters. 2012;520(2):165-73.

176.	Martin S. Why using "consciousness" in psychotherapy? Insight, metacognition and self-consciousness. New Ideas in Psychology. 2023;70:101015.

177.	Magill M, Kiluk BD, Ray LA. Efficacy of cognitive behavioral therapy for alcohol and other drug use disorders: is a one-size-fits-all approach appropriate? Substance Abuse and Rehabilitation. 2023:1-11.

178.	Bright B. The Psychological Foundations of Transpersonal Coaching: Why and How it Works. 2020.

179.	Kornfield J. A path with heart: the classic guide through the perils and promises of spiritual life: Random House; 2008.

180.	Poland B. Coming Back to Our True Nature: What Is the Inner Work That Supports Transition? 2020.

181.   Bergmann U. Consciousness Examined: An Introduction to the Foundations of Neurobiology for EMDR. J EMDR Prac Res. 2012(3):87-91.
182.   Siegel IR. EMDR as a transpersonal therapy: A trauma-focused approach to awakening consciousness. Journal of EMDR Practice and Research. 2018;12(1):24-43.
183.   Corrigan FM. Mindfulness, dissociation, EMDR and the anterior cingulate cortex: A hypothesis. Wiley Online Library; 2002.
184.   Morris H. Hamilton's Pharmacopeia. 2016.
185.   Gibney A, Pollan M. How to Change Your Mind:. 2022.
186.   Schwartzberg L. Fantastic Fungi. 2019.
187.   Storkel P. Psychedelia. 2020.
188.   Chandler T. Dosed. 2019.
189.   White R. Ayahuasca: Vine of the Soul. 2010.
190.   Boonman R. The Last Shaman. 2016.
191.   Strassman R, Burroughs M, Dean M. DMT: The spirit molecule. 2010.
192.   Guerra C. Embrace of the Serpent. 2015.
193.   Niles M. The reality of truth. 2016.
194.   Nir G, Eitan M. Trip of Compassion. 2017.
195.   González D, Cantillo J, Pérez I, Farré M, Feilding A, Obiols JE, et al. Therapeutic potential of ayahuasca in grief: a prospective, observational study. PSYCHOPHARMACOLOGY. 2020;237(4):1171-82.
196.   Soler J, Elices M, Franquesa A, Barker S, Friedlander P, Feilding A, et al. Exploring the therapeutic potential of Ayahuasca: acute intake increases mindfulness-related capacities. Psychopharmacology. 2016;233(5):823-9.
197.   Weiss B, Wingert A, Erritzoe D, Campbell WK. Prevalence and therapeutic impact of adverse life event reexperiencing under ceremonial ayahuasca. Scientific reports. 2023;13(1):9438-.
198.   Kiraga MK, Mason NL, Uthaug M, van Oorsouw KIM, Toennes SW, Ramaekers JG, et al. Persisting Effects of Ayahuasca on Empathy, Creative Thinking, Decentering, Personality, and Well-Being. Frontiers in pharmacology. 2021;12:721537-.
199.   Shah FI, Shehzadi S, Akram F, Haq IU, Javed B, Sabir S, et al. Unveiling the Psychedelic Journey: An Appraisal of Psilocybin as a Profound Antidepressant Therapy. Molecular biotechnology. 2023.
200.   Ross S, Bossis A, Guss J, Agin-Liebes G, Malone T, Cohen B, et al. Rapid and sustained symptom reduction following psilocybin treatment for anxiety and depression in patients with life-threatening cancer: a randomized controlled trial. Journal of psychopharmacology (Oxford). 2016;30(12):1165-80.
201.   van der Meer PB, Fuentes JJ, Kaptein AA, Schoones JW, de Waal MM, Goudriaan AE, et al. Therapeutic effect of psilocybin in addiction: A systematic review. Frontiers in psychiatry. 2023;14:1134454.
202.   Davis AK, Barrett FS, May DG, Cosimano MP, Sepeda ND, Johnson MW, et al. Effects of psilocybin-assisted therapy on major depressive disorder: a randomized clinical trial. JAMA Psychiatry. 2021;78(5):481-9.
203.   Doesburg-van Kleffens M, Zimmermann-Klemd AM, Gründemann C. An Overview on the Hallucinogenic Peyote and Its Alkaloid Mescaline: The

Importance of Context, Ceremony and Culture. Molecules (Basel, Switzerland). 2023;28(24):7942.

204.   Bohn A, Kiggen MHH, Uthaug MV, van Oorsouw KIM, Ramaekers JG, van Schie HT. Altered States of Consciousness During Ceremonial San Pedro Use. The International journal for the psychology of religion. 2022;ahead-of-print(ahead-of-print):1-23.

205.   Heink A, Katsikas S, Lange-Altman T. Examination of the Phenomenology of the Ibogaine Treatment Experience: Role of Altered States of Consciousness and Psychedelic Experiences. Journal of Psychoactive Drugs. 2017;49(3):201-8.

206.   Heaven R. Shamanic Quest for the Spirit of Salvia: The Divinatory, Visionary, and Healing Powers of the Sage of the Seers: Inner Traditions/Bear; 2013.

207.   Brito-da-Costa AM, Dias-da-Silva D, Gomes NG, Dinis-Oliveira RJ, Madureira-Carvalho Á. Pharmacokinetics and pharmacodynamics of salvinorin a and salvia divinorum: Clinical and forensic aspects. Pharmaceuticals. 2021;14(2):116.

208.   National Academies of Sciences E, Medicine. Therapeutic effects of cannabis and cannabinoids. The health effects of cannabis and cannabinoids: The current state of evidence and recommendations for research: National Academies Press (US); 2017.

209.   Sandoval G, Most A. Bufo alvarius and the entheogenic experience: Lunaria Ediciones; 2020.

210.   Fuentes JJ, Fonseca F, Elices M, Farré M, Torrens M. Therapeutic use of LSD in psychiatry: a systematic review of randomized-controlled clinical trials. Frontiers in psychiatry. 2020;10:494327.

211.   Bayne T, Carter O. Dimensions of consciousness and the psychedelic state. Neuroscience of Consciousness. 2018;2018(1).

212.   Levine J, Ludwig AM. Alterations in consciousness produced by combinations of LSD, hypnosis and psychotherapy. Psychopharmacologia. 1965;7(2):123-37.

213.   Liechti ME, Dolder PC, Schmid Y. Alterations of consciousness and mystical-type experiences after acute LSD in humans. Psychopharmacology. 2017;234:1499-510.

214.   Ramos CS, Thornburg M, Long K, Sharma K, Roth J, Lacatusu D, et al. The therapeutic effects of ketamine in mental health disorders: A narrative review. Cureus. 2022;14(3).

215.   Strong CE, Kabbaj M. Neural mechanisms underlying the rewarding and therapeutic effects of ketamine as a treatment for alcohol use disorder. Frontiers in Behavioral Neuroscience. 2020;14:593860.

216.   Vlisides P, Bel-Bahar T, Nelson A, Chilton K, Smith E, Janke E, et al. Subanaesthetic ketamine and altered states of consciousness in humans. British journal of anaesthesia. 2018;121(1):249-59.

217.   Passie T, Adams H-A, Logemann F, Brandt SD, Wiese B, Karst M. Comparative effects of (S)-ketamine and racemic (R/S)-ketamine on

psychopathology, state of consciousness and neurocognitive performance in healthy volunteers. European Neuropsychopharmacology. 2021;44:92-104.

218.   Zhang X, Hack L, Heifets B, Suppes T, van Roessel P, Yesavage J, et al. 43. Acute Effects of MDMA on Intrinsic Functional Connectomes Associated With Altered States of Consciousness and Defensiveness. Biological Psychiatry. 2023;93(9):S87-S8.

219.   van der Kolk BA, Wang JB, Yehuda R, Bedrosian L, Coker AR, Harrison C, et al. Effects of MDMA-assisted therapy for PTSD on self-experience. PLoS One. 2024;19(1):e0295926.

220.   Raichle ME, MacLeod AM, Snyder AZ, Powers WJ, Gusnard DA, Shulman GL. A default mode of brain function. Proceedings of the National Academy of Sciences. 2001;98(2):676-82.

221.   Tseng J, Poppenk J. Brain meta-state transitions demarcate thoughts across task contexts exposing the mental noise of trait neuroticism. Nature Communications. 2020;11(1):3480.

222.   Stephens GJ, Silbert LJ, Hasson U. Speaker–listener neural coupling underlies successful communication. Proceedings of the national academy of sciences. 2010;107(32):14425-30.

223.   Shamay-Tsoory SG, Tibi-Elhanany Y, Aharon-Peretz J. The ventromedial prefrontal cortex is involved in understanding affective but not cognitive theory of mind stories. Social neuroscience. 2006;1(3-4):149-66.

224.   Hasson U, Landesman O, Knappmeyer B, Vallines I, Rubin N, Heeger DJ. Neurocinematics: The Neuroscience of Film. Projections. 2008;2(1):1-26.

225.   Oatley K. Fiction: Simulation of Social Worlds. TRENDS IN COGNITIVE SCIENCES. 2016;20(8):618-28.

226.   Iacoboni M, Molnar-Szakacs I, Gallese V, Buccino G, Mazziotta JC, Rizzolatti G. Grasping the Intentions of Others with One's Own Mirror Neuron System. PLoS biology. 2005;3(3):e79-e.

227.   Gerrig RJ. Experiencing narrative worlds: On the psychological activities of reading. New Haven, CT, US: Yale University Press; 1993. xi, 273-xi, p.

228.   Green MC, Brock TC. The Role of Transportation in the Persuasiveness of Public Narratives. Journal of Personality and Social Psychology. 2000;79(5):701-21.

229.   Barraza JA, Alexander V, Beavin LE, Terris ET, Zak PJ. The heart of the story: Peripheral physiology during narrative exposure predicts charitable giving. Biological Psychology. 2015;105:138-43.

230.   Zak PJ, Stanton AA, Ahmadi S. Oxytocin increases generosity in human. PLos ONE. 2007;11:1-5.

231.   Lederman LC, Menegatos LM. Sustainable recovery: The self-transformative power of storytelling in Alcoholics Anonymous. Journal of Groups in Addiction & Recovery. 2011;6(3):206-27.

232.   Arminen I. Second stories: the salience of interpersonal communication for mutual help in Alcoholics Anonymous. Journal of Pragmatics. 2004;36(2):319-47.

233. Reiff CM, Richman EE, Nemeroff CB, Carpenter LL, Widge AS, Rodriguez CI, et al. Psychedelics and Psychedelic-Assisted Psychotherapy. American Journal of Psychiatry. 2020;177(5):391-410.

234. Modlin NL, Stubley J, Maggio C, Rucker JJ. On Redescribing the Indescribable: Trauma, Psychoanalysis and Psychedelic Therapy. British journal of psychotherapy. 2023;39(3):551-72.

235. Frecska E, Bokor P, Winkelman M. The Therapeutic Potentials of Ayahuasca: Possible Effects against Various Diseases of Civilization. Frontiers in pharmacology. 2016;7:35-.

236. Ahmad F. Personality traits as predictor of cognitive biases: moderating role of risk-attitude. Qualitative Research in Financial Markets. 2020;12(4):465-84.

237. Roberts BW. A revised sociogenomic model of personality traits. Journal of personality. 2018;86(1):23-35.

238. Kandler C, Bratko D, Butković A, Hlupić TV, Tybur JM, Wesseldijk LW, et al. How genetic and environmental variance in personality traits shift across the life span: Evidence from a cross-national twin study. Journal of Personality and Social Psychology. 2021;121(5):1079.

239. Gescher DM, Kahl KG, Hillemacher T, Frieling H, Kuhn J, Frodl T. Epigenetics in Personality Disorders: Today's Insights. Frontiers in psychiatry. 2018;9:579-.

240. Grof S, Grob C, Bravo G, Roger Walsh M. Birthing the transpersonal. Journal of Transpersonal Psychology. 2008;40(2):155.

241. Vernon PA, Villani VC, Vickers LC, Harris JA. A behavioral genetic investigation of the Dark Triad and the Big 5. Personality and individual differences. 2008;44(2):445-52.

242. Wright ZE, Pahlen S, Krueger RF. Genetic and Environmental Influences on Diagnostic and Statistical Manual of Mental Disorders-Fifth Edition (DSM-5) Maladaptive Personality Traits and Their Connections With Normative Personality Traits. Journal of abnormal psychology (1965). 2017;126(4):416-28.

243. Abu Raya M, Ogunyemi AO, Broder J, Carstensen VR, Illanes-Manrique M, Rankin KP. The neurobiology of openness as a personality trait. Frontiers in Neurology. 2023;14:1235345.

244. Bland AM. The Enneagram: A review of the empirical and transformational literature. The Journal of Humanistic Counseling, Education and Development. 2010;49(1):16-31.

245. McCarter BG. Application of Theory. Leadership in Chaordic Organizations. 1 ed. United Kingdom: CRC Press; 2013. p. 147-91.

246. Jarrett C. Be who you want: Unlocking the science of personality change: Simon and Schuster; 2021.

247. Bi G-Q, Poo M-M. Synaptic modification by correlated activity : Hebb's postulate revisited. Annual review of neuroscience. 2001;24(1):139-66.

248. Duckworth RA. Neuroendocrine mechanisms underlying behavioral stability: implications for the evolutionary origin of personality. Annals of the New York Academy of Sciences. 2015;1360(1):54-74.

249.   Pedrosa LRR, Coimbra GDS, Corrêa MG, Dias IA, Bahia CP. Time Window of the Critical Period for Neuroplasticity in S1, V1, and A1 Sensory Areas of Small Rodents: A Systematic Review. Frontiers in neuroanatomy. 2022;16:763245-.

250.   Lepow L, Morishita H, Yehuda R. Critical Period Plasticity as a Framework for Psychedelic-Assisted Psychotherapy. Front Neurosci. 2021;15:710004.

251.   Nardou R, Sawyer E, Song YJ, Wilkinson M, Padovan-Hernandez Y, de Deus JL, et al. Psychedelics reopen the social reward learning critical period. Nature; London. 2023;618(7966):790-8.

252.   Compton WC, Hoffman E. Positive Psychology: The Science of Happiness and Flourishing: SAGE Publications; 2019.

253.   Bodri B. Sport Visualization for the Elite Athlete: Build Mental Imagery Skills to Enhance Athletic Performance: Top Shape Publishing, LLC; 2018.

254.   Atkinson WW. Thought vibration: Or, the law of attraction in the thought world: New Thought Publishing Company; 1906.

255.   Chopra D. The spontaneous fulfillment of desire : harnessing the infinite power of coincidence. New York: Harmony Books; 2003.

256.   Beauregard M. Expanding Reality: The Emergence of Postmaterialist Science: Iff Books; 2021.

257.   Lipton B. The biology of belief: Unleashing the power of consciousness, matter and miracles.: Elite Books; 2005.

258.   Radin D, Lund N, Emoto M, Kizu T. Effects of Distant Intention on Water Crystal Formation: A Triple-Blind Replication. Journal of scientific exploration. 2010;22(4).

259.   Colloca L. Neurobiology of the Placebo Effect, Part I: Elsevier Science; 2018.

260.   Kienle GS, Kiene H. The Powerful Placebo Effect: Fact or Fiction? Journal of clinical epidemiology. 1997;50(12):1311-8.

261.   Freed CR, Breeze RE, Fahn S. Placebo Surgery in Trials of Therapy for Parkinson's Disease. The New England journal of medicine. 2000;342(5):353-5.

262.   Jones BDM, Razza LB, Weissman CR, Karbi J, Vine T, Mulsant LS, et al. Magnitude of the Placebo Response Across Treatment Modalities Used for Treatment-Resistant Depression in Adults A Systematic Review and Meta-analysis. JAMA network open. 2021;4(9):e2125531.

263.   Moseley JB, O'Malley K, Petersen NJ, Menke TJ, Brody BA, Kuykendall DH, et al. A Controlled Trial of Arthroscopic Surgery for Osteoarthritis of the Knee. New England Journal of Medicine. 2002;347(2):81-8.

264.   Hobson N, Bonk D, Inzlicht M. Rituals decrease the neural response to performance failure. PeerJ. 2017.

265.   Hobson N, Risen J, Inzlicht M. The Psychology of Rituals: An Integrative Review and Process-Based Framework. SSRN Electronic Journal. 2017.

266.   El-Sayes J, Harasym D, Turco CV, Locke MB, Nelson AJ. Exercise-induced neuroplasticity: a mechanistic model and prospects for promoting plasticity. The Neuroscientist. 2019;25(1):65-85.

267.   Hötting K, Röder B. Beneficial effects of physical exercise on neuroplasticity and cognition. Neuroscience & Biobehavioral Reviews. 2013;37(9):2243-57.

268.   Tolahunase MR, Sagar R, Faiq M, Dada R. Yoga-and meditation-based lifestyle intervention increases neuroplasticity and reduces severity of major depressive disorder: A randomized controlled trial. Restorative neurology and neuroscience. 2018;36(3):423-42.

269.   Froeliger B, Garland EL, McClernon FJ. Yoga meditation practitioners exhibit greater gray matter volume and fewer reported cognitive failures: results of a preliminary voxel-based morphometric analysis. Evidence-Based Complementary and Alternative Medicine. 2012;2012.

270.   Krause-Sorio B, Siddarth P, Kilpatrick L, Milillo MM, Aguilar-Faustino Y, Ercoli L, et al. Yoga prevents gray matter atrophy in women at risk for Alzheimer's disease: A randomized controlled trial. Journal of Alzheimer's Disease. 2022;87(2):569-81.

271.   Teixeira-Machado L, Arida RM, de Jesus Mari J. Dance for neuroplasticity: A descriptive systematic review. Neuroscience & Biobehavioral Reviews. 2019;96:232-40.

272.   Müller P, Rehfeld K, Schmicker M, Hökelmann A, Dordevic M, Lessmann V, et al. Evolution of neuroplasticity in response to physical activity in old age: the case for dancing. Frontiers in aging neuroscience. 2017;9:56.

273.   Vivar C, Potter MC, van Praag H. All about running: synaptic plasticity, growth factors and adult hippocampal neurogenesis. Neurogenesis and neural plasticity. 2013:189-210.

274.   Lin T-W, Shih Y-H, Chen S-J, Lien C-H, Chang C-Y, Huang T-Y, et al. Running exercise delays neurodegeneration in amygdala and hippocampus of Alzheimer's disease (APP/PS1) transgenic mice. Neurobiology of learning and memory. 2015;118:189-97.

275.   Andre C. Looking at Mindfulness: 25 Ways to Live in the Moment Through Art: Blue Rider Press; 2015.

276.   Doll A, Holzel Britta KH, Bratec SM, Boucard CC, Xie X, Wohlschlager Afra MW, et al. Mindful attention to breath regulates emotions via increased amygdala–prefrontal cortex connectivity. NeuroImage. 2016;134:305.

277.   Zaccaro A, Piarulli A, Laurino M, Garbella E, Menicucci D, Neri B, et al. How Breath-Control Can Change Your Life: A Systematic Review on Psycho-Physiological Correlates of Slow Breathing. Frontiers in human neuroscience. 2018;12:353-.

278.   Borge CR, Hagen KB, Mengshoel AM, Omenaas E, Moum T, Wahl AK. Effects of controlled breathing exercises and respiratory muscle training in people with chronic obstructive pulmonary disease: results from evaluating the quality of evidence in systematic reviews. BMC Pulm Med. 2014;14:184.

279.   Gorgoni M, D'Atri A, Lauri G, Rossini PM, Ferlazzo F, De Gennaro L. Is sleep essential for neural plasticity in humans, and how does it affect motor and cognitive recovery? Neural Plast. 2013;2013:103949.

280.   Weiss JT, Donlea JM. Roles for Sleep in Neural and Behavioral Plasticity: Reviewing Variation in the Consequences of Sleep Loss. Front Behav Neurosci. 2021;15:777799.

281.   Pickersgill JW, Turco CV, Ramdeo K, Rehsi RS, Foglia SD, Nelson AJ. The combined influences of exercise, diet and sleep on neuroplasticity. Frontiers in psychology. 2022;13:831819.

282.   Iranzo A, Santamaria J. Sleep in neurodegenerative diseases. Sleep Medicine: A Comprehensive Guide to Its Development, Clinical Milestones, and Advances in Treatment. 2015:271-83.

283.   Jessen NA, Munk ASF, Lundgaard I, Nedergaard M. The glymphatic system: a beginner's guide. Neurochemical research. 2015;40:2583-99.

284.   Zwilling CE, Wu J, Barbey AK. Investigating nutrient biomarkers of healthy brain aging: a multimodal brain imaging study. npj Aging. 2024;10(1):27.

285.   Preissner CE, Oenema A, de Vries H. Examining socio-cognitive factors and beliefs about mindful eating in healthy adults with differing practice experience: a cross-sectional study. BMC Psychology. 2022;10(1):268.

286.   Cherpak CE. Mindful Eating: A Review Of How The Stress-Digestion-Mindfulness Triad May Modulate And Improve Gastrointestinal And Digestive Function. Integr Med (Encinitas). 2019;18(4):48-53.

287.   Ohly H, White M, Wheeler B, Bethel A, Ukoumunne O, Nikolaou V, et al. Attention Restoration Theory: A Systematic Review of the Attention Restoration Potential of Exposure to Natural Environments. Journal of Toxicology and Environmental Health, Part B. 2016;19:1-39.

288.   Stenfors CUD, Van Hedger SC, Schertz KE, Meyer FAC, Smith KEL, Norman GJ, et al. Positive Effects of Nature on Cognitive Performance Across Multiple Experiments: Test Order but Not Affect Modulates the Cognitive Effects. Frontiers in psychology. 2019;10.

289.   Bratman GN, Hamilton JP, Hahn KS, Daily GC, Gross JJ. Nature experience reduces rumination and subgenual prefrontal cortex activation. Proceedings of the National Academy of Sciences - PNAS. 2015;112(28):8567-72.

290.   Polheber JP, Matchock RL. The presence of a dog attenuates cortisol and heart rate in the Trier Social Stress Test compared to human friends. Journal of behavioral medicine. 2014;37(5):860-7.

291.   Barker SB, Wolen AR. The Benefits of Human–Companion Animal Interaction: A Review. Journal of Veterinary Medical Education. 2008;35(4):487-95.

292.   McCardle P, McCune S, Griffin J, Esposito L, Freund L, Grandin T. Animals in our lives: human-animal interaction in family, community, and therapeutic settings. Middletown: American Library Association CHOICE; 2011. p. 400.

293.   Friedmann E, Gee NR, Simonsick EM, Kitner-Triolo MH, Resnick B, Adesanya I, et al. Pet ownership and maintenance of cognitive function in community-residing older adults: evidence from the Baltimore Longitudinal Study of Aging (BLSA). Scientific reports. 2023;13(1):14738-.

294.   Curl AL, Bibbo J, Johnson RA. Neighborhood Engagement, Dogs, and Life Satisfaction in Older Adulthood. Journal of applied gerontology. 2021;40(12):1706-14.

295.   Brooks K, Brooks K. Enhancing sports performance through the use of music. Journal of exercise physiology online. 2010;13(2).

296.   Thoma MV, La Marca R, Broennimann R, Finkel L, Ehlert U, Nater UM. The Effect of Music on the Human Stress Response. PloS one. 2013;8(8):e70156-e.

297.   Chanda ML, Levitin DJ. The neurochemistry of music. TRENDS IN COGNITIVE SCIENCES. 2013;17(4):179-93.

298.   T. Habib P. COVID-19 symphony: A review of possible music therapy effect in supporting the immune system of COVID-19 patient. Highlights in BioScience. 2021;4:bs202105.

299.   Molnar-Szakacs I, Overy K. Music and mirror neurons: from motion to 'e'motion. Social cognitive and affective neuroscience. 2006;1(3):235-41.

300.   Scataglini S, Van Dyck Z, Declercq V, Van Cleemput G, Struyf N, Truijen S. Effect of Music Based Therapy Rhythmic Auditory Stimulation (RAS) Using Wearable Device in Rehabilitation of Neurological Patients: A Systematic Review. Sensors (Basel, Switzerland). 2023;23(13):5933.

301.   Balbag MA, Pedersen NL, Gatz M. Playing a Musical Instrument as a Protective Factor against Dementia and Cognitive Impairment: A Population-Based Twin Study. International journal of Alzheimer's disease. 2014;2014:836748-6.

302.   Schlaug G. The brain of musicians. A model for functional and structural adaptation. Annals of the New York Academy of Sciences. 2001;930(1):281-99.

303.   Hudziak JJ, Albaugh MD, Ducharme S, Karama S, Spottswood M, Crehan E, et al. Cortical thickness maturation and duration of music training: Health-promoting activities shape brain development. Journal of the American Academy of Child & Adolescent Psychiatry. 2014;53(11):1153-61. e2.

304.   Choi U-S, Sung Y-W, Hong S, Chung J-Y, Ogawa S. Structural and functional plasticity specific to musical training with wind instruments. Frontiers in human neuroscience. 2015;9:597-.

305.   Kappen PR, van den Brink J, Jeekel J, Dirven CMF, Klimek M, Donders-Kamphuis M, et al. The effect of musicality on language recovery after awake glioma surgery. Frontiers in human neuroscience. 2023;16:1028897-.

306.   Provine RR. Laughter as a scientific problem: An adventure in sidewalk neuroscience. Journal of Comparative Neurology. 2015;524(8).

307.   Landoni AM. A laughing matter: Transforming trauma through therapeutic humor and expressive arts therapy. 2019.

308.   Foglia L, Wilson RA. Embodied cognition. Wiley Interdisciplinary Reviews: Cognitive Science. 2013;4(3):319-25.

309.   Caruana F, Palagi E, de Waal FBM. Cracking the laugh code: laughter through the lens of biology, psychology and neuroscience. Philosophical transactions of the Royal Society of London Series B Biological sciences. 2022;377(1863):20220159-.

310.   Kramer CA. As If: Connecting Phenomenology, Mirror Neurons, Empathy, and Laughter. PhaenEx. 2012;7(1):275-308.

311.   Berk L, Lee J, Mali D, Lohman E, Bains G, Daher N, et al. Humor associated mirthful laughter increases the intensity of power spectral density (μV 2) EEG gamma wave band frequency (31–40Hz) which is associated with neuronal synchronization, memory, recall, enhanced cognitive processing and other brain health benefits when compared to distress. The FASEB journal. 2016;30(S1).

312.   Meer JNvd, Breakspear M, Chang LJ, Sonkusare S, Cocchi L. Movie viewing elicits rich and reliable brain state dynamics. Nature Communications. 2020;11(1):5004.

313.   Carpe I, editor The alchemy of animation: A neuroplastic art media of communication and transformation. EDULEARN17 Proceedings; 2017: IATED.

314.   Maffei A. Spectrally resolved EEG intersubject correlation reveals distinct cortical oscillatory patterns during free-viewing of affective scenes. Psychophysiology. 2020;57(11):e13652-n/a.

315.   Lentoor AG, Motsamai TB, Nxiweni T, Mdletshe B, Mdingi S. Protocol for a systematic review of the effects of gardening physical activity on neuroplasticity and cognitive function. AIMS neuroscience. 2023;10(2):118-29.

316.   Van Den Berg AE, Custers MH. Gardening promotes neuroendocrine and affective restoration from stress. J Health Psychol. 2011;16(1):3-11.

317.   Farmer N, Touchton-Leonard K, Ross A. Psychosocial Benefits of Cooking Interventions: A Systematic Review. Health Educ Behav. 2018;45(2):167-80.

318.   Manera V, Petit P-D, Derreumaux A, Orvieto I, Romagnoli M, Lyttle G, et al. 'Kitchen and cooking,' a serious game for mild cognitive impairment and Alzheimer's disease: a pilot study. Frontiers in Aging Neuroscience. 2015;7.

319.   Lee B, Park J-Y, Jung WH, Kim HS, Oh JS, Choi C-H, et al. White matter neuroplastic changes in long-term trained players of the game of "Baduk"(GO): a voxel-based diffusion-tensor imaging study. Neuroimage. 2010;52(1):9-19.

320.   Gauthier A, Kato PM, Bul KCM, Dunwell I, Walker-Clarke A, Lameras P. Board Games for Health: A Systematic Literature Review and Meta-Analysis. Games Health J. 2019;8(2):85-100.

321.   Lillo-Crespo M, Forner-Ruiz M, Riquelme-Galindo J, Ruiz-Fernandez D, Garcia-Sanjuan S. Chess Practice as a Protective Factor in Dementia. International journal of environmental research and public health. 2019;16(12):2116.

322.   Clemenson GD, Stark CE. Virtual environmental enrichment through video games improves hippocampal-associated memory. Journal of Neuroscience. 2015;35(49):16116-25.

323.   Anguera JA, Boccanfuso J, Rintoul JL, Al-Hashimi O, Faraji F, Janowich J, et al. Video game training enhances cognitive control in older adults. Nature; London. 2013;501(7465):97-101.

324.   Cameron HA, Glover LR. Adult Neurogenesis: Beyond Learning and Memory. Annual review of psychology. 2015;66(1):53-81.

325.   Maguire EA, Gadian DG, Johnsrude IS, Good CD, Ashburner J, Richard SJF, et al. Navigation-Related Structural Change in the Hippocampi of Taxi Drivers. Proceedings of the National Academy of Sciences - PNAS. 2000;97(8):4398-403.

326.   Cabib S, Campus P, Conversi D, Orsini C, Puglisi-Allegra S. Functional and Dysfunctional Neuroplasticity in Learning to Cope with Stress. Brain sciences. 2020;10(2):127.

327.   Niblett M, Beuret K, Pasternak C. Why Travel?: Understanding Our Need to Move and How It Shapes Our Lives
Why Travel?: Bristol University Press; 2021 01 Jul. 2021. 13-32 p.

328.   Schlegel A, Alexander P, Fogelson SV, Li X, Lu Z, Kohler PJ, et al. The artist emerges: Visual art learning alters neural structure and function. NeuroImage (Orlando, Fla). 2015;105:440-51.

329.   Bolwerk A, Mack-Andrick J, Lang FR, Doerfler A, Maihoefner C. How Art Changes Your Brain: Differential Effects of Visual Art Production and Cognitive Art Evaluation on Functional Brain Connectivity. PloS one. 2014;9(7):e101035-e.

330.   Zaidel DW. Creativity, brain, and art: biological and neurological considerations. Frontiers in human neuroscience. 2014;8:389.

331.   Guidotti R, Del Gratta C, Perrucci MG, Romani GL, Raffone A. Neuroplasticity within and between functional brain networks in mental training based on long-term meditation. Brain Sciences. 2021;11(8):1086.

332.   Arumugam K. Neural correlates of religious and spiritual experiences 2015.

333.   Newberg A, Wintering N, Morgan D, Waldman M. The measurement of regional cerebral blood flow during glossolalia: A preliminary SPECT study. Psychiatry research. 2006;148(1):67-71.

334.   Inzlicht M, McGregor I, Hirsh JB, Nash K. Neural Markers of Religious Conviction. Psychological science. 2009;20(3):385-92.

335.   Schjoedt U, Stødkilde-Jørgensen H, Geertz AW, Roepstorff A. Highly religious participants recruit areas of social cognition in personal prayer. Social cognitive and affective neuroscience. 2009;4(2):199-207.

336.   Kapogiannis D, Barbey AK, Su M, Krueger F, Grafman J. Neuroanatomical Variability of Religiosity. PloS one. 2009;4(9):e7180-e.

337.   Walker MP, Stickgold R. Sleep, memory, and plasticity. Annu Rev Psychol. 2006;57:139-66.

338.   Nielsen T, Lara-Carrasco J. Nightmares, dreaming, and emotion regulation: A review. The new science of dreaming: Volume 2 Content, recall, and personality correlates. Praeger perspectives. Westport, CT, US: Praeger Publishers/Greenwood Publishing Group; 2007. p. 253-84.

339.   Horikawa T, Tamaki M, Miyawaki Y, Kamitani Y. Neural decoding of visual imagery during sleep. Science. 2013;340(6132):639-42.
340.   Perogamvros L, Schwartz S. The roles of the reward system in sleep and dreaming. Neurosci Biobehav Rev. 2012;36(8):1934-51.
341.   Aizenstat S. Dream tending: Spring Journal New Orleans, LA; 2009.
342.   Holzinger B, Mayer L. Lucid Dreaming Brain Network Based on Tholey's 7 Klartraum Criteria. Frontiers in psychology. 2020;11:1885-.
343.   Tononi G, Cirelli C. Sleep and synaptic homeostasis: A hypothesis. Brain Research Bulletin. 2003;62(2):143-50.
344.   Stellar JE, Gordon AM, Piff PK, Cordaro D, Anderson CL, Bai Y, et al. Self-Transcendent Emotions and Their Social Functions: Compassion, Gratitude, and Awe Bind Us to Others Through Prosociality. Emotion Review. 2017;9(3):200-7.
345.   Ly C, Greb AC, Cameron LP, Wong JM, Barragan EV, Wilson PC, et al. Psychedelics Promote Structural and Functional Neural Plasticity. Cell Rep. 2018;23(11):3170-82.
346.   Vollenweider FX, Preller KH. Psychedelic drugs: neurobiology and potential for treatment of psychiatric disorders. Nat Rev Neurosci. 2020;21(11):611-24.
347.   Gattuso JJ, Perkins D, Ruffell S, Lawrence AJ, Hoyer D, Jacobson LH, et al. Default Mode Network Modulation by Psychedelics: A Systematic Review. International Journal of Neuropsychopharmacology. 2022;26(3):155-88.
348.   Berkovich-Ohana A, Wilf M, Kahana R, Ariely A, Malach R. Repetitive speech elicits widespread deactivation in the human cortex: the "Mantra" effect? Brain and Behavior 2015;5(7).
349.   Farb NA, Anderson AK, Segal ZV. The mindful brain and emotion regulation in mood disorders. Can J Psychiatry. 2012;57(2):70-7.
350.   Csikszentmihalyi M, Csikszentmihalyi M, Abuhamdeh S, Nakamura J. Flow. Flow and the foundations of positive psychology: The collected works of Mihaly Csikszentmihalyi. 2014:227-38.
351.   Dietrich A. Functional neuroanatomy of altered states of consciousness: The transient hypofrontality hypothesis. Conscious Cogn. 2003;12(2):231-56.
352.   Pi Y-L, Wu X-H, Wang F-J, Liu K, Wu Y, Zhu H, et al. Motor skill learning induces brain network plasticity: A diffusion-tensor imaging study. PLOS ONE. 2019;14(2):e0210015.
353.   dos Santos RG, Bouso JC, Alcázar-Córcoles MÁ, Hallak JEC. Efficacy, tolerability, and safety of serotonergic psychedelics for the management of mood, anxiety, and substance-use disorders: a systematic review of systematic reviews. Expert Review of Clinical Pharmacology. 2018;11(9):889-902.
354.   Osório FdL, Sanches RF, Macedo LR, Dos Santos RG, Maia-de-Oliveira JP, Wichert-Ana L, et al. Antidepressant effects of a single dose of ayahuasca in patients with recurrent depression: a preliminary report. Brazilian Journal of Psychiatry. 2015;37:13-20.
355.   Bouso JC, Palhano-Fontes F, Rodríguez-Fornells A, Ribeiro S, Sanches R, Crippa JAS, et al. Long-term use of psychedelic drugs is associated with

differences in brain structure and personality in humans. European Neuropsychopharmacology. 2015;25(4):483-92.

356.	Palhano-Fontes F, Andrade KC, Tofoli LF, Santos AC, Crippa JAS, Hallak JEC, et al. The Psychedelic State Induced by Ayahuasca Modulates the Activity and Connectivity of the Default Mode Network. PLOS ONE. 2015;10(2):e0118143.

357.	Shao L-X, Liao C, Gregg I, Davoudian PA, Savalia NK, Delagarza K, et al. Psilocybin induces rapid and persistent growth of dendritic spines in frontal cortex in vivo. Neuron. 2021;109(16):2535-44. e4.

358.	Catlow BJ, Song S, Paredes DA, Kirstein CL, Sanchez-Ramos J. Effects of psilocybin on hippocampal neurogenesis and extinction of trace fear conditioning. Experimental brain research. 2013;228:481-91.

359.	Nichols DE. Psychedelics. Pharmacological reviews. 2016;68(2):264-355.

360.	Halberstadt AL. Recent advances in the neuropsychopharmacology of serotonergic hallucinogens. Behavioural brain research. 2015;277:99-120.

361.	He DY, Ron D, He DY, Ron D. Autoregulation of glial cell line-derived neurotrophic factor expression: implications for the long-lasting actions of the anti-addiction drug, Ibogaine. The FASEB journal. 2006;20(13):2420-2.

362.	Alper KR, Lotsof HS, Kaplan CD. The ibogaine medical subculture. Journal of ethnopharmacology. 2008;115(1):9-24.

363.	C. Mash D. Breaking the cycle of opioid use disorder with ibogaine. The American Journal of Drug and Alcohol Abuse. 2018;44(1):1-3.

364.	Addy PH. Acute and post-acute behavioral and psychological effects of salvinorin A in humans. Psychopharmacology. 2012;220:195-204.

365.	Atigari DV, Uprety R, Pasternak GW, Majumdar S, Kivell BM. MP1104, a mixed kappa-delta opioid receptor agonist has anti-cocaine properties with reduced side-effects in rats. Neuropharmacology. 2019;150:217-28.

366.	Lu H-C, Mackie K. An introduction to the endogenous cannabinoid system. Biological psychiatry. 2016;79(7):516-25.

367.	Fogaça MV, Campos AC, Coelho LD, Duman RS, Guimarães FS. The anxiolytic effects of cannabidiol in chronically stressed mice are mediated by the endocannabinoid system: role of neurogenesis and dendritic remodeling. Neuropharmacology. 2018;135:22-33.

368.	Szabo A. Psychedelics and immunomodulation: novel approaches and therapeutic opportunities. Frontiers in immunology. 2015;6:128584.

369.	Dakic V, Nascimento JM, Sartore RC, de Moraes Maciel R, de Araujo DB, Ribeiro S, et al. Short term changes in the proteome of human cerebral organoids induced by 5-methoxy-N, N-dimethyltryptamine.

370.	Vollenweider FX, Kometer M. The neurobiology of psychedelic drugs: implications for the treatment of mood disorders. Nature Reviews Neuroscience. 2010;11(9):642-51.

371.	Carhart-Harris RL, Muthukumaraswamy S, Roseman L, Kaelen M, Droog W, Murphy K, et al. Neural correlates of the LSD experience revealed by multimodal neuroimaging. Proceedings of the National Academy of Sciences. 2016;113(17):4853-8.

372.   Gasser P, Holstein D, Michel Y, Doblin R, Yazar-Klosinski B, Passie T, et al. Safety and efficacy of lysergic acid diethylamide-assisted psychotherapy for anxiety associated with life-threatening diseases. The Journal of nervous and mental disease. 2014;202(7):513-20.

373.   Duman RS, Aghajanian GK. Synaptic dysfunction in depression: potential therapeutic targets. science. 2012;338(6103):68-72.

374.   Autry AE, Monteggia LM. Brain-derived neurotrophic factor and neuropsychiatric disorders. Pharmacological reviews. 2012;64(2):238-58.

375.   Zarate CA, Singh JB, Carlson PJ, Brutsche NE, Ameli R, Luckenbaugh DA, et al. A randomized trial of an N-methyl-D-aspartate antagonist in treatment-resistant major depression. Archives of general psychiatry. 2006;63(8):856-64.

376.   Parrott AC. MDMA, serotonergic neurotoxicity, and the diverse functional deficits of recreational 'Ecstasy'users. Neuroscience & Biobehavioral Reviews. 2013;37(8):1466-84.

377.   Thompson M, Callaghan PD, Hunt GE, Cornish JL, McGregor IS. A role for oxytocin and 5-HT1A receptors in the prosocial effects of 3, 4 methylenedioxymethamphetamine ("ecstasy"). Neuroscience. 2007;146(2):509-14.

378.   Carhart-Harris RL, Wall MB, Erritzoe D, Kaelen M, Ferguson B, De Meer I, et al. The effect of acutely administered MDMA on subjective and BOLD-fMRI responses to favourite and worst autobiographical memories. International Journal of Neuropsychopharmacology. 2014;17(4):527-40.

FIGURES AND TABLES

INDEX

www.ingramcontent.com/pod-product-compliance
Lightning Source LLC
Chambersburg PA
CBHW050642270326
41927CB00012B/2840